International Series in Operations Research & Management Science

Volume 260

More information about this series at http://www.springer.com/series/6161

International Series in Operations Research & Management Science

Series Editor

Camille C. Price
Stephen F. Austin State University, TX, USA

Associate Series Editor

Joe Zhu
Worcester Polytechnic Institute, MA, USA

Founding Series Editor

Frederick S. Hillier
Stanford University, CA, USA

Paul Tae-Woo Lee • Zaili Yang
Editors

Multi-Criteria Decision Making in Maritime Studies and Logistics

Applications and Cases

Springer

Editors
Paul Tae-Woo Lee
Institute of Maritime Logistics
Ocean College
Zhejiang University
Zhoushan, Zhejiang Province, China

Zaili Yang
School of Maritime and Mechanical Engineering
Liverpool John Moores University
Liverpool, UK

ISSN 0884-8289 ISSN 2214-7934 (electronic)
International Series in Operations Research & Management Science
ISBN 978-3-319-62336-8 ISBN 978-3-319-62338-2 (eBook)
DOI 10.1007/978-3-319-62338-2

Library of Congress Control Number: 2017955023

Printed on acid-free paper

This Springer imprint is published by Springer Nature
The registered company is Springer International Publishing AG
The registered company address is: Gewerbestrasse 11, 6330 Cham, Switzerland

This book is dedicated to the family members of the Editors, who have supported their academic life.

Paul Tae-Woo Lee and Zaili Yang

Contents

About the Editors

Paul Tae-Woo Lee is a Professor of Maritime Transport and Logistics and Director of Institute of Maritime Logistics, Ocean College, Zhejiang University in Zhoushan, China. He received his PhD from Cardiff University in the UK. He was a Visiting Scholar at, among others, the Faculty of Economics and Politics in Cambridge, University of Plymouth, Hong Kong Polytechnic University, and Dalian Maritime University. His research interests include maritime economics and maritime logistics. He is Editor-in-Chief of *Journal of International of Logistics and Trade* and an Associate Editor of *Transportation Research Part E* and *Journal of Shipping and Trade*.

Zaili Yang is a Professor of Maritime Transport at Liverpool John Moores University (LJMU), UK. His research interests are system safety, security and risk-based decision making modelling, especially their applications in marine and supply chain systems. His research findings have been published in more than 170 refereed papers in risk and supply chain areas, including over 80 (60 SCI/SSCI-cited) journal papers. He is an editorial board member of Transportation Research Part E: Logistics and Transport Review and an associate editor of IMA Journal of Management Mathematics.

About the Editors

About the Authors

Hanna Barbara Rasmussen is a postdoctoral staff at the Centre for Maritime Health and Society, University of Southern Denmark. Trained as a sociologist, in 2013 Dr. Rasmussen defended her PhD within accident prevention in the Danish oil and gas industry. Hanna's research area apart from safety in the oil and gas industries is accident prevention in the fishing and shipping industries and situation awareness in connection to energy efficiency in minor working vessels.

Zhan Bian is an Assistant Professor at Capital University of Economics and Business, China. Dr. Bian's research interests are logistics system planning, transportation management, and supply chain design and management. She has published over 15 papers, and secured one project from National Natural Science Foundation of China (NSFC).

Steve Bonsall currently retired but is active in nautical associations. He is a member of the Nautical Institute, Honourable Company of Master Mariners and sits on their Education and Training Committee and Technical Committee and Merseyside Master Mariners Committee. He was formally Head of Maritime at Liverpool John Moores University (LJMU).

Emrah Bulut is an Associate Professor of Logistics at Yildiz Technical University, Turkey. His research interests include shipping business, forecasting, and decision making. He received his PhD degree on "The Economic Analysis on Ship Investment and Management Strategy of Dry Bulk Shipping" from the Graduate School of Maritime Sciences, Kobe University. Dr. Bulut contributes several studies related to the market entry-exit time on shipping, shipping investment strategy decision and ship valuation. He has published papers in scholarly journals and conferences such as *International Journal of Shipping and Transport Logistics, Applied Mathematical Modelling, Quality and Quantity, Expert Systems with Applications,* and *Applied Soft Computing.*

Yu-Wang Chen is currently a Senior Lecturer in Decision Sciences at Alliance Manchester Business School, The University of Manchester, UK. His research interests are mainly in the areas of Decision and Risk Analysis, Operational Research, and Data Analytics. He has published more than 30 refereed articles in leading academic journals, such as *Computers & Operation Research*, *Information Sciences*, *Knowledge-Based Systems*, *EJOR*, and *IEEE T-SMC*, and three book chapters. He also acts as Associate Editor of the *Decision Analytics Journal* (Springer) and Editorial Board Member of *Web Intelligence: An International Journal* and *International Journal of Productivity and Performance Management* (Emerald).

Okan Duru is an Assistant Professor of Maritime Studies at Nanyang Technological University Singapore. His major research interests are maritime economics, economic pluralism, maritime policies, ship investment and finance. Dr. Duru received his PhD degree on "The Long-term Econometric Analysis of Dry Bulk Shipping" from the Graduate School of Maritime Sciences, Kobe University. He has published papers in various journals and conferences as well as reviewing papers.

Hang Fang is a Logistics Engineer in Dongfeng Peugeot Citroën Automobile Company Ltd. (DPCA), P.R. China. He received his MSc and BSc in Logistics Engineering from Wuhan University of Technology in 2016 and 2014, respectively. His major research interests include risk assessment of maritime transportation, logistics risk modelling, and simulation of logistics systems.

Min-Ho Ha is a research fellow at Nanyang Technological University, Singapore. He received his BSc in Shipping Management from Korea Maritime University in 2007, and MSc in Transport & Maritime Management from University of Antwerp, Belgium, in 2009 and in Shipping Management from Korea Maritime University in 2009, respectively, and PhD degree in Transport Logistics from Liverpool John Moores University, UK, in 2017. He has extensive experience on research project bids and management gained from a number of national competitive research projects funded by the Ministry of Oceans and Fisheries, Port Authorities, and Korea Shipping Association in South Korea.

Zhihong Jin is a Professor and Dean of College of Transportation Management, Dalian Maritime University, China. Prof. Jin's research interests are logistics system planning and management, transportation planning and management technology, and supply chain design and management. Professor Jin has published four books and more than 160 papers, and obtained five projects from National Natural Science Foundation of China (NSFC), seven projects granted by Ministry of Education of China, Ministry of Transport of China, and Liaoning Province, and two international cooperation projects. He is a member of Teaching Guiding Committee of Ministry of Education, China. He is an editorial board member of

International Journal of Shipping and Transport Logistics and Associate Editor-in-Chief of *Journal of Dalian Maritime University.*

Na Li is an Assistant Professor at Transportation Management College, Dalian Maritime University, China, and a Visiting Scholar at Maritime Institute, Strome College of Business in Old Dominion University, USA. Dr. Li's research interests are container terminal operation and optimization. She has published some papers about quay crane scheduling, berth allocation, and truck appointments. She has secured four projects from Ministry of Education of China and Liaoning Province.

Cheng-Wei Lin received his MSc and PhD degrees in traffic and transportation management from National Chiao Tung University, Taiwan, in 1994 and 2004, respectively. Since 2004, he has been affiliated with Department of Logistics and Shipping Management, Kainan University, Taiwan. Before joining Kainan University, he was a research fellow in the Energy Research Group at National Chiao Tung University. His current research interests include multiple criteria decision analysis, fuzzy set theory, rough set theory, maritime policy and management, logistic management, and so on.

Marie Lützen has many years of experience within the areas of maritime transport and technology. Dr. Lützen has an education as master mariner (1989), naval architect (1998) and has a PhD (2002) in Naval Architecture and Ocean Engineering from The Technical University of Denmark. Marie has a large experience – both practical and theoretical – in ship operation and maritime technology and has very broad knowledge of the maritime sector. She has participated in several projects economically supported and funded by the industry, the government, and private foundations.

Feng Ma is an Associate Professor in Intelligent Transport System Research Centre, National Engineering Research Centre of Water Transportation Safety (WTS), Wuhan University of Technology, P. R. China. His research interests are mainly in the areas of maritime safety, artificial intelligence, and unmanned surface vessel. He has published over ten research articles in leading journals, such as *Transportation Research Part C: Emerging Technologies*, *Reliability Engineering and System Safety*, and *Ocean Engineering.*

Lefteris Maistralis is currently serving in the Oil and Gas Industry heading Ocean Rig's, QHSE's department, system safety, and risk-based decision-making processes for both offshore and onshore operations. Dr. Maistralis was previously serving as Director of Europe/Middle East/Africa for American Bureau of Shipping (ABS) Nautical Systems Operations as well as Head of Europe for Risk and Reliability Services for ABS consulting. Graduated as a Naval Architect/Marine Engineer, Lefteris held various technical positions in the maritime industry.

Adolf K.Y. Ng is a Professor of Transportation and Supply Chain Management at the Asper School of Business of the University of Manitoba (Canada). He is also the Director of the Transport Institute of the University of Manitoba. He obtained his DPhil from University of Oxford (UK), and excels in the research and teaching of maritime transportation management and geography, climate adaptation planning and management, and global supply chains. He is the Associate Editor of *Maritime Policy & Management* and the Co-editor of *Journal of Transport Literature*.

John Psarras is a Professor and Director of the Decision Support Systems (DSS) Laboratory of the School of Electrical and Computer Engineering of the National Technical University of Athens, Greece. His research interests focus on the area of decision support systems with emphasis on the management of energy, environment, and information systems. He has more than 35 years of experience in managerial, consulting, and technical positions in the energy sector. He has been Team Coordinator in complex, multinational research, and development projects funded by the EC in the areas of energy policy and markets, clean energy technologies, capacity building and training, decision support systems, and project and programme development and management. He has extensive experience in networking activities and in facilitation of international co-operation, acquired through the management of multi-cultural teams in the frame of multi-country projects and in communication and collaboration with high-level EC officials. He has published more than 200 papers in international journals and conference proceedings.

Zhuohua Qu is a Senior Lecturer in Liverpool Business School at Liverpool John Moore University, UK. Dr. Qu is specialized in quality control and process excellence of logistics systems. She has an expertise in both experimental studies and modelling over different supply chains systems including food, energy and healthcare, etc. She has published more than ten academic papers and two PhD students under her direct supervision.

Jingzheng Ren is an Assistant Professor of Operations Research of The Hong Kong Polytechnic University and an honorary Associate Professor of Energy Strategic Management at the University of Southern Denmark. Dr. Ren also serves as the Associate Editor of SCI-indexed journal – *Journal of Renewable and Sustainable Energy*, Associate Editor of *Renewable Energy & Sustainable Development*, and Guest Editor of several leading SCI-indexed international Journals. He has also co-authored more than 70 papers in the leading refereed journals, and many of his papers have been identified and highlighted as the Key Scientific Article contributing to the excellence in energy research.

Sung-Ho Shin holds a PhD degree from Department of Logistics and Maritime Studies, The Hong Kong Polytechnic University, Hong Kong China. He has a research experience involving in various governmental projects at the Korea Maritime Institute (KMI), which is a government think-tank specializing in the area of shipping, ports, and marine issues in Korea. His research interests are in maritime transport and logistics, ship accidents, and risk analysis.

Eleftherios Siskos is a Chemical Engineer, and in the final stage of his PhD candidature in the Decision Support Systems Laboratory of the School of Electrical and Computer Engineering of the National Technical University of Athens. Through his work, he has gained experience in modelling and solving real-world optimization problems and in simulating and evaluating risky decisions under uncertainty. His research interests fall into the areas of multi-criteria decision analysis, decision support systems, multi-objective mathematical programming, and robustness analysis of forecasting and decision models. Eleftherios has significant experience in e-government, energy planning, climate policy making, and environmental management towards sustainable development. He has published more than 25 papers in international journals, book chapters, and conference proceedings.

Dimitrios I. Stavrou is a superintendent engineer in a shipping company. Dr. Stavrou was retired from the Hellenic Navy as a Lieut. Commander Officer after his 23-year service. He is a Naval Architecture and Marine Engineer of the National Technical University of Athens (NTUA), Greece. His field of expertise includes maritime safety and security, risk analysis and assessment, human factor, multi-criteria analysis, and decision support systems. He is the author of more than 20 technical papers for journals and international conferences. For his PhD studies, he was supported by the "IKY fellowships of excellence for postgraduate studies in Greece-Siemens program".

Nikolaos P. Ventikos is an Associate Professor in the Laboratory for Maritime Transport of the School of Naval Architecture and Marine Engineering at the National Technical University of Athens (NTUA), Greece. Dr. Ventikos's expertise lies with marine safety; risk engineering; (oil) marine pollution; human factors; crew-centered design; salvage engineering; maritime security; lifecycle analysis and design; port operations; ship recycling; maritime sustainability; ship-generated waste management, blue growth, and statistical/probabilistic modelling. Nikolaos has published his work in peer-reviewed journals and to more than 100 conferences and symposia.

Chengpeng Wan is a joint PhD candidate between Wuhan University of Technology, China, and Liverpool John Moores University (LJMU), UK. He is currently doing research on the risk assessment of container maritime logistics. He received his bachelor's degree in Marine Engineering from Wuhan University of Technology in 2012. Wan's major research interests include performance evaluation, uncertainty treatment in risk modelling and assessment, risk-based multiple criteria decision making, and resilience management of transportation systems.

Jin Wang is a Professor of Marine Technology at Liverpool John Moores University (LJMU), UK. Following just less than five years' research as a Research Associate at Newcastle University, UK, he joined LJMU as a Lecturer in 1995, and was promoted as Reader in Marine Engineering and Professor of Marine Technology in 1999 and 2002, respectively. He has been involved in safety and reliability research of large engineering systems with significant financial support from the UK research councils, EU, etc.

Ying Wang is an Assistant Professor in the School of Economics and Administration at the Yantai University, China. She received her ME and PhD in the Graduate School of Logistics at the Incheon National University, South Korea, in 2012 and 2015, respectively. Ying's specific academic field is maritime transportation management, fuzzy sets theory, decision science, and forecasting issues. She has published a lot of research articles in international academic journals such as *Tourism Management*, *Journal of Air Transport Management*, *Transport Policy*, *Maritime Policy & Management*, and *Maritime Economics and Logistics*.

Qi Xu is a Lecturer in College of Transportation Management, Dalian Maritime University, China. Dr. Xu's research interests are container multimodal transport system optimization, logistics system optimization and simulation, and supply chain design and management. She has published more than ten papers related to her research interests, and secured three projects which are granted by the National Natural Science Foundation of China (NSFC), Natural Science Foundation of Liaoning Province in China, and Research Foundation of Post Doctor in China.

Gi-Tae Yeo is a Professor at the Graduate School of Logistics in the Incheon National University, South Korea. He obtained his PhD in Shipping and Logistics from University of Plymouth, UK. He was an Editor-in-Chief of the *Asian Journal of Shipping Logistics*. He is the President of Korean Association of Shipping and Logistics and Dean of Graduate School of Logistics in Incheon National University. He has completed supervision of 50 master/PhD researchers in research areas of maritime economics and logistics.

Di Zhang is the Vice Director of and an Associate Professor at National Engineering Research Center for Water Transport Safety, Wuhan University of Technology (WUT), China. He received his PhD in Vehicle Operation Engineering from WUT in 2011. With the financial support from the China Scholarship Council, he was a full-time researcher at Liverpool John Moores University from October 2010 to September 2011. Di's major research interests include risk assessment and decision science applied in marine systems. He has authored three book chapters, 25 refereed journal papers, and over 30 refereed conference papers. He is an associate fellow of Royal Institute of Navigation.

Contributors

Zhan Bian College of Business Administration, Capital University of Economics and Business, Beijing, China

Stephen Bonsall School of Engineering, Liverpool John Moores University, Liverpool, UK

Emrah Bulut Department of Business Administration, Yildiz Technical University, Istanbul, Turkey

Yu-Wang Chen Decision and Cognitive Sciences Research Centre, The University of Manchester, Manchester, UK

Okan Duru School of Civil and Environmental Engineering, Nanyang Technological University, Singapore, Singapore

Hang Fang National Engineering Research Center for Water Transport Safety (WTSC), Wuhan University of Technology, Wuhan, China

Min-Ho Ha Liverpool Logistics Offshore and Marine Research Institute (LOOM), Liverpool John Moores University, Liverpool, UK

Zhihong Jin College of Transportation Management, Dalian Maritime University, Dalian, China

Paul Tae-Woo Lee Ocean College, Zhejiang University, Zhoushan, Zhejiang Province, China

Na Li College of Transportation Management, Dalian Maritime University, Dalian, China

Cheng-Wei Lin Department of Logistics and Shipping Management, Kainan University, Taoyuan City, Taiwan

Marie Lützen Mechanical Engineering, Department of Technology and Innovation, University of Southern Denmark, Odense, Denmark

Feng Ma Intelligent Transport System Research Center, Wuhan University of Technology, Wuhan, People's Republic of China

National Engineering Research Center of Water Transportation Safety (WTS), Wuhan, People's Republic of China

Lefteris Maistralis Risk & Reliability Services, ABS Maritime Services (Hellas) LLC, Piraeus, Greece

Adolf K.Y. Ng Department of Supply Chain Management, Asper School of Business, University of Manitoba, Winnipeg, MB, Canada

Transport Institute, Asper School of Business, University of Manitoba, Winnipeg, MB, Canada

John E. Psarras School of Electrical and Computer Engineering, National Technical University of Athens, Athens, Greece

Qi Xu College of Transportation Management, Dalian Maritime University, Dalian, China

Zhuohua Qu Liverpool Business School, Liverpool John Moores University, Liverpool, UK

Hanna Barbara Rasmussen Centre of Maritime Health and Society, University of Southern Denmark, Odense, Denmark

Jingzheng Ren Department of Industrial and Systems Engineering, The Hong Kong Polytechnic University, Hong Kong, SAR, China

Sung-Ho Shin Department of Logistics and Maritime Studies, The Hong Kong Polytechnic University, Kowloon, SAR, China

Eleftherios Y. Siskos School of Electrical and Computer Engineering, National Technical University of Athens, Athens, Greece

Dimitrios I. Stavrou School of Naval Architecture and Marine Engineering, National Technical University of Athens, Athens, Greece

Nikolaos P. Ventikos School of Naval Architecture and Marine Engineering, National Technical University of Athens, Athens, Greece

Chengpeng Wan Intelligent Transportation Systems Research Center (ITSC), Wuhan University of Technology, Wuhan, China

National Engineering Research Center for Water Transport Safety (WTSC), Wuhan University of Technology, Wuhan, China

Liverpool Logistics, Offshore and Marine (LOOM) Research Institute, Liverpool John Moores University, Liverpool, UK

Jin Wang School of Engineering, Liverpool John Moores University, Liverpool, UK

Ying Wang School of Economics and Management, Yantai University, Yantai, China

Zaili Yang Liverpool Logistics Offshore and Marine Research Institute (LOOM), Liverpool John Moores University, Liverpool, UK

School of Engineering, Liverpool John Moores University, Liverpool, UK

Gi-Tae Yeo Graduate School of Logistics, Incheon National University, Incheon, South Korea

Di Zhang Intelligent Transportation Systems Research Center (ITSC), Wuhan University of Technology, Wuhan, China

National Engineering Research Center for Water Transport Safety (WTSC), Wuhan University of Technology, Wuhan, China

Chapter 1
Introduction

Paul Tae-Woo Lee and Zaili Yang

After a careful literature survey of the publications on Multi-Criteria Decision Making (MCDM) (e.g., Chen and Hwang 1992; Figueira et al. 2005; Hwang and Masud 1979: Hwang and Yoon 1981; Lai and Hwang 1994; Tanino et al. 2003; Zopounidis and Pardalos 2010), we found that few books have been published to address the theoretical demands on the use of advanced survey techniques and MCDM methods in a complementary way. Furthermore, there are increasing practical concerns on the scanty real cases of using MCDM methods in shipping, port, and logistics, available from the International Series in Operations Research and Business Management by Springer, and/or other sources/publishers such as Ishizaka and Nemery (2013), Triantaphyllou (2000) and Tzeng and Huang (2011). To fulfill this gap, we have edited this book entitled Multi-Criteria Decision Making in Maritime Studies and Logistics, which has peculiar advantages and characteristics, demonstrated by a few highlights and research challenges as follows.

- This book applies MCDM to real case studies in a wide range of areas relating to the maritime subject including shipping, port, maritime logistics, cruise ports, waterfront developments, and shipping finance, etc. In such areas, researchers, students and industrialists, in general, felt struggling to apply MCDM to find the solutions to their real problems in practice.
- More than four thousand papers involving MCDM methods have been published in international journals since 2000, according to Web of Science. However to present the most important and concise information in these journal papers, their

P.T.-W. Lee (✉)
Ocean College, Zhejiang University, Zhoushan, Zhejiang Province, China
e-mail: paultaewoo.lee@zju.edu.cn

Z. Yang
Liverpool Logistics, Offshore and Marine Research Institute (LOOM), Liverpool John Moores University, Liverpool, UK

P.T.-W. Lee, Z. Yang (eds.), *Multi-Criteria Decision Making in Maritime Studies and Logistics*, International Series in Operations Research & Management Science 260,
DOI 10.1007/978-3-319-62338-2_1

authors do not often describe the calculation processes of using the MCDM methods, leading to limited access to the relevant detailed information by students and early stage researchers. This book having a focus on the in-depth step-by-step applications of the most popular MDCM methodologies will be able to address this challenge.

- Our teaching experience confirms that students can easily learn the principles and theoretical backgrounds of MCDM but feel struggling to apply them to real cases, because the applications of MCDM require well designed questionnaires to collect data from respondents. However there are few publications, showing real samples of the questionnaires applied in solving MCDM problems. This book discloses some samples of the questionnaires with reference to the applications of MCDM to real cases.
- The real cases described in this book also address the emerging issues in the maritime context such as green shipping, port, and logistics as well as security and safety issues, revealing new hybrid MDCM solutions in complex dynamic decision making environments.

This book brings together an eclectic collection of twelve chapters which seek to respond to the above challenges. The first contribution by Zhuohua Qu, Chengpeng Wan, Zaili Yang, and Paul Tae-Woo Lee (Chap. 2) is an overview of major MCDM techniques. This chapter describes the detailed mathematical steps of well-established MCDM methodologies such as Analytic Hierarchy Process (AHP), TOPSIS, VIKOR, ELECTRE, and PROMETHEE in order to provide a holistic knowledge. It also analyses the advantages and disadvantages of the methods so as to serve as a foundation to their applications in the ensuing chapters.

Recognizing that AHP is very functional and popular in both academia and professional life, Emrah Bulut and Okan Duru (Chap. 3) claim that there is various biases and misuse of the method due to lack of the full understanding of its comprehensive theoretical basis. Their work discusses, first, the theory of AHP in detail with references to the strong assumptions inherently adjunct to the method and their practical impacts on the AHP analysis. Secondly, it investigates a fuzzy AHP (FAHP) approach and its capability and rationale in dealing with decision problems of ambiguous information. Last, empirical applications including dry port location and shipping asset selection have been conducted to demonstrate their feasibility and to provide effective solutions to maritime and logistics problems.

Green shipping has become a focal issue aiming to mitigate the negative environmental impacts caused by maritime transportation in conjunction with the effort from the International Maritime Organization (IMO). Chapter 4 by Jingzheng Ren, Marie Lützen, and Hanna Barbara Rasmussen proposes a generic methodology to identify the key factors influencing green shipping and to establish an evaluation system for the assessment of shipping greenness. The authors employ Analytic Network Process (ANP) to determine the relative importance of the identified factors in green shipping with the consideration of their interdependences and interactions for realizing precise evaluation of shipping greenness. In addition, Chap. 4 adopts Interpretative Structuring Modeling (ISM) to analyze the cause-effect relationships among the measures of and solutions to the greenness of shipping. Consequently, it contributes to analyzing the influential factors of green shipping and studying the strategic measures for enhancing the greenness of shipping in a hybrid approach.

Chapter 5 by Zaili Yang, Lefteris Mastralis, Stephen Bonsall, and Jin Wang proposes a new function of fuzzy Evidential Reasoning (ER) to improve the vessel selection process in which multiple criteria with insufficient and ambiguous information are evaluated and synthesised. By doing so, a numerical case study of selecting an oil tanker based on a voyage charter party is presented to demonstrate the proposed method. Chapter 5 contributes to overcoming the difficulty and complexity in selecting vessels that the stakeholders of conflicting interests encounter, to helping analysts or decision makers derive rational decisions from uncertain and incomplete data contained in different quantitative and qualitative forms. In the decision making process, the Window-based software tool called Intelligent Decision System (IDS) via ER (Yang and Xu 2000) is used to build up the model, define alternatives and criteria, and perform different assessments according to the decision makers' requirements.

Feng Ma and Yu-wang Chen propose, in Chap. 6, a novel methodology by using an Artificial Potential Field (APF) model and the ER approach to estimate the collision probabilities of monitoring targets for coastal radar surveillance. Initially, the probability of a monitoring target being a real moving vessel is estimated using the records of manual operations and the ER rule. Subsequently, the bridges, piers and other encountering vessels in a waterway are characterized as collision potential fields using an APF model. As a result, the positional collision potential of any monitoring vessel can be obtained through overlapping all the collision potential fields together. The probabilities of authenticity and the collision potential are further formulated as two pieces of evidence on which the Dempster's rule of combination is used to reason the collision probability of a monitoring target. The vessels associated with high collision probabilities can be highlighted for supervisors' decision on risk avoidance, as they potentially pose high risks to safety.

Based on the literature review of port performance evaluation and brainstorms with domain experts, Chap. 7 by Chengpeng Wan, Di Zhang, and Hang Fang attempts to develop the inland port performance assessment model (IPPAM) using AHP and ER and a utility function. IPPAM is a dynamic complex system involving many indicators from four main perspectives, namely, infrastructure, operations and management, financial status, and development potential with a case study by evaluating the performance of the Port of Wuhan in China from 2007 to 2013. This chapter contributes to developing a hierarchical model for the evaluation of inland port performance by using AHP and ER in a complementary way in which AHP is employed to calculate the relative importance of the relevant qualitative and quantitative criteria, while ER is hired to deal with synthesis in order to achieve the estimation of inland port performance. The novel model and flexible method presented in this chapter could be applied for evaluating performance of inland ports in other areas in order to formulate corresponding measures to improve their management and development.

Ship-to-ship (STS) transfers which can cause the adverse effects of a potential accident raise great concerns to the stakeholders, who require capable risk decision techniques for risk avoidance. Chapter 8 by Dimitrios I. Stavrou, Eleftherios Y. Siskos, Nikolaos P. Ventikos, and John E. Psarras proposes a challenging approach to address the aforementioned risk and uncertainty, by utilizing a STOCHASTIC Utility Additives (UTA) method with the help of the philosophy of aggregation–disaggregation coupled with a robustness control procedure. It can therefore contribute to evaluating the risks of STS transfer of cargo operations. The applied methodology in Chap. 8 provides the STS transfer operators with an alternative way to conduct risk assessment using experts' knowledge. Moreover, this chapter highlights the addition of the bipolar robustness control procedure to manage robustness of the model which can eliminate the weaknesses of the classical STOCHASTIC UTA approach and improve the effectiveness of the extracted results.

In Chap. 9, Paul Tae-Woo Lee, Cheng-Wei Lin, and Sung-Ho Shin apply Entropy and Grey Relation Analysis (GRA) to analyze the relative weights of financial ratios and the major four shipping companies in Korea and Taiwan: Evergreen, Yang Ming, Hanjin and Hyundai Merchant Marine. The four companies are ranked according to the real values of their financial ratios taken and calculated from the financial statements audited by certified public accountants in the period of 1999–2009. The authors explore a practical procedure of financial ratio analysis to identify the various features relating to financial crisis by investigating their financial statements. The findings in this chapter help shipping managers develop their business policy to mitigate the impacts of financial crisis on their companies, in particular at the scene of the bankruptcy of Hanjin Shipping Company in 2017.

Chapter 10 by Zhihong Jin, Na Li, Xu Qi, and Zhan Bian systemically applies modern heuristics to solving MCDM problems in the fields of operation optimisation in container terminals. A container terminal consists of three geographically interrelated core areas: container terminal, anchorage ground, and gateway. With respect to these three key areas, three MCDM optimisation models are developed to deal with the container loading sequence problem in the terminal, the tugboat scheduling problem in the anchorage ground and resource deployment for truck appointment system at the gateway, respectively. More specifically, in terms of container loading sequence problem, the authors develop a two-phase hybrid dynamic algorithm aiming to generate an optimal movement sequence for the crane to retrieve all the containers from a given yard to a ship. As far as the tugboat scheduling problem is concerned, a hybrid simulated annealing algorithm is proposed to solve the addressed problem. The factors of multi-anchorage bases and three stages of operations (berthing/shifting-berth/unberthing) are considered and the objective is to minimize the total operation time for all tugboats and the waste of the tugboats horsepower in use at the same time. Finally, when the truck appointment system at the gateway is considered, a bi-objective model is set up to minimize resource input and balance workloads. Modern heuristics method based on non-dominated genetic algorithm is proposed to solve difficulties of simultaneous optimization of resource input and appointment quotas. Numerical experiments are undertaken to evaluate the effectiveness of the proposed algorithms and show the efficiency of the proposed algorithm.

Arguing that bunkering port selection is typically a multi-criteria group decision problem and, in many practical situations, decision makers cannot form proper judgments using incomplete and uncertain information in an environment with exact and crisp values, Ying Wang, Gi-Tae Yeo, and Adolf K.Y. Ng propose a hybrid Fuzzy-Delphi-TOPSIS based methodology with a sensitivity analysis in Chap. 11. This chapter contributes to developing a benchmarking framework for choosing optimal bunkering ports for liner shipping companies along a regular liner route by evaluating the bunkering ports' performance. The proposed framework can enable decision makers to better understand the complex relationships of the relevant key performance factors and assist managers in comprehending the present strengths and weaknesses of their strategies.

Port performance indicators (PPIs) can interact with each other (outer dependency) and/or feedback themselves (inter dependency). Given the fact that previous studies have done little on the analysis of interdependency among the PPIs, Min-Ho Ha and Zaili Yang propose a new conceptual PPIs' interdependency model using a hybrid approach of a fuzzy logic based evidential reasoning (FER), DEMATEL and AHP in Chap. 12. The combined approach of DEMATEL and AHP is applied to calculate the weights of dependent PPIs which are used as a part of the FER model to measure and analyze port performance. The framework proposed in Chap. 12 has been successfully implemented in dealing with both objective data and subjective data in a unified manner to incorporate multiple objectives of key stakeholders, which is validated through the case study of six container terminals in South Korea. From the case study results, the new conceptual PPIs' interdependency model offers diagnostic instruments to decision makers to identify the strengths and weaknesses of the terminals. Accordingly, decision makers in the terminals can identify the particular areas for improvement to enhance their competitiveness based on any necessary comparisons.

References

Chen SJ, Hwang CL (1992) Fuzzy multiple attribute decision making: methods and applications. Springer, Berlin

Figueira J, Greco S, Ehrogott M (eds) (2005) Multiple criteria decision analysis: state of the art surveys. Springer, New York

Hwang CL, Masud AS (1979) Multiple objective decision making. Springer, Berlin

Hwang CL, Yoon K (1981) Multiple attribute decision making methods and applications: a state-of-the-art survey. Lecture notes in economics and mathematical systems, vol. 186. Springer, Berlin

Ishizaka A, Nemery P (2013) Multi-criteria decision analysis: methods and software. Wiley, Chicheste

Lai YJ, Hwang CL (1994) Fuzzy multiple objective decision making: methods and applications. Springer, Berlin

Tanino T, Tanaka T, Inuiguchi M (eds) (2003) Multi-objective programming and goal-programming: theory and applications. Springer, Berlin/Heidelberg

Triantaphyllou E (2000) Multi-criteria decision making methods: a comparative study. Kluwer, Dordrecht

Tzeng GH, Huang JJ (2011) Multi criteria decision making analysis: methods and applications. CRC Press, Boca Raton

Yang JB, Xu DL (2000) "An introduction to an intelligent decision system: IDS", Proceedings of the 42nd Annual Conference of the UK Operational Research Society, Sep. 12–14, Swansea, UK

Zopounidis C, Pardalos PM (eds) (2010) Handbook of multicriteria analysis. Springer, New York

Chapter 2
A Discourse of Multi-criteria Decision Making (MCDM) Approaches

Zhuohua Qu, Chengpeng Wan, Zaili Yang, and Paul Tae-Woo Lee

Abstract During a decision making process, decision makers often need to handle large amount of information in order to reach a rational decision. Such information can be incomplete, uncertain and even contradictory to each other. Multi-criteria decision making (MCDM) methods provide effective and popular solutions to aid decisions under uncertainty. Well-established MCDM methodologies such as AHP, TOPSIS, VIKOR, ELECTRE, and PROMETHEE are reviewed with particular reference to their standard frameworks in this chapter to provide a holistic knowledge base on their applications individually and/or collectively in the other chapters in this book.

Keywords MCDM • AHP • TOPSIS • VIKOR • ELECTRE • PROMETHEE • Fuzzy set • Evidential Reasoning

Z. Qu
Liverpool Business School, Liverpool John Moores University, Liverpool, UK

C. Wan
Intelligent Transportation Systems Research Center (ITSC), Wuhan University of Technology, Wuhan, China

National Engineering Research Center for Water Transport Safety (WTSC), Wuhan University of Technology, Wuhan, China

Liverpool Logistics, Offshore and Marine (LOOM) Research Institute, Liverpool John Moores University, Liverpool, UK

Z. Yang
Liverpool Logistics, Offshore and Marine Research Institute (LOOM), Liverpool John Moores University, Liverpool, UK

P.T.-W. Lee (✉)
Ocean College, Zhejiang University, Zhoushan 316021, Zhejiang Province, China
e-mail: paultaewoo.lee@zju.edu.cn

P.T.-W. Lee, Z. Yang (eds.), *Multi-Criteria Decision Making in Maritime Studies and Logistics*, International Series in Operations Research & Management Science 260,
DOI 10.1007/978-3-319-62338-2_2

1 Introduction

MCDM methods can be defined as structured frameworks that deal with the process of making decisions in the presence of multiple objectives (Pohekar and Ramachandran 2004). They are used to find the best opinion from all of the feasible alternatives in the presence of multiple, usually conflicting, decision criteria (Pomerol and Romero 2000).

A MCDM process typically defines objectives, chooses the criteria to measure the objectives, specifies alternatives, transforms the criterion scales into commensurable units, assigns weights to the criteria that reflect their relative importance, selects, and applies a mathematical algorithm for ranking and choosing an alternative (Ananda and Herath 2009). During a decision making process, objectives can be uncertain, complex and even conflicting; criteria can be cardinal or ordinal; and information can be exact or fuzzy. MCDM methods are considered to be an important tool as they allow decision-makers to select a solution by tackling all above mentioned difficulties and complexity.

Uncertainty is an important aspect to be addressed in MCDM. It is defined as a situation in which a person does not have appropriate quantitative and qualitative information to describe, prescribe or predict deterministically and numerically a system, its behaviour or other characteristics (Zimmermann 2000). Uncertainty in principle is originated from failures, assumptions, unavailability or incompleteness of data.

There are a large number of MCDM methods established in the literature. They differ from each other in terms of the required quality and quantity of additional information, the methodologies, the user-friendliness of the methods and their associated software, the sensitivity tools used, and the mathematical properties (Zavadskas and Turskis 2011). However, it is noteworthy that none of them is considered suitable under all MCDM environments and therefore hybrid approaches are often developed to deal with complex scenarios involving different types of uncertainties.

2 Analytical Hierarchy Process (AHP)

AHP was developed by Saaty (1980). As one of the most widely used MDCM approaches, AHP is capable of assisting criteria selection, criteria importance analysis and alternative evaluation. The best decision can be made when qualitative and quantitative aspects of a decision are included (Saaty 1990). AHP uses the concept of pair-wise comparisons to improve the efficiency of synthesising qualitative and quantitative evaluations in a decision process. It contains different alternatives and criteria for judging the alternatives. The approach allows decision makers to express their opinions by comparing two alternatives at a time rather than simultaneously on all the alternatives. It simplifies and expedites a decision making process on complex issues. The visibility and easiness characteristics of AHP contribute to its popularity across different industries. Vaidya and Kumar (2006) revealed that the AHP method has been used in nearly 150 applications. Examples of using AHP in the shipping and maritime sectors include, port choice and competitiveness evaluation (Lam and Dai 2012; Yeo et al. 2010; Yeo et al. 2014), vessel selection (Yang et al. 2009b; Xie et al. 2008), port allocation (Carlos Perez-Mesa et al. 2012), risk estimation of ship operations (Ung et al. 2006), design support evaluation for the offshore industry (Sii and Wang 2003), port service quality ranking (Ugboma et al. 2004), maritime regulation implementation (Karahalios et al. 2011), ship operational energy efficiency (Beşikçi et al. 2016), choice of ship flag (Chou and Ding 2016), assessment of the maritime labour convention compliance (Akyuz et al. 2015) and marine accident analysis (Sahin and Senol 2015).

AHP uses a mathematical process to handle subjective judgements of an individual or a group in a decision making process. It consists of four steps: (1) establishing the hierarchy of criteria and alternatives, (2) making pair-wise comparisons of the criteria, and estimating the weights of the criteria and the relative performance values of the alternatives with respect to each criterion, (3) aggregating the weights and performance values for alternative priority, and (4) checking the consistency of the judgements to verify the result.

Step 1 Establish the hierarchy of criteria and alternatives

Hierarchy is the base of AHP. In order to conduct an AHP study, a hierarchy of clear criteria and alternatives need to be constructed. Figure 2.1 shows an example of hierarchy with defined criteria and alternatives.

Step 2 Make a pair-wise comparison decision matrix (M).

A pair-wise comparison matrix (M) is constructed for all the criteria (Eq. (2.1). a in the matrix represents a quantified judgement on a pair of criteria (e.g. a_{12} represents the importance of *Criterial 1(C1)* over *Criterial 2 (C2)*). A scale of "1" to "9" is adopted to conduct non-quantitative pair-wise comparisons of two elements (Saaty 1980). Judgements are given verbally as indicated in Table 2.1 before corresponding score is allocated.

$$M = \begin{array}{c} \\ C_1 \\ C_2 \\ \vdots \\ C_j \\ \vdots \\ C_n \end{array} \begin{array}{c} \begin{array}{cccccc} C_1 & C_2 & \dots & C_i & \dots & C_n \end{array} \\ \begin{bmatrix} a_{11} & a_{21} & \dots & a_{i1} & \dots & a_{n1} \\ a_{12} & a_{22} & \dots & a_{i2} & \dots & a_{n2} \\ \vdots & \vdots & & \vdots & & \vdots \\ a_{1j} & a_{2j} & \dots & a_{ij} & \dots & a_{nj} \\ \vdots & \vdots & & \vdots & & \vdots \\ a_{1n} & a_{2n} & \dots & a_{in} & \dots & a_{nn} \end{bmatrix} \end{array} \tag{2.1}$$

Step 3 Normalize the decision matrix and calculate the priorities of this matrix to obtain the weights of criteria $w_1, w_2, \dots$ *and* w_n.

In order to calculate the weight of each criterion, the comparison matrix has to be normalized. This can be done by summing each set of column values; then each value is divided by its corresponding summed value. The relative weight of the kth criteria is obtained through averaging the values of the kth row in the matrix. This can be presented by using Eq. (2.2).

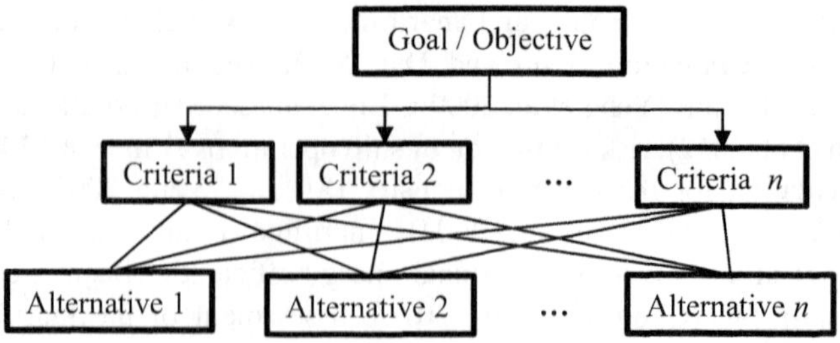

Fig. 2.1 Hierarchy of criterial and alternatives

Table 2.1 Judgement scores in AHP (Saaty 1980)

Score	Judgment	Explanation
1	Equally	Two activities contribute equally to the objective
3	Moderately	Experience and judgment slightly favor one activity over another
5	Strongly	Experience and judgment strongly favor one activity over another
7	Very strongly	An activity is strongly favored and its dominance demonstrated in practice
9	Extremely	The evidence favoring one activity over another is of the highest possible order of affirmation
2,4,6,8	Intermediate values between two adjacent judgments	When compromise is needed

$$w_k = \frac{1}{n}\sum_{j=1}^{n}\frac{a_{kj}}{\sum_{i=1}^{n} a_{ij}}, \quad k = (1, 2, \ldots n) \tag{2.2}$$

where, a_{ij} is the entry of row i and column j in a comparison matrix of order n and w_k is the weight of a specific criterion k in the pairwise comparison matrix.

Step 4 Check consistency of the judgements to verify the result

In order to derive meaningful weights, a minimal consistency is required and a test must be done. The consistency of the comparison matrices is tracked by a Consistency Ratio (CR). CR index in AHP is used in order to maintain consistency in decision making of the responders. CR can be defined as follows:

$$CR = \frac{CI}{RI} \tag{2.3}$$

CI is the consistency index and RI is the random index. CI can be defined as follows:

$$CI = \frac{\lambda_{max} - n}{n - 1} \tag{2.4}$$

where λ_{max} defined as the maximum eigenvalue can be approximately calculated in Eq. (2.5).

$$\lambda_{max} = \frac{\sum_{j=1}^{n} \frac{\sum_{k=1}^{n} w_k a_{jk}}{w_j}}{n} \quad j = (1, 2, \ldots n), k = (1, 2, \ldots n) \tag{2.5}$$

where w_j and w_k are the weights of criteria obtained in Step 2.

According to Saaty (1995), CR should be less than or equal to 10% to be acceptable. Higher CR value indicates the need of adjustment of the judgements.

The rating for each alternatives against each criterion can be obtained by following a similar procedure. The decision alternatives can then be priorities by using the weighted average rating.

3 Analytical Network Process (ANP)

The AHP has some advantages over other methods because of its simplicity and its ability to rank parts of a multi-criteria problem in a hierarchical structure (Chen and Lin 2006). However it lacks the ability to model the interdependencies among the criteria, which constrains its applications in complex systems such as transport networks. Analytical Network Process (ANP) (Saaty 1990) was developed to complement the AHP in a way that the criteria are presented in a network (instead of hierarchy) structure.

ANP, being capable of modelling interdependency among the decision factors, becomes a useful MCDM tool since its development. It is an extension of AHP and allows the consideration of interdependence among and between levels of criteria and alternatives. ANP uses a network without the need to specify levels as in hierarchy. It provides a logical way of dealing dependency. Networks in ANP include clusters of elements that may influence each other. A pairwise comparison matrix is established for all elements (Eq. (2.6). The respondents need to answer the questions such as: "Given an element and its upper level objective, which of the two elements influences the given element more with respect to the upper level objective, and how much more influence it has than another element?" The responses are presented numerically, scaled on the basis of Saaty's 1–9 scale (see Table 2.1), where 1 presents indifference between the two elements and 9 stands for overwhelming dominance of the element under consideration (in the row of the matrix) over the comparison element (in the column of the matrix). The local weights for all the elements are then generated by using Eq. (2.7. The local weights derived from the pairwise comparison matrices become a part of the inputs of a supermatrix. A supermatrix with its general entry matrices is shown as Eqs. (2.8) and (2.9). The weighted supermatrix is obtained through multiplying the priority vectors of each element in the un-weighted supermatrix with the priority vectors of the corresponding clusters. The global weights are then obtained through raising the weight supermatrix to limiting power.

$$A = a_{(ij)} = \begin{array}{c} \\ e_1 \\ e_2 \\ \vdots \\ e_i \\ \vdots \\ e_m \end{array} \begin{array}{c} \begin{array}{ccccccc} e_1 & e_2 & \cdots & e_j & \cdots & e_m \end{array} \\ \begin{bmatrix} a_{11} & a_{12} & \cdots & a_{1j} & \cdots & a_{1m} \\ a_{21} & a_{22} & \cdots & a_{2j} & \cdots & a_{2m} \\ \vdots & \vdots & \vdots & \vdots & \vdots & \vdots \\ a_{i1} & a_{i2} & \cdots & a_{ij} & \cdots & a_{im} \\ \vdots & \vdots & \vdots & \vdots & \ddots & \vdots \\ a_{m1} & a_{m2} & \cdots & a_{mj} & \cdots & a_{mm} \end{bmatrix} \end{array} \tag{2.6}$$

where $a_{(ij)} = 1$ if $i = j$; $a_{(ij)} = \frac{1}{a_{ji}}$, $i = (1, 2, \ldots m)$ and $j = (1, 2, \ldots m)$.

Suppose there are m elements to be compared in a matrix, let $e_1, e_2, \ldots, e_m$ denote the different elements, where $a_{(ij)}$ represents the level of influences that the respondent believes when element e_i is compared with e_j. When scoring is conducted for a pair, a reciprocal value is automatically assigned to the reverse comparison within the matrix.

$$w_k = \frac{1}{m} \sum_{j=1}^{m} \frac{a_{kj}}{\sum_{i=1}^{n} a_{ij}} \qquad k = (1, 2, \ldots m) \tag{2.7}$$

where, w_k is the priority vector of the kth element in the pairwise comparison matrix. Its value is a part of the supermatrix in the later steps. Literately, it repeats the ANP procedure in Sect. 3.

$$\begin{array}{ccccc} C_1 & \cdots & C_j & \cdots & C_n \\ e_{11} \cdots e_{1m_1} & \cdots & e_{j1} \cdots e_{jm_j} & \cdots & e_{n1} \cdots e_{nm_n} \end{array}$$

$$W = \begin{array}{c} \\ C_1 \\ \\ \vdots \\ \\ C_i \\ \\ \vdots \\ \\ C_n \\ \\ \end{array} \begin{array}{c} e_{11} \\ e_{12} \\ \vdots \\ e_{1m_1} \\ e_{i1} \\ e_{i2} \\ \vdots \\ e_{im_i} \\ e_{n1} \\ e_{n2} \\ \vdots \\ e_{nm_n} \end{array} \begin{bmatrix} W_{11} & \cdots & W_{1j} & \cdots & W_{1n} \\ \vdots & & \vdots & & \vdots \\ W_{i1} & \cdots & W_{ij} & \cdots & W_{in} \\ \vdots & & \vdots & & \vdots \\ W_{n1} & \cdots & W_{nj} & \cdots & W_{nn} \end{bmatrix} \tag{2.8}$$

where C_n denotes the nth cluster (top level objective), e_{nm} represents the mth element in the nth cluster, and W_{ij} is the principal eigenvector of the influence of the elements in the jth cluster compared to the i_{th} cluster. If the jth cluster has no influence on the jth cluster, then $W_{ij}=0$. W_{ij} (Eq. (2.9) represents the values of priority vectors of elements from the cluster C_i in relation to elements from the cluster C_j.

$$W_{ij} = \begin{vmatrix} w_{i1}^{j1} & w_{i1}^{j2} & \cdots & w_{i1}^{jm_j} \\ w_{i2}^{j1} & w_{i2}^{j2} & \cdots & w_{i2}^{jm_j} \\ \vdots & \vdots & \vdots & \vdots \\ w_{im_i}^{j1} & w_{im_i}^{j2} & \cdots & w_{im_i}^{jm_j} \end{vmatrix} \tag{2.9}$$

ANP can be used to model a problem that needs to be presented by a hierarchic or a network structure and then establish a pairwise comparison relationship within the structure. It allows for dependence and includes independence. It has the ability to prioritize groups or clusters of elements. It can handle interdependence better than AHP and "can support a complex, networked decision-making with various intangible criteria" (Tsai et al, 2010, p. 3884). However, ANP has two disadvantages: firstly, it is difficult to provide correct network structure among criteria even for experts, and different structures lead to different results. Secondly, to form a supermatrix all criteria have to be pair-wise compared with regard to all other criteria, which is difficult and also unnatural (Yu and Tzeng 2006).

4 Technique for Order Preference by Similarity to an Ideal Solution (TOPSIS)

TOPSIS, as one of the classical decision making methods for solving MCDM problems, was developed by Hwang and Yoon (1981). The method is based on the principle that the chosen alternative should have the farthest Euclidean distance from the negative ideal solution (NIS), and the shortest from the positive ideal solution (PIS). More specifically the solution that maximizes the benefit criteria and minimizes the cost criteria will be selected as the best (Zouggari and Benyoucef 2012). TOPSIS can be sometimes used to replace AHP in the process of ranking the alternatives. In other words, it is often the case that the AHP is used to assign the weight of the selection criteria while the TOPSIS is applied to prioritise the selection alternatives. The procedure of the TOPSIS method contains six steps.

Step1 Identify alternatives and criteria to establish a decision making matrix

A decision matrix D can be established to record data and it can be expressed as below:

$$D = \begin{array}{c} \\ A_1 \\ \vdots \\ A_i \\ \vdots \\ A_m \end{array} \begin{array}{c} \begin{array}{ccccc} C_1 & \cdots & C_j & \cdots & C_n \end{array} \\ \begin{bmatrix} x_{11} & \cdots & x_{12} & \cdots & x_{1n} \\ \vdots & & \vdots & & \vdots \\ x_{i1} & \cdots & x_{ij} & \cdots & x_{in} \\ \vdots & & \vdots & & \vdots \\ x_{m1} & \cdots & x_{mj}\cdots & & x_{mn} \end{bmatrix} \end{array} \tag{2.10}$$

$$W = \left[w_1 \ldots w_j \ldots w_n\right]$$

where each A_i represent alternative i considered; C_j is the criterion used to measure the performance of each alternative; and x_{ij} is the rating of the ith alternative with respect to the jth criterion. w_j is the subjective importance estimation of the jth criterion which is defined by the decision makers.

Step 2 Normalize the decision making matrix

The decision making matrix can be normalized through Eq. (2.11).

$$R_{ij} = \overline{x_{ij}} \Big/ \sqrt{\sum_{k=1}^{n} x_{ik}^2}, (i = 1, 2, \ldots m), (k = 1, 2, ..j..n) \tag{2.11}$$

Step 3 Construct weighted normalized fuzzy decision matrix

The weighted normalized decision matrix can be constructed by multiplying the normalized decision matrix R_{ij} with the associated weights w_j as follows:

$$V_{ij} = w_j R_{ij} \tag{2.12}$$

Step 4 Determine PIS and NIS

The PIS and NIS can be expressed as:

$$\begin{aligned} V^* &= \left\{ \left(\sum \max V_{ij}, j \in B \right), \left(\sum \min V_{ij}, j \in C \right) \right\} \\ &= \left\{ V_1^*, V_2^*, \ldots V_n^* \right\} \\ V^- &= \left\{ \left(\sum \min V_{ij}, j \in B \right), \left(\sum \max V_{ij}, j \in C \right) \right\} \\ &= \left\{ V_1^-, V_2^-, \ldots V_n^- \right\} \end{aligned} \tag{2.13}$$

where B and C indicate the sets of benefit and cost criteria respectively. V^* stands for the values of PIS, whereas V^- is for NIS.

Step 5 Obtain the separation measures

The separation of each alternative from the PIS and NIS can be represented by the Euclidean distance using the following equations.

$$\begin{aligned} S^* &= \sqrt{\sum_{j=1}^{n} \left(V_{ij} - V_j^* \right)^2} \qquad (i = 1, 2, \ldots, m) \\ S^- &= \sqrt{\sum_{j=1}^{n} \left(V_{ij} - V_j^- \right)^2} \qquad (i = 1, 2, \ldots, m) \end{aligned} \tag{2.14}$$

Step 6 Determine the relative closeness of the investigated alternatives to the PIS

$$P_i = S^- / (S^* + S^-) \tag{2.15}$$

Once the values for relative closeness of all alternatives are obtained, they can then be ranked based on the P_i in descending order. The higher the P_i value, the closer an alternative is to the PIS.

TOPSIS is a popular method due to its simplicity and the ability to identify the best alternative quickly. It is also useful for both qualitative and quantitative data. The output can be a preferential ranking using both negative and positive criteria (Aly et al. 2013). However there are some disadvantages of the traditional TOPSIS method. For example, the Euclidean distance algorithm it uses does not consider the correlation of attributes. The weight coefficients acquired by expert judgements have arguably subjective bias (Wang et al. 2015).

5 VIsekriterijumska optimizacija i KOmpromisno Resenje (VIKOR)

VIKOR[1] stands for **VI**sekriterijumska optimizacija i **KO**mpromisno **R**esenje in Serbian language developed by Serafim Opricovic, who called it the compromise ranking method. VIKOR is an outranking method for a finite set of alternative actions to be ranked and selected among criteria and solves a discrete multi-criteria problem with non-commensurable and conflicting criteria. As Opricovic and Tzeng (2003: p.228) and Opricovic (2011, p.12984) stated, the origin of the VIKOR method should be credited to Duckstein and Opricovic (1980) who started to develop it with the help of L_p-metric that Yu (1973) introduced in the compromise programming method (Zeleny 1973). In other words, it focuses on asking and selecting the best from a set of alternatives, and determines compromise solutions to a problem with conflicting criteria, which can help the decision makers to reach a final decision. The compromise solution is a feasible solution which is the closest to the ideal. However, most literature of VIKOR indicated that the method is rooted in Opricovic's book (1998) published in Serbian (Visekriterijumska optimizacija sistema u gradjevinarstvu), implying that it was written in English by quoting it as 'Multicriteria Optimization of Civil Engineering Systems" in their publications. As a result, it causes as if the book would be easily available in English for latecomers' reference. VIKOR as one of the MCDA methods has been further popularized by Opricovic and Tzeng (2004 and 2007). An extension of VIKOR to determine fuzzy compromise solution for multicriteria is presented in Opricovic (2007).

Both the VIKOR and the TOPSIS methods are distance-based. However, a compromise solution in VIKOR is determined based on mutual concessions, while, in TOPSIS, the best solution is determined by the shortest distance from the PIS and the farthest distance from the NIS, without the consideration of relative importance of these distances (Opricovic and Tzeng 2007). A detailed comparative analysis of TOPSIS and VIKOR can be found in Opricovic and Tzeng (2004). Assuming that there are m alternatives, denoted as $A_1, A_2, \ldots, A_m$. For an alternative A_i, the merit of the jth aspect is denoted by $x_{ij}, j = 1,2,\ldots,n$. Then, the procedure of traditional VIKOR for compromise-ranking can be described as the following steps:

Step 1 Determine the best x_j^* and the worst x_j^- values of all criterion functions, where $j = 1,2,\ldots,n$;

[1]Not a few of its acronyms and full names in the existing literature have typos. The last author of this chapter has confirmed them from Opricovic by email.

$$\begin{array}{c} x_j^* = \max\left[(x_{ij})|j = 1,2,\ldots,n\right],\ x_j^- = \min\left[(x_{ij})|j = 1,2,\ldots,n\right], \\ \text{if the } j^{th} \text{ criterion represents a benefit;} \\ x_j^* = \min\left[(x_{ij})|j = 1,2,\ldots,n\right], x_j^- = \max\left[(x_{ij})|j = 1,2,\ldots,n\right], \\ \text{if the } j^{th} \text{ criterion represents a cost.} \end{array} \tag{2.16}$$

Step 2 Compute the values of S_i and R_i, where, $i = 1,2,\ldots,m$, by the relations

$$S_i = \sum_{j=1}^{n} w_j \frac{x_j^* - x_{ij}}{x_j^* - x_j^-} \tag{2.17}$$

$$R_i = \max\left[w_j \frac{x_j^* - x_{ij}}{x_j^* - x_j^-} |j = 1,2,\ldots,n\right] \tag{2.18}$$

where, w_j is the weight of the jth criterion. S_i and R_i denote the utility measure and the regret measure, respectively, for the alternative A_i.

Step 3 Compute the value Q_i, where, $i = 1,2,\ldots,m$, by the relations

$$Q_i = v\left(\frac{S_i - S^*}{S^- - S^*}\right) + (1 - v)\left(\frac{R_i - R^*}{R^- - R^*}\right) \tag{2.19}$$

$$\begin{array}{c} S^* = \min[(S_i)|i = 1,2,\ldots,m], S^- = \max[(S_i)|i = 1,2,\ldots,m]; \\ R^* = \min[(R_i)|i = 1,2,\ldots,m], R^- = \max[(R_i)|i = 1,2,\ldots,m] \end{array} \tag{2.20}$$

where, v is the weight for the strategy of maximum group utility and 1-v is the weight of the individual regret. v is usually set to 0.5 (Opricovic 1998).

Step 4 Rank the alternatives, sorting by the S, R and Q values in a decreasing order. The results are three ranking lists. The less the value of Q_i is, the better decision of the alternatives A_i is.

Step 5 Propose a compromise solution, the alternative (A') which is ranked the best by the minimum value of Q, if the following two conditions are satisfied:

C1. "Acceptable advantage": $Q(A'') - Q(A') \geq D$, where A'' is the alternative with second position in the ranking list by Q; $DQ = 1/(m - 1)$ and m is the number of alternatives.

C2. "Acceptable stability in decision making": The alternative A' should also be the best in terms of S and/or R value (The lower the value of S/R is, the better).

If one of these conditions is not satisfied, it is not possible to select directly the best solution of the set but a subset of preferable options can be defined, which consists of (Opricovic and Tzeng 2007):

1. Alternatives A' and A'' if only the condition C2 is not satisfied, or
2. Alternatives A', A'',..., $A^{(M)}$ if the condition C1 is not satisfied, where $A^{(M)}$ is determined by the relation $Q(A^M) - Q(A') < DQ$ for maximum M.

The VIKOR calculates the ratio of positive and negative ideal solution, and thus proposes a compromise solution with an advantage rate. Therefore, it is particular helpful in a situation where the decision maker is not able to express their preference at the beginning of decision-making process. It has been applied for dealing with MCDM problems in various fields including design and manufacturing, marketing, supply chain management, and risk management, to name just a few (Yazdani and Graeml 2014).

6 Elimination Et Choix Traduisant la REalité (ELECTRE)

The acronym of ELECTRE stands for ELimination Et Choix Traduisant la REalité (ELimination and Choice Expressing the REality) (Benayoun et al. 1966). As a family member of MCDM methods, the ELECTRE method was originated in the mid-1960s at the European consultancy company SEMA for commercial reasons (Sevkli 2010). It was initially devised for choosing the best action from a given set of alternatives, and was later referred to as ELECTRE I. This approach has evolved into a number of variants, such as ELECTRE II (Roy and Bertier 1973), ELECTRE III (Roy 1978), and ELECTRE TRI (Yu 1992), for the purpose of different types of problems being addressed, such as choosing, ranking or sorting. Figueira et al. (2005) stated more detailed introduction of different ELECTRE methods, as well as their history, developments, and main features. ELECTRE is based on the study of outranking relations between alternatives, taking two at a time. Concordance and discordance indexes are used to analyse such relations, which can be viewed as the measures of satisfaction and dissatisfaction of a decision maker when choosing one alternative over the other. Assume there are m alternatives and n decision criteria for a MCDM problem, and each alternative is evaluated with respect to n criteria. The decision matrix can be donated as the same by Eq. (2.10) (in Sect. 4). Then, the ELECTRE method can be summarised as follows.

The decision matrix $D=(x_{ij})_{m\times n}$ is firstly normalised through Eq. (2.11). A weighted normalized decision matrix $V=(v_{ij})_{m\times n}$ can then be constructed by multiplying the normalized one with the associated weights using Eq. (2.12). The procedures for normalizing and weighting a decision matrix exactly follow the first three steps in the TOPSIS method, detailed information on how to obtain a weighted normalized decision matrix is no longer repeated in this section.

After that, concordance and discordance sets are determined. For each pair of alternatives A_p and A_q ($p,q=1,2,\ldots$,m and,$p\neq q$), the set of criteria is divided into two distinct subsets. In terms of the criteria against which alternative A_p is preferred to alternative A_q, the concordance set is composed as

$$C(p,q)=\{j|v_{pj}>v_{qj}\} \tag{2.21}$$

$C(p,q)$ is the collection of attributes where A_p is better than, or equal, to A_q. On completing $C(p,q)$, the discordance set $D(p,q)$, as a complement of $C(p,q)$, is obtained by investigating the criteria against which A_p is better than A_q. it can be presented as

$$D(p,q) = \{j|v_{pj} < v_{qj}\} \tag{2.22}$$

The concordance index of $C(p,q)$ is generated by adding the values of weights of concordance set elements, defined as

$$C_{pq} = \sum_{j^*} w_{j^*} \tag{2.23}$$

where j^* are the attributes contained in the concordance set $C(p,q)$. The discordance index $D(p,q)$ represents the degree of disagreement in $A_p \to A_q$ (it means that alternative A_p outranks A_q, which indicates a situation where performance values of A_p are better or at least equal than those offered by A_q in respect of the majority of criteria) in Eq. (2.24):

$$D_{pq} = \frac{\sum_{j^+} | v_{pj^+} - v_{qj^+} |}{\sum_{j} | v_{pj} - v_{qj} |} \tag{2.24}$$

where j^+ are the attributes contained in the discordance set $D(p,q)$. This method implies that A_p outranks A_q when $C_{pq} \geq \bar{C}$ and $D_{pq} \leq \bar{D}$, where, threshold values $\bar{C}$ and $\bar{D}$ are usually set by decision makers (Sevkli 2010).

One main weakness of ELECTRE methods is that threshold values for the concordance and discordance indices are usually decided according to decision makers' opinion, which brings in subjectivity. However, as important members belonging to family of outranking methods, ELECTRE methods are still popular despite its existence for more than four decades, and its application can be seen in, for example, energy management (Mousavi et al. 2017), supply chain management (Fahmi et al. 2016), and risk assessment (Govindan and Jepsen 2016).

7 Preference Ranking Organization METHods for Enrichment Evaluation (PROMETHEE)

ROMETHEE is a popular MCDM outranking method dealing with the evaluation problems, originally introduced by Brans (1982). Brans et al. (1984) elaborated the method as a new family member of outranking methods in multi-criteria analysis.

Brans and Vincke (1985) further developed it with sophisticated mathematical reasoning and published their work in *Management Science*. Brans and Mareschal (1994) introduced a decision support system and visual software, named as PROMCALC & GAIA, showing some examples and requisites to demonstrate the applications of PROMETHEE in reality.

The PROMETHEE method contains PROMETHEE I for partial ranking of alternatives, PROMETHEE II for complete ranking, PROMETHEE III for ranking based on interval, and PROMETHEE IV for ranking in continuous viable solutions. Other members include PROMETHEE V (Brans and Mareschal 1992), PROMETHEE VI (Brans and Mareschal 1995), PROMETHEE GDSS (Macharis et al. 1998), and visual interactive module GAIA (Geometrical Analysis for Interactive Aid) for graphical representation (Brans and Mareschal 2005), and etc. On a review paper on PROMETHEE (Behzadian et al. 2010), 217 papers using the method were investigated from 100 journals in the period of 1985–2009.

The core idea of PROMETHEE is the pairwise comparison of alternatives along each recognized criterion, taking the inner relationships of each evaluation facts into account. It derives a (partial or complete) ranking of a finite set of feasible alternatives based on a positive outranking flow, a negative outranking flow and a net outranking flow. PROMTHEE is capable of addressing decision-makers' evaluation problems through reasonable normalization, thus avoiding inconsistent ranking results with the characteristic functions, and providing them with visual software so as to easily deal with the evaluation problems and sensitive analysis. Given its structure, this method allows a direct operation on the variables included in a decision matrix, without requiring any normalization. PROMETHEE II, which presents the fundamental to the implementation of other PROMETHEE methods, consists the following steps (Behzadian et al. 2010):

Step 1 Evaluate the alternatives with respect to the criteria (assuming there m alternatives and n criteria), and determine the deviations based on pair-wise comparisons:

$$d_j(a,b) = g_j(a) - g_j(b) \tag{2.25}$$

where, a and b are two alternatives, and $d_j(a,b)$ denotes the difference between the evaluations of a and b on each criterion.

Step 2 Calculate the preference between the alternatives a and b via function:

$$P_j(a,b) = F_j\left[d_j(a,b)\right],\ j = 1,2,\ \ldots,n \tag{2.26}$$

where $P_j(a,b)$ denotes the reference of alternative a with regard to alternative b against the jth criterion, as a function of $d_j(a,b)$. F_j is a preference function, which translates the difference between the evaluations of alternatives a and b on the jth criterion into a preference degree ranging from 0 to 1. There are six basic types of preference functions as proposed by Brans and Vincke (1985) including usual criterion, U-shape criterion, V-shape criterion, level criterion, V-shape with indifference criterion and Gaussian criterion.

Step 3 Calculate the overall or global preference index

$$\pi(a,b) = \sum_{j=1}^{n} P_j(a,b)w_j \tag{2.27}$$

where, w_j is the weight of the jth criterion, and $\pi(a,b)$ of a over b is defined as the weighted sum of $p(a,b)$ for each criterion.

Step 4 Calculated the positive and negative outranking flows

$$\phi^+(\mathrm{a}) = \frac{1}{\mathrm{m}-1}\sum_{\mathrm{x}\in\mathrm{A}} \pi(\mathrm{a},\mathrm{x}) \tag{2.28}$$

$$\phi^-(a) = \frac{1}{m-1}\sum_{x\in A} \pi(x,a) \tag{2.29}$$

where A is a collection of alternatives. The partial outranking can be obtained from the two ranks induced by ϕ^+ (positive outranking flow) and ϕ^-(negative outranking flow). a outranks b if $\phi^+(a) \geq \phi^+(b)$ and $\phi^-(a) \leq \phi^-(b)$. Otherwise, it will result to an indifference relation or incomparability of the two alternatives.

Step 5 Calculate the net outranking flow and the complete ranking.

$$\phi(a) = \phi^+(a) - \phi^-(a) \tag{2.30}$$

where $\phi(a)$ denotes the net outranking flow for each alternative. The higher the net flow is, the better the alternative.

PROMETHEE does not provide the possibility to really structure a decision problem, which may increase the difficulty for the decision maker to obtain a clear view of the targeted problem. However, it has unique advantages when important elements of the decision are difficult to quantify or compare, as criteria scores can be expressed in their own units. Moreover, it needs much less inputs compared to other MCDM methods (Gavade 2014). Its extensive application in fields such as environment management, business management, and logistics and transportation has been discussed by Behzadian et al. (2010).

8 Evidential Reasoning (ER)

The theory of evidence was first generated by Dempster (1967) and further developed by Shafer (1976). The Dempster-Shafer theory of evidence or D-S theory was originally used for information aggregation in expert systems as an approximate reasoning tool (Buchanan and Shortliffe 1984; Mantaras 1990) and then used in decision making under uncertainty and risk in contrast to Bayes decision theory (Yager 1992; Yager 1995). ER is developed on the basis of the D-S theory. The use of ER as a decision making tool has been widely reported in the literature. An important achievement of applying ER to decision analysis is to incorporate it into traditional MCDM methods for addressing the degree of belief associated with subjective judgements. The lack of data, the inability of assessors to provide precise judgements, or the failures of some assessors to provide judgements in group decision-making can result in an incomplete assessment (Yang and Xu 2002). An ER based decision making approach for MCDM problems with both qualitative and quantitative criteria under uncertainty was developed in the early 1990's (Yang and Singh 1994; Yang and Sen 1994). The kernel of such an approach is an ER algorithm, which was generated by Yang and Singh (1994), later updated by Yang and Sen (1994) and further modified by Yang (2001) and Yang and Xu (2002). ER is applied for ranking alternatives or selecting the best compromise alternative in a process, in which both quantitative and qualitative attributes are simultaneously satisfied as much as possible (Yang and Singh 1994). Several applications of this approach are addressed in the maritime related literature (Yang 2001; Sii et al. 2001; Yang et al. 2005; Yang et al. 2009a; Yang et al. 2014).

The utilization of the ER algorithm with belief structure can be explained as follows.

Suppose there is a two level hierarchy structure, a top level attribute E consists of L attributes at the lower level which include all the attributes influencing the assessment of the E. Lower level attributes can be represented as:

$$E = (e_1, e_2, \ldots e_i, \ldots e_L).$$

The weight of each attributes can be expressed as $w = (w_1, w_2, \ldots w_i \ldots w_L)$. w_i is the normalized relative weight for the ith attribute (e_i) where $0 \leq w_i \leq 1$.

Suppose there are N evaluation grades, each H_n ($n = 1, 2, \ldots N$) is a standard grade for assessing an attribute. Without loss of generality, it is assumed that $H_{(n+1)}$ is preferred to H_n. A given assessment for e_i ($i = 1, 2, \ldots L$) of an alternative can be represented as:

$$S(e_i) = \left\{ \left(H_n, \beta_{n,i} \right), n = 1, 2, \ldots N \right\} \tag{2.31}$$

where $\beta_{n,i} \geq 0$ and denotes to the degree of belief associated with the evaluation grade H_n for the attribute e_i. An assessment $S(e_i)$ is complete if $\sum_{n=1}^{N} \beta_{n,i} = 1$ and incomplete if $\sum_{n=1}^{N} \beta_{n,i} < 1$.

The basic probability assignments for each attribute can then be calculated as below.

Let $m_{n,i}$ be a basic probability mass representing the degree to which the ith attribute supports the hypothesis that the top level attribute is assessed to the nth evaluation grade. Then $m_{n,i}$ can be obtained as follows:

$$m_{n,i} = w_i \beta_{n,i} \qquad (n = 1, 2, \ldots N) \tag{2.32}$$

Let $m_{H,i}$ be the remaining probability mass unassigned to any individual grade after e_i has been assessed. $m_{H,i}$ can be calculated as follows:

$$m_{H,i} = 1 - \sum_{n=1,}^{N} m_{n,i} = 1 - w_i \sum_{n=1}^{N} \beta_{n,i} \quad (i = 1, 2 \ldots L) \tag{2.33}$$

$m_{H,i}$ contains $\bar{m}_{H,i}$ and $\tilde{m}_{H,i}$, where $\bar{m}_{H,i}$ represents the remaining probability mass that other attributes (apart from the ith attribute) contribute in the assessment. $\tilde{m}_{H,i}$ is the unassigned probability mass due to the possible incompleteness in the assessment. They can be expressed as follows:

$$\bar{m}_{H,i} = 1 - w_i \text{ and } \tilde{m}_{H,i} = w_i \left(1 - \sum\nolimits_{n=1}^{N} \beta_{n,i}\right) \tag{2.34}$$

$m_{H,i}$ can therefore be presented as:

$$m_{H,i} = \bar{m}_{H,i} + \tilde{m}_{H,i} \tag{2.35}$$

The probability assignment for an attribute can be combined as follows.

Let $m_{n,I(1)} = m_{n,1}$ $(n = 1, 2, \ldots N)$, $\bar{m}_{H,I(1)} = \bar{m}_{H,1}$, $\tilde{m}_{H,I(1)} = \tilde{m}_{H,1}$ and $m_{H,I(1)} = m_{H,1}$. A factor $K_{I(i+1)}$ is used to normalize $m_{n,I(i+1)}$ and $m_{H,I(i+1)}$ so that $\sum_{n=1}^{N} m_{n,i} + m_{H,I(i+1)} = 1$.

$$K_{I(i+1)} = \left[1 - \sum\nolimits_{t=1}^{N} \sum\nolimits_{\substack{j=1 \\ j \neq t}}^{N} m_{t,I(i)} m_{j,i+1}\right]^{-1} (i = 1, 2 \ldots L - 1) \tag{2.36}$$

The combined probability assignment $m_{n,I(L)}$ $(n = 1, 2, \ldots N)$, $\bar{m}_{H,I(L)}$, $\tilde{m}_{H,I(L)}$ and $m_{H,I(L)}$ can be generated as follows.

$$\{H_n\} : m_{n,I(i+1)} = K_{I(i+1)} \left[m_{n,I(i)} m_{n,i+1} + m_{H,I(i)} m_{n,i+1} + m_{n,I(i)} m_{H,i+1}\right] \tag{2.37}$$

$$\begin{aligned} \{H\}: m_{H,I(i)} &= \tilde{m}_{H,I(i)} + \bar{m}_{H,I(i)} \\ \tilde{m}_{H,I(i+1)} &= K_{I(i+1)}\left[\tilde{m}_{H,I(i)}\tilde{m}_{H,i+1} + \bar{m}_{H,I(i)}\tilde{m}_{H,i+1} + \tilde{m}_{H,I(i)}\bar{m}_{H,i+1}\right] \\ \bar{m}_{H,I(i+1)} &= K_{I(i+1)}\left[\bar{m}_{H,I(i)}\bar{m}_{H,i+1}\right] \end{aligned} \tag{2.38}$$

The combined degrees of belief of all the lower level attributes for the assessment of the top level attribute can then be calculated. Let β_n denote a degree of belief that the top level attribute is assessed to the grade H_n, which is generate by combining the assessments for all the associated attribute e_i,$(i = 1, 2 \ldots L)$. β_n can be calculated by:

$$\begin{aligned} \{H_n\}: \beta_n &= \frac{m_{n,I(L)}}{1 - \bar{m}_{H,I(L)}} \quad (n = 1, 2, \ldots N) \\ \{H\}: \beta_H &= \frac{\tilde{m}_{H,I(L)}}{1 - \bar{m}_{H,I(L)}} \end{aligned} \tag{2.39}$$

The overall assessment for the top level attribute E can be represented by Eq. (6.22).

$$S(E) = (H_n, \beta_n) \quad (n = 1, 2, \ldots N) \tag{2.40}$$

ER is capable of dealing with problems with both quantitative and qualitative criteria. It introduces the concepts of belief structure and belief decision matrix, which makes it possible to model uncertainties of various types of nature in a unified format for further analysis without resorting to sensitivity analysis. However there are some criticisms in the application of ER. Processing the data in a belief decision matrix by hand is rather difficult. But this issue is largely addressed through the development of the IDS software. In addition, interpreting the outcome represented by a belief structure is not as straight forward as interpreting a simple score (Xu 2012).

9 Fuzzy Logic

Fuzzy logic was first conceptualized by Zadeh in 1965 (Zadeh 1965). Recognizing the reality that many criteria involved in a decision making process are far from precise or clear, the idea of applying fuzzy logic into MCDM has been widely discussed for more than two decades. Fuzzy logic is a superset of conventional Boolean logic with extensions to account for imprecise information. Instead of crisp membership of a set, its membership is fuzzy or imprecise. Fuzzy logic permits vague information, knowledge and concepts to be used in an exact mathematical manner. Linguistic variables such as “definite”, “likely”, “average”, “unlikely” and “impossible” are necessary media used to describe continuous and overlapping states. This enables qualitative and imprecise reasoning statements to be incorporated with fuzzy algorithms or fuzzy rule bases producing simpler, more intuitive and better-behaved models. Fuzzy logic is based on the principle that every crisp value belongs to all relevant fuzzy sets to various extents, called the degrees of membership. Pure fuzzy logic has extremely limited applications in business (the only popularised application is the Sony Palmtop) and the main use of fuzzy logic is as an underlying logic system for fuzzy expert decision making systems (Pai et al. 2003). It has been successfully applied for a wide range of single and MCDM problems. For instance, Chou (2010) proposes a fuzzy MCDM methodology for solving the container transhipment hub port selection dilemma under fuzzy environment. Wang and Lee (2010) utilize a fuzzy MCDM method for evaluating the financial performance of container shipping companies based on extended fuzzy preference relation and using linguistic weights. The application of fuzzy logic in MCDM becomes more compelling when being combined with AHP, (Bulut et al. 2012; Hsu 2012; Chao and Lin 2011), TOPSIS (e.g. Yeh and Chang 2009; Durbach and Stewart, 2012; Yang and Wang 2013; Kannan et al. 2014), VIKOR (e.g. Kaya and Kahraman 2010; Shemshadi et al. 2011), ELECTRE (e.g. Sevkli 2010; Chen and Xu 2015), and PROMETHEE (e.g. Shirinfar and Haleh 2011; Gupta et al. 2012; Tavakoli et al. 2013).

10 Conclusion

This chapter has introduced eight most popular MCDM methods, which rank a finite set of alternatives with respect to their frameworks, algorithms, and advantages and disadvantages. It has also presented their applications in the literature mainly within the context of shipping, port and logistics. The readers can therefore have a better understanding of their own applicability and suitability. Having said that, this chapter has laid down a platform for the following chapters in this book, which focus on new applications of MCDM methods as well as their hybrid approach in maritime and logistics areas.

Acknowledgements The authors would like to thank the EU FP7 Marie Curie IRSES project "ENRICH" (612546) for its financial support to this research.

References

Akyuz E, Karahalios H, Celik M (2015) Assessment of the maritime labour convention compliance using balanced scorecard and analytic hierarchy process approach. Marit Policy Manag 42(2):145–162

Aly MF, Attia HA, Mohammed AM (2013) Integrated fuzzy (GMM)-TOPSIS model for best design concept and material selection process. Int J Innov Res Sci Eng Technol 2:6464–6486

Ananda J, Herath G (2009) A critical review of multi-criteria decision making methods with special reference to forest management and planning. Ecol Econ 68(10):2535–2548

Beşikçi EB, Kececi T, Arslan O, Turan O (2016) An application of fuzzy-AHP to ship operational energy efficiency measures. Ocean Eng 121:392–402
Buchanan BG, Shortliffe EH (1984) Rule based expert systems: the mycin experiments of the stanford heuristic programming project. Addison-Wesley, Reading
Bulut E, Duru O, Kececi T, Yoshida S (2012) Use of consistency index, expert prioritization and direct numerical inputs for generic fuzzy-AHP modeling: a process model for shipping asset management. Expert Syst Appl 39(2):1911–1923
Behzadian M, Kazemzadeh RB, Albadvi A, Aghdasi M (2010) PROMETHEE: a comprehensive literature review on methodologies and applications. Eur J Oper Res 200(1):198–215
Benayoun R, Roy B, Sussman B(1996) ELECTRE: Une méthode pour guider le choix en présence de points de vue multiples. Note de travail 49, SEMAMETRA International, Direction Scientifique
Brans JP (1982) L'ingenierie de la decision. Elaboration dinstruments daide a la decision. Methode PROMETHEE. In: Nadeau R, Landry M (eds) Laide a la Decision: Nature, Instrument set Perspectives Davenir. Presses de Universite Laval, Quebec, pp 183–214
Brans JP, Mareschal B (1992) PROMETHEE V: MCDM problems with segmentation constraints. INFOR: Inf Syst Oper Res 30(2):85–96
Brans JP, Mareschal B (1994) The PROMCALC and GAIA decision support system for MCDA. Decis Support Syst 12:297–310
Brans JP, Mareschal B (1995) The PROMETHEE VI procedure: how to differentiate hard from soft multicriteria problems. J Decis Syst 4(3):213–223
Brans JP, Mareschal B (2005) PROMETHEE methods. In: Multiple criteria decision analysis: state of the art surveys. Springer, New York, pp 163–196
Brans JP, Mareschal B, Vincke P (1984) PROMETHEE: a new family of outranking methods in multicriteria analysis. In: Brans JP (ed) Operational research, IFORS 84. North-Holland, Amsterdam, pp 477–490
Brans JP, Vincke P (1985) A preference ranking organization method (the PROMETHEE method for multiple criteria decision making). Manag Sci 31(6):647–656
Carlos Perez-Mesa J, Galdeano-Gomez E, Salinas Andujar JA (2012) Logistics network and externalities for short sea transport: an analysis of horticultural exports from southeast Spain. Transp Policy 24:188–198
Chao S-L, Lin Y Jr (2011) Evaluating advanced quay cranes in container terminals. Transp Res Part E-Logist Transp Rev 47(4):432–445
Chen N, Xu Z (2015) Hesitant fuzzy ELECTRE II approach: a new way to handle multi-criteria decision making problems. Inf Sci 292:175–197
Chen TL, Lin KL (2006) Complementing AHP with habitual domains theory to identify key performance indicators for service industry. In: IEEE international conference on service operations and logistics and informatics, pp 84–89
Chou C-C (2010) Application of FMCDM model to selecting the hub location in the marine transportation: a case study in southeastern Asia. Math Comput Model 51(5–6):791–801
Chou CC, Ding JF (2016) An AHP Model for the choice of ship flag: a case study of Tawanese shipowners. Int J Marit Eng 185:A61–A68
Dempster AP (1967) Upper and lower probabilities induced by a multivalued mapping. Ann Math Stat 38(2):325–339
Duckstein L, Opricovic S (1980) Multiobjective optimization in river basin development. Water Res J 16(1):14–20
Durbach IN, Stewart TJ (2012) Modeling uncertainty in multi-criteria decision analysis. Eur J Oper Res 223(1):1–14
Fahmi A, Kahraman C, Bilen Ü (2016) ELECTRE I Method using hesitant linguistic term sets: an application to supplier selection. Int J Comput Intell Syst 9(1):153–167
Figueira J, Mousseau V, Roy B (2005) ELECTRE methods. In: Multiple criteria decision analysis: state of the art surveys. Springer, New York, pp 133–153

Gavade RK (2014) Multi-criteria decision making: an overview of different selection problems and methods. Int J Comput Sci Inf Technol 5(4):5643–5646

Govindan K, Jepsen MB (2016) Supplier risk assessment based on trapezoidal intuitionistic fuzzy numbers and ELECTRE TRI-C: a case illustration involving service suppliers. J Oper Res Soc 67(2):339–376

Gupta R, Sachdeva A, Bhardwaj A (2012) Selection of logistic service provider using fuzzy PROMETHEE for a cement industry. J Manuf Technol Manag 23(7):899–921

Hwang C-L, Yoon K (1981) Multiple attributes decision making methods and applications. Springer, Berlin

Hsu W-KK (2012) Ports' service attributes for ship navigation safety. Saf Sci 50(2):244–252

Kannan D, Lopes De Sousa Jabbour AB, Chiappetta Jabbour CJ (2014) Selecting green suppliers based on GSCM practices: using fuzzy TOPSIS applied to a Brazilian electronics company. Eur J Oper Res 233(2):432–447

Karahalios H, Yang ZL, Williams V, Wang J (2011) A proposed system of hierarchical scorecards to assess the implementation of maritime regulations. Saf Sci 49(3):450–462

Kaya T, Kahraman C (2010) Multicriteria renewable energy planning using an integrated fuzzy VIKOR & AHP methodology: the case of Istanbul. Energy 35(6):2517–2527

Lam JSL, Dai J (2012) A decision support system for port selection. Transp Plan Technol 35 (4):509–524

Macharis C, Brans JP, Mareschal B (1998) The GDSS PROMETHEE procedure – a PROMETHEE–GAIA based procedure for group decision support. J Decis Syst 7:283–307

Mantaras RLD (1990) Approximate reasoning models. Ellis Harwood Ltd., Chichester

Mousavi M, Gitinavard H, Mousavi SM (2017) A soft computing based-modified ELECTRE model for renewable energy policy selection with unknown information. Renew Sust Energ Rev 68:774–787

Opricovic S (1998) Multicriteria optimization of civil engineering systems (in Serbian, Visekriterijumska optimizacija sistema u gradjevinarstvu). Faculty of Civil Engineering, Belgrade

Opricovic S (2007) A fuzzy compromise solution for multicriteria problems. Int J Uncertain Fuzziness Knowledge-based Syst 15(3):363–380

Opricovic S (2011) Fuzzy VIKOR with an application to water resources planning. Expert Syst Appl 38:12983–12990

Opricovic S, Tzeng GH (2003) Multicriteria expansion of a competence set using genetic algorithm. In: Tanino T, Tanaka T, Inuiguchi M (eds) Multi-objective programming and goal-programming: theory and applications. Springer, Berlin/Heidelberg, pp 221–226

Opricovic S, Tzeng GH (2004) Compromise solution by MCDM methods: a comparative analysis of VIKOR and TOPSIS. Eur J Oper Res 156(2):445–455

Opricovic S, Tzeng GH (2007) Extended VIKOR method in comparison with outranking methods. Eur J Oper Res 178(2):514–529

Pai RR, Kaflepalh VR, Caudill RJ, Zhou MC, Ieee I (2003) Methods toward supply chain risk analysis. In: 2003 I.E. international conference on systems, man and cybernetics, vols 1–5, conference proceedings, pp 4560–4565

Pohekar SD, Ramachandran M (2004) Application of multi-criteria decision making to sustainable energy planning—a review. Renew Sust Energ Rev 8(4):365–381

Pomerol JC, Romero SB (2000) Multicriterion decision in management: principles and practice. 1st ed. (2000) (trans: James C from French). Kluwer, Norwell

Roy B (1978) ELECTRE III: un algorithme de classements fonde sur une representation floue des preference en presence de criteres multiples. Cahiers de CERO 20(1):3–24

Roy B, Bertier P (1973) La methode ELECTRE II: une methode au media-planning. In: Ross M (ed) Operational research. North-Holland, Amsterdam, p 1972

Saaty TL (1980) The analytic hierarchy process. McGraw-Hill, New York

Saaty TL (1990) How to make a decision – the analytical hierarchy process. Eur J Oper Res 48 (1):9–26

Saaty TL (1995) Transport planning with multiple criteria: the analytic hierarchy process applications and progress review. J Adv Transp 29(1):81–126
Sahin B, Senol YE (2015) A novel process model of marine accident analysis by using generic fuzzy-AHP algorithm. J Navig 68:162–183
Sevkli M (2010) An application of the fuzzy ELECTRE method for supplier selection. Int J Prod Res 48(12):3393–3405
Shafer G (1976) A mathematical theory of evidence. Princeton University Press, Princeton
Shemshadi A, Shirazi H, Toreihi M, Tarokh MJ (2011) A fuzzy VIKOR method for supplier selection based on entropy measure for objective weighting. Expert Syst Appl 38 (10):12160–12167
Shirinfar M, Haleh H (2011) Supplier selection and evaluation by Fuzzy multi-criteria decision making methodology. Int J Ind Eng Prod Res 22(4):271–280
Sii HS, Ruxton T, Wang J (2001) A Fuzzy-logic-based approach to subjective safety modelling for maritime products. Reliab Eng Syst Saf 73:19–34
Sii HS, Wang J (2003) A design-decision support framework for evaluation of design options/proposals using a composite structure methodology based on the approximate reasoning approach and the evidential reasoning method. Proc Inst of Mech Eng Part E-J Process Mech Eng 217(E1):59–76
Tavakoli M, Tabriz AA, Farahani R, Rezapour E (2013) Application of fuzzy goal programming and F-PROMETHEE approaches in evaluating and selecting the best suppliers in supply chain. J Basic Appl Sci Res 3(2):1115–1127
Tsai W, Leu J, Liu J, Lin S, Shaw M (2010) A MCDM approach for sourcing strategy mix decision in IT projects. Expert Syst Appl 37(5):3870–3886
Ugboma C, Ibe C, Ogwude IC (2004) Service quality measurements in ports of a developing economy: Nigerian ports survey. Manag Serv Qual Int J 14(6):487–495
Ung ST, Williams V, Chen HS, Bonsall S, Wang J (2006) Human error assessment and management in port operations using fuzzy AHP. Mar Technol Soc J 40(1):73–86
Vaidya OS, Kumar S (2006) Analytic hierarchy process: an overview of applications. Eur J Oper Res 169(1):1–29
Wang P., Li, Y., Wang, Y.H. and Zhuo, Z.Q. 2015. A New method based on TOPSIS and response surface method for MCDM problems with interval numbers. Math Probl Eng 2015, 11, http://dx.doi.org/10.1155/2015/938535
Wang YJ, Lee HS (2010) Evaluating financial performance of Taiwan container shipping companies by strength and weakness indices. Int J Comput Math 87:38–52
Xie X, Xu D-L, Yang J-B, Wang J, Ren J, Yu S (2008) Ship selection using a multiple-criteria synthesis approach. J Mar Sci Technol 13(1):50–62
Xu DL (2012) An introduction and survey of the evidential reasoning approach for multiple criteria decision analysis. Ann Oper Res 195(1):163–187
Yager RR (1992) On the specificity of a possibility distribution. Fuzzy Sets Syst 50(3):279–292
Yager RR (1995) An approach to ordinal decision-making. Int J Approx Reason 12(3–4):237–261
Yang JB, Sen P (1994) A general multi-level evaluation process for hybrid MADM with uncertainty. IEEE Trans Syst Man Cybern 34(10):1458–1473
Yang JB, Singh MG (1994) An evidential reasoning approach for multiple-attribute decision-making with uncertainty. IEEE Trans Syst Man Cybern 24(1):1–18
Yang JB, Xu DL (2002) On the evidential reasoning algorithm for multiple attribute decision analysis under uncertainty. IEEE Trans Syst Man Cybern Part a-Syst Hum 32(3):289–304
Yang JB (2001) Rule and utility based evidential reasoning approach for multiattribute decision analysis under uncertainties. Eur J Oper Res 131(1):31–61
Yang W-E, Wang J-Q (2013) Multi-criteria semantic dominance: a linguistic decision aiding technique based on incomplete preference information. Eur J Oper Res 231(1):171–181
Yang ZL, Mastralis L, Bonsall S, Wang J (2009a) Incorporating uncertainty and multiple criteria in vessel selection. Proc Inst Mech Eng Part M-J Eng Marit Environ 223(M2):177–188

Yang ZL, Ng AKY, Wang J (2014) Incorporating quantitative risk analysis in port facility security assessment. Transp Res A Policy Pract 59:72–90
Yang ZL, Wang J, Bonsall S, Fang QG (2009b) Use of fuzzy evidential reasoning in maritime security assessment. Risk Anal 29(1):95–120
Yang Z, Wang J, Bonsall S, Yang J-B, Fang Q-G (2005) A subjective risk analysis approach for container supply chains. Int J Autom Comput 2(1):85–92
Yazdani M, Graeml FR (2014) VIKOR and its applications: a state-of-the-art survey. Int J Strateg Decision Sci (IJSDS) 5(2):56–83
Yeh C-H, Chang Y-H (2009) Modeling subjective evaluation for fuzzy group multicriteria decision making. Eur J Oper Res 194(2):464–473
Yeo G-T, Ng AKY, Lee PT-W, Yang Z (2014) Modelling port choice in an uncertain environment. Marit Policy Manag 41(3):251–267
Yeo GT, Song DW, Dinwoodie J, Roe M (2010) Weighting the competitiveness factors for container ports under conflicting interests. J Oper Res Soc 61(8):1249–1257
Yu PL (1973) A class of solutions for group decision problems. Manag Sci 19(8):936–946
Yu R, Tzeng GH (2006) A soft computing method for multi-criteria decision making with dependence and feedback. Appl Math Comput 180:63–75
Yu W (1992) ELECTRE TRI (aspects méthodologiques et manuel d'utilisation). Document-Université de Paris-Dauphine, LAMSADE
Zadeh LA (1965) Fuzzy sets. Inf Control 8(3):338–353
Zavadskas EK, Turskis Z (2011) Multiple criteria decision making (MCDM) methods in economics: an overview. Technol Econ Dev Econ 17(2):397–427
Zeleny M (1973) Compromise programming. In: Cochrane JL, Zeleny M (eds) Multiple criteria decision making. University of South Carolina Press, Columbia, pp 262–375
Zimmermann HJ (2000) An application-oriented view of modeling uncertainty. Eur J Oper Res 122(2):190–198
Zouggari A, Benyoucef L (2012) Simulation based fuzzy TOPSIS approach for group multi-criteria supplier selection problem. Eng Appl Artif Intell 25(3):507–519

Chapter 3
Analytic Hierarchy Process (AHP) in Maritime Logistics: Theory, Application and Fuzzy Set Integration

Emrah Bulut and Okan Duru

Abstract In the last few decades, there is a growing interest in using Analytic Hierarchy Process (henceforth AHP), and it is frequently employed in solving the maritime industry problem since the 2000s. The AHP method is a powerful instrument to decompose complex decision-making problems and to simplify (facilitate) decision makers' cognitive burden. In contrast to its predecessors, AHP is capable of executing both hard and soft information (i.e. numerical data/input and subjective/judgemental assessment respectively) through a top-down investigation of micro aspects in each level of the hierarchy. Although AHP is very functional and popular in both academia and professional life, there are various biases and misuse of the method which are heavily based on the lack of theoretical basis. The AHP method has several underlying assumptions, and each assumption needs to be investigated and demonstrated through specific decision making problems. Ignoring these fundamentals of AHP eventually initiates various forms of inconsistencies and sometimes implicit invalidity which is difficult to detect from derived results. In this chapter, the theory of AHP will be discussed in detail with references to other theories in social sciences and its practical impacts on the AHP analysis. In addition to the conventional AHP methodology, the fuzzy set extension (Fuzzy AHP or FAHP) and its rationale in particular problems will be investigated. Empirical applications will help clarifying its capability of solving some maritime and logistics problems while developing hands-on experience with numerical examples.

Keywords Analytic Hierarchy Process • Fuzzy logic • Decision theory • Rational choice theory • Consistency control

---

E. Bulut
Department of Business Administration, Yildiz Technical University, Istanbul, Turkey

O. Duru (✉)
School of Civil and Environmental Engineering, Nanyang Technological University, Singapore, Singapore
e-mail: okanduru@ntu.edu.sg

P.T.-W. Lee, Z. Yang (eds.), *Multi-Criteria Decision Making in Maritime Studies and Logistics*, International Series in Operations Research & Management Science 260,
DOI 10.1007/978-3-319-62338-2_3

1 Introduction

The outranking approach has been developed and applied in MCDM problems for several decades, and its pure numerical framework was the only instrument till expert judgement and soft knowledge orientation has been pioneered by fathers of modern MCDM literature. The key difference between the era of outranking and multi-attribute utility analysis relies on the uncertainty and quantification of expertise. For many engineering problems with numerical inputs and tangible features, outranking methods such as Preference Ranking Organization METHod for Enrichment of Evaluations (PROMETHEE) or ELimination Et Choix Traduisant la REalité (ELimination and Choice Expressing Reality, ELECTRE) would be the most suitable and practical solutions. However, the most of real world problems need to be investigated with soft information and immeasurable expert engagement. Subjective attributes such as colour, the perception of convenience and design can only be derived by expert/user consultation and surveys. Although the problem has been well-known for centuries, a parametric approach capable of simplifying complex decision making environment (e.g. lots of attributes, interlocking features) as well as reducing the cognitive load on decision makers was not developed till the 1980s. With the growing academic interest and customer oriented business philosophies, collecting data about user experiences and perception has become an integral part in MCDM literature.

In the mid-1980s, Thomas L. Saaty proposed the AHP after studying on queuing theory and linear optimisation for two decades (Saaty 1980). If we have a look his academic background and biography, decision making problems become one of his academic interests while he worked for the U.S. Arms Control and Disarmament Agency and dealt with conflict resolution. His engagement with complex decision problems seems as a sparking point and key motivation. Conflict resolution is a term used for processes resolving conflicts through negotiation, arbitration, and diplomacy etc. Conflicts usually consist of gains and losses for different regimes and solutions. Therefore, conflict resolution can be classified as a multi-attribute utility problem.

The AHP is based on three major principles of human behaviour and cognitive operation: (1) *understanding a complex problem by the decomposition of various elements*, (2) *comparison of features for measuring their impact*, and (3) *synthesis (gathering knowledge and creating collective arguments)*. Similar to many other decision making instruments, one of the early applications of the AHP was about a transport problem, development of a transport plan for Sudan. Quality Function Deployment (QFD) was also developed by Yoji Akao based on an identical decomposition framework, and its core instrument, House of Quality, first utilised in 1972 for the design of an oil tanker at Mitsubishi Heavy Industries Shipyard (Kobe, Japan) (Akao and Mazur 2003). After several decades, Thomas L. Saaty, the father of AHP, has been awarded by the Akao Prize of the QFD Institute.

By the introduction of the method, the number of scholarly publications addressing AHP has grown exponentially, and more than 30,000 publications just in 6 years (2010–2016) have mentioned the method according to the statistics by Google Scholar search metrics (dated May 11th, 2016). With the development of academic publishing and growing number of scholars in different fields, the AHP method has been applied to various kinds of MCDM problems and tested for various real life challenges. In contrast to its usefulness and performance, AHP is applied improperly in many studies ignoring the fundamental principles and assumptions behind the method. The huge volume of ill-defined and misrepresented AHP applications has created its own illusory literature which could in turn mislead young scholars in particular. Therefore, this chapter aims not only to illustrate relevant uses of AHP in solving maritime and logistics problems, but also to clarify its core principles to rehabilitate future research in a deceptive scholarly publishing environment.

The application process of the AHP can be classified as the following stages (Saaty 2000):

1. *Pre-survey and consultation for defining the complex decision problem,*
2. *Decomposition of complex problem into attributes,*
3. *Classification and selection of attributes,*
4. *Setting up the decision hierarchy,*
5. *Designing the survey platform and its content (representation of survey),*
6. *Defining a group of experts for the AHP survey,*
7. *Collecting responses (data) from pre-defined experts,*
8. *Consistency Control Loop:*

   Monitoring the reliability of survey responses (consistency control),
   Redesign and reapplication of the AHP survey if consistency check is failed,

9. *Post-survey solution of the hierarchy (calculating relative weights),*
10. *Representation of final results, interpretation and final decision making.*

Before approaching to the application stage, it would be timely to have a look and review theoretical basis of the AHP and potential failures caused by ignoring these theoretical background. According to our literature review, vast number of the AHP studies violate or ignore theoretical foundations of the AHP which eventually conclude impractical or irrational academic practices. Therefore, we strongly encourage scholars and readers of this chapter to review the theory of AHP and its linkages to practical applications. Hence, the theory and essential prerequisites of an AHP application are discussed in the following section (Sect. 2). In Sect. 3, the traditional AHP method is investigated through its algorithm and an application to the dry port location selection problem. The dry port location problem also introduces one of motivations behind the Fuzzy extended AHP approach. Increasing number of criteria and alternatives also means that an extreme number of pairwise comparison at the precision of Saaty scale. Section 4 investigates the Fuzzy AHP (FAHP) and its application to the shipping asset selection problem. There are various fuzzification approaches in the AHP, but Chang (1996)'s synthetic extent analysis is preferred in this chapter since it is the most cited version of the FAHP varieties. Finally, a holistic assessment of the AHP and FAHP algorithms and their effectiveness is presented in the conclusion (Sect. 5).

2 Theory of AHP

The major contribution of Thomas Saaty with the discovery of AHP is the decomposition of the complex problems. While previous techniques like ELECTRE or The Technique for Order of Preference by Similarity to Ideal Solution (TOPSIS) also decompose objective problems, AHP improved the approach one step forward by integrating intangible aspects. AHP also proposed a hierarchy design which classifies key attributes and their sub-attributes as well as alternatives in a single top-down illustration (See Fig. 3.1). A typical AHP hierarchy begins with the objective at the top, then a number of criteria/attribute is placed below the objective. A number of sub-attribute may be defined below a certain criterion. Finally, alternatives for ranking or selection are listed below the entire hierarchy, and each alternative is connected to the bottom of each vertical flow of attributes. Among the entire AHP analysis process, structuring and designing the hierarchy are probably the most critical stages since the AHP hierarchy figures out the whole problem. Posterior interpretations are developed based on this initial hierarchy of attributes.

In AHP applications, a designer (scholar) frequently pay little attention to the structure of the hierarchy and the selection of relevant attributes. On the other hand, the most critical and costly stage is actually building of the hierarchy. Once a designer defines the hierarchy, significant time and effort are spent on empirical stages, and any inconsistency at these stages will result in loss of time and money for redesign and reapplication of the entire experiment. Therefore, we would like to pay our most attention to this stage of the AHP and emphasize the theoretical underpinnings of the method.

The AHP method relies on a number of theories which validate the empirical work, and their absence completely invalidates all efforts spent for the analysis. Among these theories, *the Rational Choice Theory* (Gilboa 2010) plays a central role in the method. Without understanding the principles of rational choice theory, one may easily overlook the problem and ignore some fundamentals. The six pillars (axioms) of the rational choice theory can be classified as follows[1]:

- *Completeness*: An individual must be able to choose among these possibilities of

  (i) *a_i is preferred to a_j,*
  (ii) *a_j is preferred to a_i,*
  (iii) *or individual is indifferent between alternatives.*

  where a_i and a_j are two independent alternatives for a choice problem.

[1]More details on the rational choice theory may be found at Gilboa (2010).

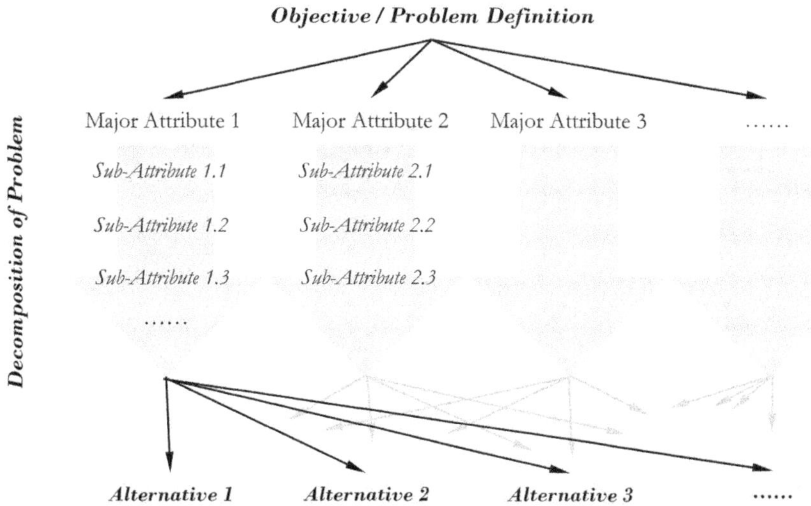

Fig. 3.1 Hierarchy design for the AHP analysis

In other words, individuals are decisive among alternatives. There is no indecision, and individuals can define their choices (*Reflexivity* is a sub-form of this axiom, and therefore it is ignored here). '*Decline to state*' is not a relevant option. In a mathematical term, there is no pairwise disjoint among subsets in a preference relation matrix.

- *Transitivity*: Preferences must be internally consistent. If a_i is preferred to a_j, and a_j is preferred to a_k, then a_i is preferred to a_k (a_i, a_j and a_k are three independent alternatives). In more precise format, the level of preference must be consistent too. For example, *A* is three times better than *B*, and *B* is two times better than *C*, then *A* must be six times (2 × 3) better than *C* (*A*, *B* and *C* are three independent alternatives)
- *No Decoy Effect* or *Robustness of Preferences*: When transitivity is ensured, the ranking of alternatives must not be changed when an alternative is excluded. Considering the example used to illustrate transitivity, *A* must be preferred to *B* even when *C* is excluded. If individual prefers *B* to *A* when only *A* and *B* is presented, then a decoy effect is present in the problem. A popular example of decoy effect in marketing may be found at the subscription offers by *The Economist Magazine* (See Fig. 3.2). With the given bundle of options, no reader is expected to choose print-only or digital-only offers. The presentation of two irrational options (print-only and digital-only) even at the same prices (Fig. 3.2a) is just for rationalizing the 'best value' option. In Singapore sales (Fig. 3.2b), the decoy effect is much clearer with a slight price difference. For example, excluding the digital-only option (Fig. 3.2b) may significantly change customers' choice (Print-only at S$55 versus Print + Digital at S$70).

Fig. 3.2 Subscription offers for *The Economist Magazine* in U.S.A. (**a**) and Singapore (**b**) *(Source: https://subscriptions.economist.com/ – Dated: Mar 28th, 2017)*

- *Freedom of Choice* and *No Externality*: Choices are freely made without any kind of disruptive incentives or external (third party) effects. For example, the presentation and framing of questions in an AHP survey should not affect the independent choice of individuals.
- *Full Information*: Experts should be really experts with full information about the context and the problem.

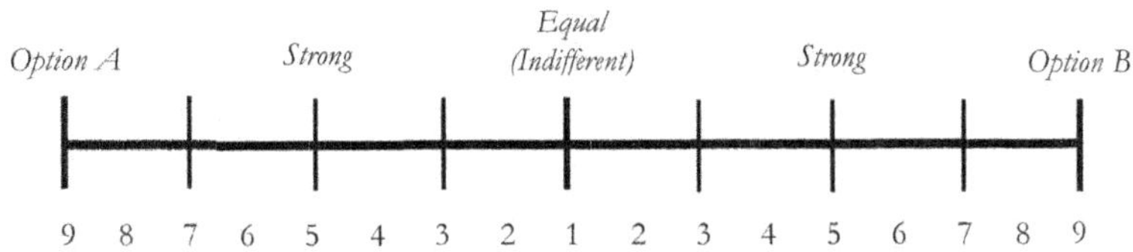

Fig. 3.3 Pairwise comparison format based on 9-point Likert Scale (i.e. *the Saaty Scale*) (Saaty 1977)

- *Non-satiation*: Individual does not satiate at any level of satisfaction and does not ignore remaining preferential judgements. For example, if the expert has three alternatives to compare (e.g. assume A > B > C) and feel satiated with the second best alternative, then the first two alternatives may be indifferent irrationally (e.g. A = B). If an individual never experienced a first class seat at any airline company, then any first class seat in any airline company may be indifferent and as good as any other first class seat. Such an expert is not relevant for given problem.

Each of these axioms has reflections in the AHP application and analysis. Among these axioms, completeness is the most passive assumption behind the AHP. In the traditional format, individuals are given a pairwise comparison survey, and therefore, they are asked to prefer the level of choice (any of the options) or indifference (Fig. 3.3). By completing the survey, an individual ensures the completeness of preferences. However, remaining assumptions behind the AHP analysis need to be investigated thoughtfully for eliminating biased results.

2.1 Transitivity Axiom, Scale Selection Problem and Rank Reversal

The transitivity axiom is the leading challenge of the AHP approach, and it is frequently addressed that the conventional AHP matrices may not validate the axiom. The essential problem behind the challenge arises from the measurement instrument, i.e. Likert Scale. Transitivity axiom obviously requires unbounded scale since any bounded scale will have spillovers out of the boundary to ensure the axiom itself. For example, *A* may be *five* times better than *B* (selecting Likert Scale-5), and *B* may also be *five* times better than *C*, then *A* must be *25* times (5×5) better than *C* which is not in line with 9-point bounded Likert Scale. Therefore, the AHP method suffers from scale intransitivity. A significant volume of the literature dealt with the scale intransitivity problem (Hanbin and Nuanchen 2010; Ji and Jiang 2003; Lootsma 1993; Ma and Zheng 1991; Salo and Hämäläinen 1997) while there is still no unique scale that can completely eliminate the intransitivity problem. The scale selection problem is not only connected with the scale intransitivity, but it should also comply with the '*no decoy effect*' axiom.

Another key discussion about AHP in the literature is originated from the *Rank Reversal* problem (Saaty 1987; Millet and Saaty 2000; Schenkerman 1994; Wang and Elhag 2006). In the traditional form of AHP, a preference matrix (a_{ij}) is composed of individuals' pairwise comparisons. The ranking of preferences with a_{ij} is expected to be robust even when an alternative is excluded from the problem (e.g. $a_{i-1,j-1}$). We call this problem as decoy effect (also known as the attraction effect) in marketing research (Dhar and Simonson 2003). However, the rank reversal problem is not a result of decoy or attraction, but it is probably initiated by sub-grouping of the entire preference matrix and limits of memory and execution of human brain. When individuals (experts) respond to AHP surveys, each pairwise comparison is evaluated separately, and they do not need to worry about logical consistency of their responses. Therefore, an individual may have different conclusions about the same problem in different sub-sections of it. An average AHP survey has 20–30 pairwise comparisons, and one may easily be mentally disconnected through survey responses. The downside of pairwise comparison surveys is individuals need to respond a large number of questions which requires a significant amount of mental workload and encourages irrational choices (one of rationales of FAHP is the mental workload. Fuzzification process is a clustering approach to the Saaty scale which in turn converts the scale into linguistic choices. See Sect. 4 for details).

2.2 *Mental Accounting and Irrationality*

In the last few decades, behavioural economics have developed significantly, and prospect theory plays the central role in the emergence and development of this new school of economics (Kahneman and Tversky 1979). The fundamental premise of the behavioural school is individuals (economic actors) do not perform rationally at all, but irrationally or somewhat rationally based on the context and problem. Bounded rationality concept also draws attention to several aspects and drivers of decision making task and overrules the conventional view of rational economic actor assumption which extremely idealizes the decision makers' cognitive environment (Simon 1982).

Mental accounting is an interesting topic in the behavioural school, and it tells us another story about the nature of preferences (Thaler 1985). There are a number of different scenarios to illustrate the mental accounting. In a popular example, individual needs to consider whether to purchase a product at *$20* or walk a quarter mile and buy it at *$10* in another store. In most cases, subjects prefer to walk for *$10* discount. What if an individual purchases a *$1000* product while other store sells it at *$990*? Subjects usually do not walk a quarter mile for the same amount of discount. Mental accounting definitely shows us the perception of the problem may suddenly change if the context and numbers are manipulated.

The reflection of this perception uncertainty to the AHP approach is individuals may have different perceptions in different sub-sections of the same problem based on different levels of satisfaction and/or its perception. When comparing three alternatives, one may prioritize an item since it is very close to individual's satisfaction level while undervaluing increments over this level. In 2004, A Handymax bulk carrier was sold around *$30 m* while a Panamax bulk carrier was sold at *$40 m* which means *$10 m* value for an extra *20,000-ton* cargo capacity. In the same year, a Capesize bulker (180 k dwt) was sold at *$65 m* which meant *$25 m* value for extra *100,000-ton* cargo capacity comparing to a Panamax bulker. The additional cargo capacity grows five times while its price does not triple. In 2015, the context is absolutely different (*$13,5 m, $14 m* and *$25 m* for Handymax, Panamax and Capesize bulkers respectively). There is certainly a market effect composed of idle tonnages, parcel size trends etc. Here the point is even physical circumstances may lead individuals perceive and respond differently.

Therefore, the pitfalls of perception-led inconsistencies may be tolerated as long as the preference matrices ensure a certain level of consistency in numerical resolution (to the best of our knowledge). On the other hand, workload-led inconsistency may amplify the bias, and the AHP moderators (i.e. researchers) should pay particular attention to the design and presentation of surveys to eliminate imprecise and lax responses. Otherwise, surveys may pass the consistency control (Cronbach alpha or Saaty's traditional consistency algorithm) while final results may be implicitly biased. When an AHP hierarchy has roughly 5–6 criteria and 5–6 alternatives, that means around a 100 pairwise comparisons. For a common person, this level of mental workload would be unfeasible to expose preferences satisfactorily.

Thomas Saaty (2006), the inventor of AHP method, proposed an alternative approach, *rank from ratings* which does not require pairwise comparisons. Saaty (2006) classifies two ranking approaches: *Relative measurement* based on pairwise comparisons and *absolute measurement* based on ratings (scoring). Rank from ratings is an absolute measurement approach, and it is very useful when the mental workload is very high. According to Saaty (2006), rank from ratings can be employed when the best preference and indifference among preferences are known to decision makers.

2.3 Choice Architecture, Framing and Externalities

Making '*free*' choices is an essential need for any MCDM approaches. That freedom of choice does not only imply social pressure, but it also adds up to presentation of questions and choices. In cognitive science, it is usually addressed in the context of framing (Levin et al. 1988; Scheufele and Tewksbury 2007). In other words, the AHP surveys must be organized and presented fairly, and moderators should be cautious about manipulation and priming desired results. That may happen through the hierarchy building process (selection and/or omission of criteria) as well as the design of the AHP surveys (attention to conformity bias). Considering a location selection problem for logistics hub, one may have following four criteria (as an example):

- *Scale Economics (Volume of Cargo Flow, Bundling Opportunity),*
- *Transport Facilities/Infrastructure,*
- *Political Stability/Security,*
- *Network/Connectivity.*

From a holistic perspective, the second criterion, *Transport Facilities*, is expected to emphasise the quality and safety of transport infrastructure. At the given presentation and listing of criteria, individual will be first asked to compare *Scale Economics* to *Transport Facilities*. That initial comparison will probably prime the association between these two criteria, and transport facilities may be evaluated in terms of its capacity to ensure a large volume of transportation. Then, this initial association may establish a reference used in the entire AHP survey. If *Transport Facilities* is first compared to *Network/Connectivity*, then that association may lead another meaning of Transport Facilities based on its relevant connections to surrounding locations and markets. Even the order of criteria at the time of survey may manipulate individuals unexpectedly.

In many AHP surveys, individuals are usually in doubt about desired meanings of terms used in the surveys. Misconception of criteria eventually initiates irrelevant or misleading preference matrices. Therefore, terms and phrases of the AHP surveys need to be clarified by moderator to eliminate contradicting perceptions and biased responses. Moderators should pay particular attention to convey their message and desired meanings of terms (semantics) to individuals and be aware of equivocation bias. Otherwise, an implicit double-counting error may also be induced.

In addition to content related bias, individuals may be misled by external incentives and drivers. For example, in the above location selection problem, some stakeholders may prefer a closer alternative location while others tend to highlight another location closer to their business. Having such various views is not a direct problem for AHP surveys as long as the balance of subjects is considered. Desires may differ based on the context, working environment, business segment etc. Not only responses of individuals but also incentives behind them should be taken into account. Balancing incentives are expected to neutralise sided views, and aggregated outcome and/or a small group of unbiased responses (apart from subjects neutralised) may conclude the fair preference matrix (i.e. wisdom of crowd).

2.4 Full Information and Expert Competency

Selection of experts for an AHP panel is always problematic and always a black box of decision making methodologies. Both panel size (no. of experts) and level of expertise (usually the background of expert or their current employment) are difficult to judge, and even academic journals are not able to test expert-dimension of various problems except few fundamental requirements. Expert must be selected in the context of the objective problem, and that can be monitored broadly. For general topics (e.g. national elections), anyone may be classified as a relevant individual. However, in maritime transport and logistics, individuals are expected from the industry or relevant academic/public institutions (e.g. ministry of transport). The size of expert panel is quite subjective and contextual measure while it is usually required to be over ten people at least (just a rough requirement based on practice). In some experiments, that number reaches to 100. On the other hand, there is no empirical and theoretical rationale whether outcomes would be much better along with the size of expert panel. The profile of individuals may be investigated as an indicator of expertise, but the size of the panel is unavoidably ignored since there is no objective and empirical evidence opposing the dimension. Based on Cochran (1977), there is a statistical estimation of sample size while it is a general recommendation for relatively definite and non-expert population. In the MCDM studies, the size of population is frequently indefinite, and samples are usually expert groups rather than lay people (such as in political surveys).

'*Full information*' axiom is an essential component of rationality assumption which also generated the theory of asymmetric information in economics. According to the axiom, individuals are assumed having full information about the objective in a rational choice. Rejection or ignorance of the axiom will conclude that:

Result (final selection or ranking) is incomplete;
and/or
There are ignored circumstances.

Considering the sensitivity of the AHP results (e.g. rank reversal), the level of expertise may play a significant role, and contradicting rankings may be found with a different sample of experts.

Prioritisation of expertise at the aggregation stage may be a countermeasure against unstable preferences (Bulut et al. 2012). The level of expertise can be classified based on years in the profession, consistency level of individuals or other relevant indicators. Then, expertise indicator may be utilised to prioritise responses taken from different individuals. Measuring expertise needs a kind of test in fact while AHP surveys do not have such testing procedures. When it is applicable, the moderator may conduct a pre-survey test prepared for the problem area. Once that group is classified, then the same group may be asked further AHP surveys in the same context, and pre-defined expertise ratings may be utilised again.

2.5 Non-satiation

Among other axioms of the rational choice theory, non-satiation is less known, and little emphasis is put on its rationale. Econometric analysis and fundamental theory of finance also assume non-satiation of economic actors. In addition to that scarcity is known as a central idea in economics since Lionel Robbins (1932) proposed the definition of economics based on limited resources and unlimited needs (i.e. non-satiation). Experiences and expectations of individuals eventually develop a range of possibilities, tastes, and cognitive depth. When it comes to ranking or comparison, an expert is implicitly assumed that:

- *Having experiences with compared items;*
- *Having a sense of the indifference gap between compared items.*

If we were asked to compare various meals of several countries, we could not response properly if we did not taste all of these meal options. Our answer would be robust and consistent as much as the level of our experience. Therefore, non-satiation is somewhat connected with the level of expertise. As a principle of the AHP method, every individual responding a pairwise comparison survey is assumed having previous experience of alternatives and criteria or individuals should be able to reflect an indirect knowledge which is capable of interpretation and evaluation of relevant concepts and variables.

2.6 Anonymity and Iteration of AHP Surveys

The anonymity of individuals is a key attribute of another group decision making method, DELPHI. A variation of DELPHI, Estimate-Talk-Estimate (ETE) is a relaxation of anonymity rule between iterations of DELPHI. Various combinations of AHP and DELPHI are proposed and employed in the literature (Robbins 1932; Tavana et al. 1993; Prasad and Somasekhara 1990; Joshi et al. 2011). In AHP practices, anonymity is frequently ensured involuntarily to increase response rate in surveys. Similar to DELPHI method, an iterative process may be utilized in AHP surveys too. In the fundamental AHP analysis, consistency control is required to be performed for each matrix, and if consistency is not sufficient (e.g. *0.01* in Saaty's consistency control), the corresponding section of the survey (or entire survey) is needed to be repeated. If the moderator introduces data about the first survey (the first iteration) to the expert group, then it is classified as an integration of the AHP and DELPHI. In DELPHI surveys, the objective is consensus, and once consensus is ensured, no iteration is needed further.

Although anonymity is not required in principle, similar circumstances exist in AHP surveys, and it would be much better to ensure anonymity. Direct interaction between individuals may cause groupthink bias or conformity bias (Muchnik et al. 2013) (See Sect. 2.7).

2.7 Group Homogeneity Problem and Need for Heterodoxy

Another dimension of the problem lies in the fact that groups may fail if the level of consensus and cohesiveness is very high. This phenomenon is called *groupthink bias* (also the tyranny of groupthink) (Callaway and Esser 1984; McCauley 1989; Esser 1998). Under the group homogeneity, aggregated preference of survey group may be completely biased, impractical or illogical. In fact, a certain level of heterodoxy may be healthy for AHP analysis, and it may well cover different perspectives and drawbacks. Homogeneity is also connected with the selection of experts. When experts are preferred from a closer small group with similarities, then the moderator may well achieve precision (strong consensus) while ignoring the accuracy (Fig. 3.4b) whereas accuracy is the fundamental objective of the AHP study. As it is seen in Fig. 3.4c, in some AHP surveys, the moderator may not ensure high level of precision. However, averaging AHP responses may indicate an accurate outcome.

The consensus is an indicative measure in terms of robustness of AHP studies. Under weaker consensus, AHP rankings and final weights will be more sensitive and uncertain. Rank reversal problem may be found more frequently in weak consensus. Therefore, consensus (precision) is a dimension that is required for a successful AHP application while it may not be achieved strongly in various applications. When the average of expert group is consistent and robust, many problems of AHP applications will not be diagnosed numerically.

According to Davey (2017), disagreement and encouraging productive conflict would improve the accuracy of decision making than a strong consensus. False consensus can be much costly in managerial decision making (Marks and Miller 1987).

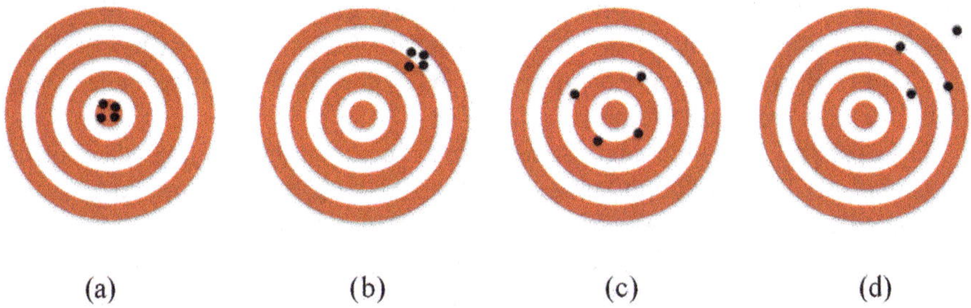

Fig. 3.4 Accuracy vs. precision: high accuracy, high precision (**a**); low accuracy, high precision (**b**); high accuracy, low precision (**c**); low accuracy, low precision (**d**)

2.8 Hierarchy Building: Independence Principle and Omission Bias

Among various academic reviews by authors of this chapter for several scholarly journals in the last decade significantly addressed a common problem in AHP applications, misconception and violation of *independence principle*. There are even many invalid AHP studies published in scholarly journals. Independence principle is an essential rule in MCDM methods, but in AHP approach particularly. One of the key questions at the beginning of an AHP analysis is which criteria (also sub-criteria) must be included in the hierarchy.

About the selection of criteria, a moderator of the AHP survey may follow two major directions:

- *Criteria may be defined based on previous literature and knowledge of scholars (including moderator);*
- *A pre-survey is conducted to define criteria having an impact on the objective.*

In both ways, a relevant and valid bundle of criteria may be defined properly. However, a substantial detail, false choice of criteria, may change results entirely. For example, an AHP survey may be conducted to select a car among alternatives. Among criteria, there may be two criteria namely *engine size* and *0–100 mph time*. An average driver may know that these criteria are strongly connected, and actually engine size is the major cause of less time for 0–100 mph. So, we have two criteria which are strongly connected with a cause-effect relationship. A similar relationship exists between engine size and *fuel consumption*. Now critical questions are what happens if moderator goes with these criteria as is and how that situation can be debiased. Having dependent criteria causes '*double-counting*' problem which overvalues a particular aspect of the objective by counting twice with two criteria. For example, if there are five criteria in the problem while two of them are dependent, the aspect represented by these two criteria will have 15% more contribution (40% in five criteria −2/5- instead of 25% in four criteria −1/4-). One fourth of the problem becomes nearly half of it. With more dependent variables, the bias grows accordingly.

The opposite of double-counting problem is no counting bias, in other words, *omission bias*. When an important aspect of the AHP objective is ignored, a certain amount of explanatory power of the application may be reduced, and it is pretty difficult to monitor such biases and their impacts. Without gathering these ignored criteria and performing an AHP analysis, the loss of explanatory power cannot be estimated and tested.

Considering the independence principle, double-counting and omission bias, the selection of criteria and hierarchy building are more of a theoretical process which strongly relies on experiences, the theory of the objective and inclusion of variables as much as possible at the first place before classification and elimination of dependency.

As it is indicated in the process of the AHP, there are four fundamental steps before AHP surveys:

1. *Pre-survey and consultation for defining the complex decision problem,*
2. *Decomposition of complex problem into attributes,*
3. *Classification and selection of attributes,*
4. *Setting up the decision hierarchy.*

AHP studies usually refer to Step 4 and occasionally Step 3. Steps 1 and 2 are thought to be performed much before the experiment, and it is taken for granted by readers as well as reviewers. On the other hand, only content, that clarifies the AHP experiment in terms of the validity of criteria selection, exists in the first two steps of AHP applications.

There are a number of debiasing strategies for AHP applications. First, moderators should pay attention to collect a large bundle of criteria about their objective. For this step, they may perform interviews, reach out previous works in the field (not only AHP applications but literature in general), conduct some short surveys preferably with open-ended questions encouraging brainstorming among experts and also reach out media and lay press to find current topics.

Second, moderators may classify their collection of criteria and develop a cause-effect diagram of these criteria to visualise possible spillovers among them. However, a delicate process comes after such organisation of criteria, the definition of dependency. It is also difficult to disconnect all criteria in various problems. Criteria are usually connected to some degree, directly or indirectly. The judgement on the level of toleration is change-making. Possibly a committee of experts may define the level of toleration for dependency, and criteria out of such toleration may be classified dependent and grouped for a single representation. The DELPHI group decision making method is usually employed at this stage of the AHP analysis. Its consensus searching mechanism helps to reach an agreement on the selection of criteria.

2.9 Scaling and Perception of Indifference at the Final Score

At the final step, an AHP analysis defines final scores (weights) for each alternative. When the objective of AHP is just ranking them, then alternatives will be found in order of their score. In most case, AHP is used for selection, and the highest final score will point out the selection of expert group. On the other hand, the difference between final scores needs to be interpreted since it may be negligible which makes some of the alternatives indifferent. Two alternatives may have final scores of *0.19* and *0.20*. Does that *0.01* difference between them make the second alternative superior to the first one? Only expert judgement may scale the difference based on the context and perception of indifference. When it comes to survival in a life threatening disease, 0.01 difference between therapies would be a great value. On the other hand, if it is about the selection of a household vehicle that may be negligible, and even that may encourage decision maker to focus only a single criterion such as fuel consumption (mpg). When two alternatives are almost same in general, a cost item may be indicative in the selection process.

About the definition of indifference, there is no global arithmetic which fits all problems. Since it is very contextual, the expert may be asked to predefine an indifference value, or they may be asked to respond an additional short survey consisting only these similar alternatives. Due to rank reversal problem (See Sect. 2.1), an additional survey with two or more similar alternatives (moderator is in doubt about a small difference among them) may come up with significantly different results; one alternative may also gain a definite majority in an additional survey.

3 Understanding the Core Mechanism of AHP and Its Algorithms

Before discussing the AHP in more detail, it would be useful to figure out its structure and macro calculations. An AHP analysis consists of n number of criteria i.e. attributes of the objective (C_1, C_2,. . ., C_n) and t number of alternatives (A_1, A_2,. . ., A_t). In some cases, there are also m number of sub-criteria defined under a certain major criterion (SC_{n1}, SC_{n2},. . ., SC_{nm}). Through a top-down approach, relative weights of major criteria are defined at the first place, then sub-criteria under the associated major criteria, and finally alternatives under the criteria and sub-criteria if a criterion is decomposed. If a criterion is not decomposed into sub-criteria, then alternatives will be compared in terms of major criterion itself. Otherwise, criterion with sub-criteria will not be evaluated directly for alternatives.

Figure 3.1 illustrated the AHP hierarchy with both single major attributes and decomposed major attributes including sub-attributes. In Fig. 3.5, connections between sub-attributes of the first criterion and alternatives are presented as an example. Having more sub-attributes eventually complicates the problem further, and that reminds us potential risks of mental accounting in a high level of task load. Large web of criteria with sub-attributes is quite impractical to evaluate with an average capacity of human judgement.

For simplification, a four-criteria and four-alternative problem is selected to illustrate high-order math in AHP calculations. In the first step, the relative weights of criteria (w_1, w_2,. . ., w_4) are defined through pairwise comparison or rank by rating, and sum of these relative weights under the objective will always sum up to 1.00 (e.g. $w_1 + w_2 + w_3 + w_4 = 1.00$).

At the next stage, each alternative will have a unique relative weight under each criterion which is again defined by either pairwise comparison or rank by rating (Fig. 3.6). Global weights of each alternative in respect to criteria is a product of relative weights of alternatives under a criterion and its own relative weight (Fig. 3.7). The sum of these global weights presents the final weights of alternatives.

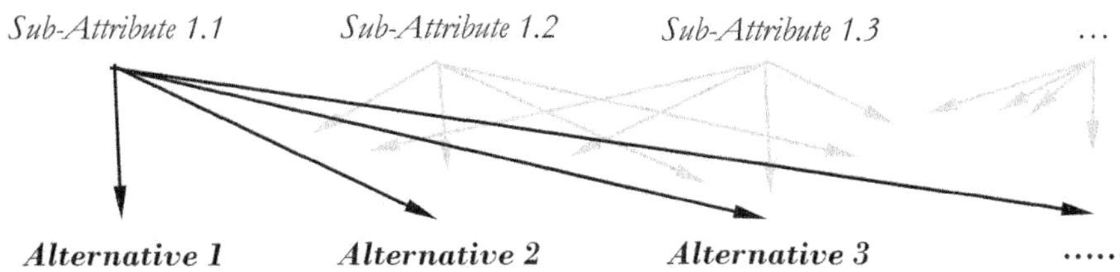

Fig. 3.5 Connection web between sub-attributes and alternatives

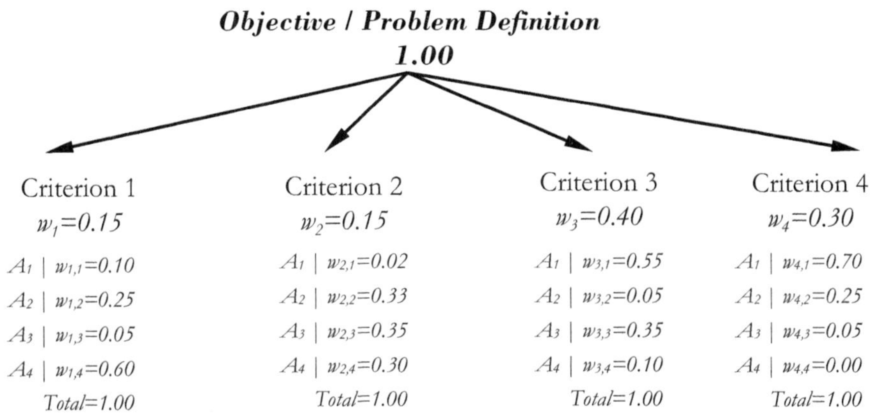

Fig. 3.6 Relative weights for alternatives with respect to each criterion

In this example, A_1 with a final score of 0.448 has the highest weight which corresponds to the first rank or the final selection.

As it is discussed in *Sect. 2.9*, the size of indifference may be worthy of consideration. Considering the results of given example, there is around 23% difference between the highest two scores (A_1 and A_2) which refer to a definite difference in the majority of AHP problems. A difference below roughly 2 ~ 3% may be reconsidered to avoid a selection failure.

3.1 The Formal Algorithm of AHP

As it is discussed previously, the aim of the AHP method is to obtain the expert's experience and knowledge to find out the best alternative with respect to each criterion and transforms the complex decision-making system into a simple elementary hierarchy system (Saaty 1977). Experts' judgement of the importance of one alternative over another can be made subjectively and usually converted to a numerical value using a scale of *1–9* (Fig. 3.3) where *1* denotes equal importance,

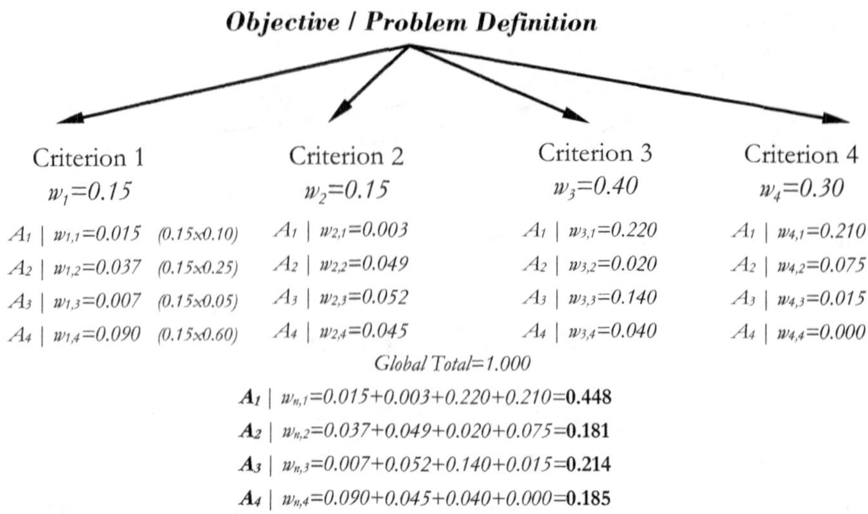

Fig. 3.7 Global weights for alternatives with respect to each criterion

and *9* denotes the highest degree of contribution/priority (Saaty 1977). Table 3.1 lists the possible judgements and their respective numerical values.

Some basic steps involved in the numerical application stage are as follows:

Step 1. Determination of the relative importance of the attributes and sub-attributes, if any.
Step 2. Determination of the relative standing (weight) of each alternative with respect to each sub-attribute, if applicable, and then successively with respect to each attribute.
Step 3. Determination of the overall priority weight (global score) of each alternative.
Step 4. Determination of consistency indicator(s) in making pairwise comparisons.
Step 5. Determination of ranking to find the best alternative.

Let's consider the $C_1, C_2, C_3,\ldots, C_i,\ldots, C_j,\ldots, C_n$, criteria at some one level in the hierarchy. In the next step, their weights of importance, $w_1, w_2, w_3,\ldots, w_i,\ldots, w_j,\ldots, w_n$, need to be computed by a researcher. Allow a_{ij} $(i, j = 1, 2, \ldots, n)$ to be the importance strength of C_i when compared to C_j. The matrix of these numbers a_{ij} is denoted A;

Table 3.1 Degree of preference between criteria and alternatives (Saaty 1977)

Linguistic preference	Numerical index
Equally important (compared items equally contribute to the objective)	*1*
Weakly important	*3*
Strongly important	*5*
Very strong important	*7*
Absolutely more important	*9*
Intermediate values between adjacent judgements	*2, 4, 6, 8*

$$A = \begin{bmatrix} a_{11} & a_{12} & \cdot & \cdot & \cdot & a_{1j} & \cdot & \cdot & \cdot & a_{1n} \\ \cdot & \cdot & & & & \cdot & & & & \cdot \\ \cdot & \cdot & & & & \cdot & & & & \cdot \\ \cdot & \cdot & & & & \cdot & & & & \cdot \\ a_{i1} & a_{i2} & \cdot & \cdot & \cdot & a_{ij} & \cdot & \cdot & \cdot & a_{in} \\ \cdot & \cdot & & & & \cdot & & & & \cdot \\ \cdot & \cdot & & & & \cdot & & & & \cdot \\ \cdot & \cdot & & & & \cdot & & & & \cdot \\ a_{n1} & a_{n2} & \cdot & \cdot & \cdot & a_{nj} & \cdot & \cdot & \cdot & a_{nn} \end{bmatrix}_{nxn}$$

where $a_{ji} = 1/a_{ij}$, and A is reciprocal. If decision maker's comparisons are logical for each criterion in the pairwise matrix, then $a_{ik} = a_{ij} * a_{jk}$ for all i, j, k and the matrix A is called perfectly consistent (Saaty 1980) (also proof for rational choice assumption).

An obvious case of a consistent matrix A is its elements;

$$a_{ij} = w_i/w_j(i,j = 1,2,\ldots,n) \tag{3.1}$$

Therefore, when the vector formed by each weighting multiplies matrix A;

$$w = (w_1, w_2, \ldots, w_n)^T \tag{3.2}$$

Finally;

$$Aw = \begin{bmatrix} w_1/w_1 & w_1/w_2 & . & . & . & w_1/w_j & . & . & . & w_1/w_n \\ w_2/w_1 & w_2/w_2 & . & . & . & w_2/w_j & . & . & . & w_2/w_n \\ . & . & & & & . & & & & . \\ . & . & & & & . & & & & . \\ & & & & & . & & & & . \\ w_i/w_1 & w_i/w_2 & . & . & . & w_i/w_j & . & . & . & w_i/w_n \\ . & . & & & & . & & & & . \\ . & . & & & & . & & & & . \\ . & . & & & & . & & & & . \\ w_n/w_1 & w_n/w_2 & . & . & . & w_n/w_j & . & . & . & w_n/w_n \end{bmatrix}_{nxn} \begin{bmatrix} w_1 \\ w_2 \\ . \\ . \\ . \\ w_j \\ . \\ . \\ . \\ w_n \end{bmatrix}_{nx1}$$

$$= n \begin{bmatrix} w_1 \\ w_2 \\ . \\ . \\ . \\ w_j \\ . \\ . \\ . \\ w_n \end{bmatrix}_{nx1} = nw$$

The relative weights obtained by decision makers from each one of n rows of matrix A;

$$A \times w = n \times w \tag{3.3}$$

where $w = (w_1, w_2, \ldots, w_n)$. W is the vector of actual relative weights, and n is the number of criterion.

Since a_{ij} subjective rate given by practitioners, Saaty (1980) proposed to use the maximum eigenvalue, $\lambda_{\max}$, to replace n to calculate actual values;

$$A \times w = \lambda_{\max} \times w \tag{3.4}$$

The normalisation of row average (NRA) is used to reveal the priority vector, and the following formula represents how to compute the weight of each criterion and/or each alternative at the final row of hierarchy;

$$w_i = \sum_{i=1}^{n} \frac{a_{ij}}{\sum_{j=1}^{n} a_{ij}}, i = 1, 2, \ldots, n. \tag{3.5}$$

3.2 Consistency Control in the Classical AHP

Saaty (1980) first presented the consistency index as a component of the AHP process. Several reasons may cause individual bias on AHP surveys and in such situations; a facilitator may decide whether the survey responses are robust. Otherwise, the survey should be replaced to provide a robust solution of the objective. Therefore, consistency control is a unique and routine part of every AHP study (Bulut et al. 2010).

The consistency ratio (*CR*) is defined as a ratio between the consistency of a given evaluation matrix (consistency index, *CI*) and the consistency of a random matrix (*RI*). The *CR* of a decision should not exceed a certain level e.g. *0.1* for a matrix larger than 4×4. Therefore, only evaluations are included which fulfil the condition $CR \leq 0.1$. *CR* can be approximated via $\lambda_{\max}$:

$$CI = \frac{\lambda_{max} - n}{n - 1} \tag{3.6}$$

where $\lambda_{\max}$ eigenvalue, and

$$CR = \frac{CI}{RI} \leq 0.1 \tag{3.7}$$

where *RI* is the average index of randomly generated weights (Table 3.2) and *n* is a number of criteria or alternatives.

3.3 Case Study: Dry Port Location

Dry port refers to a logistics centre established in an inland region directly connected by road or rail to a seaport and operating as a centre for the transhipment cargo to inland destinations. In addition to their role in cargo transhipment, dry ports may also include facilities for storage and consolidation of goods, maintenance for road or rail cargo carriers and customs clearance services. The location of these facilities at a dry port relieves competition for storage and customs space at the seaport itself. Therefore, the location for the dry port plays a critical role in the inland transportation.

In this example for the AHP application, the feasible location for the dry port is investigated. First, six criterion related to the location of the dry ports are defined by considering literature review (Ka 2011), and Table 3.3 displays a brief definition of this criterion and their abbreviations. There is obviously a number of steps that are very judgemental and somewhat subjective, and the selection of criteria is among

Table 3.2 Random consistency index (RI)

n	1	2	3	4	5	6	7	8	9	10
RI	0.00	0.00	0.58	0.90	1.12	0.24	1.32	1.41	1.45	1.49

Table 3.3 Definition of the criteria

Criteria	Abbreviation	Definition
Transportation	TR	*Transportation distance* *Quality and variety of transportation*
Natural environment	NE	*Weather conditions* *Geological conditions*
Infrastructure facilities	IS	*Existence of the logistics centre* *Security of infrastructure facilities*
Trade level	TL	*Import and export volume* *Commercial and industrial output value* *Mutual complimentary of resource*
Policy environment	PE	*Policy-oriented* *Regional cooperation environment*
Cost	CS	*Transportation cost* *Land cost*

them. The selection of criteria for this example is based on previous work, and the independence is confirmed by checking overlaps between predefined meanings of criteria. Another way of building hierarchy is to conduct a pre-survey (usually open-ended questions) and to get some initial response from experts[2].

In the second step, three different proper locations for the dry port are thought as the alternatives, A_1, A_2 and A_3. The hierarchical design of the location selection process for the dry port is presented in Fig. 3.8.

In the first step of the solution, the questionnaire needs to be sent to decision makers such as experts and academician to get a pairwise matrix for each of them. Each criterion is compared with each other through pairwise comparisons by asking "*How much importance does a criterion have compared to another criterion with respect to your preferences?*" The relative importance value can be determined using a scale in Table 3.1. Table 3.4 displays one of decision maker's pairwise

[2]The level of expertise is also usually a subjective part of the AHP applications. As it is proposed by Bulut et al. (2012), experts' valuations may be prioritised based on years spent in the industry or the level of consistency in their decision matrices. Since there is no precise and definite way of measuring expertise, the selection of experts (also prioritisation of them if needed) is left to moderators. Regarding our experiences with the AHP, we have two major groups of AHP problems and their corresponding expert definitions. For high level problems such as policy or strategy selection, researchers and policy makers may perform well and better consistency may be ensured. In practical and technical problems such as selection of an equipment or method, experts from the industry and practice would be better reflecting technical feasibility and challenges.

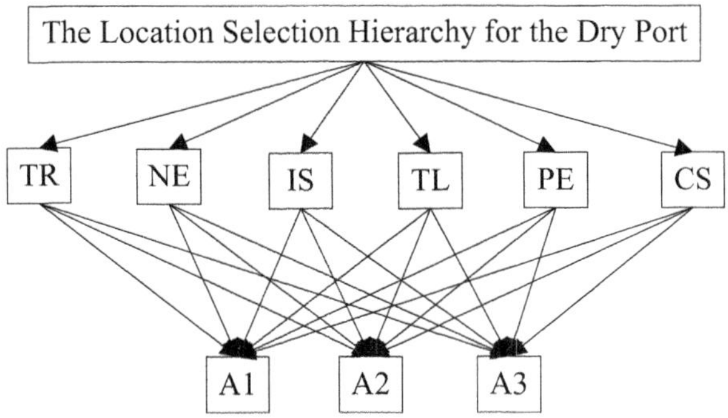

Fig. 3.8 Location selection hierarchy for the Dry port.

Table 3.4 Decision Maker's (*Decision Maker 1*) pairwise matrix (*A* vector)

	TR	*NE*	*IS*	*TL*	*PE*	*CS*
TR	1.00	0.33[a]	0.50	0.20	0.33	0.14
NE	3.00	1.00	0.33	0.33	1.00	0.20
IS	2.00	3.00	1.00	1.00	0.33	0.20
TL	5.00	3.00	1.00	1.00	3.00	0.33
PE	3.00	1.00	3.00	0.33	1.00	0.20
CS	7.00	5.00	5.00	3.00	5.00	1.00
Total	21.00	13.33	10.83	5.87	10.67	2.08

[a]*Reciprocal inputs are written in decimal format (1/3 = 0.33; 1/5 = 0.20)*

matrix (*Decision Maker 1* in Table 3.7). We clearly present how to calculate consistency ratio by considering Table 3.4.

The pairwise matrix presents the comparison of each criterion by using Saaty's scale (1977). In addition, we need to calculate the normalization matrix by using the total value of the column vector to find W vector for each criterion (Table 3.5).

After calculation w vector, we need to find $A \times w$ vector by multiplying A and w vector.

Example First row at the decision maker matrix (Table 3.4*) is*

$$a_{1j} = (1.00, 0.33, 0.50, 0.20, 0.33, 0.14)$$

and row total (w) of normalisation matrix (Table 3.5*) is*

$$w = (0.25, 0.50, 0.71, 1.17, 0.74, 2.63)$$

Then, $a_{1j} \times W$ vector is calculated as

$a_{1j} \times w = (1.00 \times 0.25,\ 0.33 \times 0.50,\ 0.50 \times 0.71,\ 0.20 \times 1.17,\ 0.33 \times 0.74,\ 0.14 \times 2.63)$

$= (0.25, 0.17, 0.36, 0.23, 0.25, 0.38)$, *as the first row at Table* 3.6.

According to the Eq. (3.4), λ_{max} can be calculated as follows (Table 3.6);

Table 3.5 Normalisation of the Decision Maker's pairwise matrix

	TR	*NE*	*IS*	*TL*	*PE*	*CS*	*Row Total (w)*
TR	0.05	0.03	0.05	0.03	0.03	0.07	0.25
NE	0.14	0.08	0.03	0.06	0.09	0.10	0.50
IS	0.10	0.23	0.09	0.17	0.03	0.10	0.71
TL	0.24	0.23	0.09	0.17	0.28	0.16	1.17
PE	0.14	0.08	0.28	0.06	0.09	0.10	0.74
CS	0.33	0.38	0.46	0.51	0.47	0.48	2.63

Example: a_{11}, 0.05 = 1.00/21.00 (inputs from Table 3.4*)*

Table 3.6 $A \times w$ vector

	TR	*NE*	*IS*	*TL*	*PE*	*CS*	*Row total ($A \times w$)*	*Aw/w*
TR	0.25	0.17	0.36	0.23	0.25	0.38	1.63	6.45
NE	0.76	0.50	0.24	0.39	0.74	0.53	3.15	6.35
IS	0.51	1.49	0.71	1.17	0.25	0.53	4.64	6.54
TL	1.26	1.49	0.71	1.17	2.23	0.88	7.73	6.62
PE	0.76	0.50	2.13	0.39	0.74	0.53	5.04	6.80
CS	1.77	2.48	3.55	3.50	3.71	2.63	17.64	6.70

$$\lambda_{\max} = \frac{6.45+6.35+6.54+6.62+6.80+6.70}{6} = 6.58 \quad \textit{(by using Eq. 3.4)}$$

$$CI = \frac{(6.58-6)}{(6-1)} = 0.12 \quad \textit{(by using Eq. 3.6)}$$

RI is 1.24 for the seven criteria (*See Table* 3.2).

$$CR = 0.12/1.24 = 0.09 \quad \textit{(by using Eq. 3.7) (below 0.1, consistent)}$$

Decision maker's pairwise matrix is found consistency since *CR* value is less than *0.1*. The above calculation and definition should be applied to each decision maker's pairwise matrix to confirm that they are consistency before calculation of the priority for all criteria.

In this example, five practitioners and scholars in logistics research participate in the questionnaire, and the aggregated matrix based on an average of their pairwise matrix need to be computed to reveal priority of each criterion. Table 3.7 shows each decision maker's pairwise matrix.

Table 3.7 Pairwise matrix for all decision makers

Decision Maker 1						
	TR	*NE*	*IS*	*TL*	*PE*	*CS*
TR	1.00	0.33	0.50	0.20	0.33	0.14
NE	3.00	1.00	0.33	0.33	1.00	0.20
IS	2.00	3.00	1.00	1.00	0.33	0.20
TL	5.00	3.00	1.00	1.00	3.00	0.33
PE	3.00	1.00	3.00	0.33	1.00	0.20
CS	7.00	5.00	5.00	3.00	5.00	1.00
Total	21.00	13.33	10.83	5.87	10.67	2.08
Decision Maker 2						
	TR	*NE*	*IS*	*TL*	*PE*	*CS*
TR	1.00	0.20	1.00	0.33	1.00	0.20
NE	5.00	1.00	3.00	1.00	5.00	0.33
IS	1.00	0.33	1.00	1.00	2.00	0.20
TL	3.00	1.00	1.00	1.00	3.00	0.33
PE	1.00	0.20	0.50	0.33	1.00	0.20
CS	5.00	3.00	5.00	3.00	5.00	1.00
Total	16.00	5.73	11.50	6.67	17.00	2.27
Decision Maker 3						
	TR	*NE*	*IS*	*TL*	*PE*	*CS*
TR	1.00	1.00	1.00	0.33	1.00	0.20
NE	1.00	1.00	1.00	0.20	1.00	0.33
IS	1.00	1.00	1.00	0.20	1.00	0.14
TL	3.00	5.00	5.00	1.00	5.00	0.33
PE	1.00	1.00	1.00	0.20	1.00	0.14
CS	5.00	3.00	7.00	3.00	7.00	1.00
Total	12.00	12.00	16.00	4.93	16.00	2.15
Decision Maker 4						
	TR	*NE*	*IS*	*TL*	*PE*	*CS*
TR	1.00	0.20	1.00	0.20	1.00	0.14
NE	5.00	1.00	5.00	1.00	5.00	0.33
IS	1.00	0.20	1.00	0.20	1.00	0.20
TL	5.00	1.00	5.00	1.00	5.00	0.50
PE	1.00	0.20	1.00	0.20	1.00	0.14
CS	7.00	3.00	5.00	2.00	7.00	1.00
Total	20.00	5.60	18.00	4.60	20.00	2.32
Decision Maker 5						
	TR	*NE*	*IS*	*TL*	*PE*	*CS*
TR	1.00	0.33	0.33	0.20	1.00	0.14
NE	3.00	1.00	1.00	0.33	3.00	0.20
IS	3.00	1.00	1.00	1.00	3.00	0.20
TL	5.00	3.00	1.00	1.00	5.00	0.33
PE	1.00	0.33	0.33	0.20	1.00	0.14
CS	7.00	5.00	5.00	3.00	7.00	1.00
Total	20.00	10.67	8.67	5.73	20.00	2.02

Table 3.8 Aggregated matrix for the criteria

	TR	*NE*	*IS*	*TL*	*PE*	*CS*
TR	1.00	0.41	0.77	0.25	0.87	0.17
NE	3.40	1.00	2.07	0.57	3.00	0.28
IS	1.60	1.11	1.00	0.68	1.47	0.19
TL	4.20	2.60	2.60	1.00	4.20	0.37
PE	1.40	0.55	1.17	0.25	1.00	0.17
CS	6.20	3.80	5.40	2.80	6.20	1.00
Total	17.80	9.47	13.00	5.56	16.73	2.17

Table 3.9 Normalisation of the aggregated matrix for the criteria and their priority

	TR	*NE*	*IS*	*TL*	*PE*	*CS*	*Priority*
TR	0.06	0.04	0.06	0.05	0.05	0.08	0.06
NE	0.19	0.11	0.16	0.10	0.18	0.13	0.14
IS	0.09	0.12	0.08	0.12	0.09	0.09	0.10
TL	0.24	0.27	0.20	0.18	0.25	0.17	0.22
PE	0.08	0.06	0.09	0.05	0.06	0.08	0.07
CS	0.35	0.40	0.42	0.50	0.37	0.46	0.42

Table 3.8 displays the aggregated matrix that based on the average of the all decision maker's matrix found consistent. After normalization of the aggregated matrix by using column vector, we calculate the row average of each criterion to find their priority as in Table 3.9.

At the second step, the same steps applied in the calculation of the priority for each criterion are also used for the calculation of the priority of each alternative. The individual pairwise matrix of decision makers for the alternatives under each criterion is analysed whether they are consistent. Table 3.10 displays the normalisation of the aggregated matrix for the alternatives by considering each criterion.

Finally, the ranking scores of alternatives are calculated by multiplying the weights of alternatives with priorities of each criterion (Table 3.11). The alternative with the highest ranking score is selected for recommendation to the location of the dry port complex (Table 3.12). In this case, the alternative *A2* is found as the best alternative.

The indifference between two best alternatives, *A2* and *A3*, may be evaluated further by decision makers. When the gap between two alternatives is small (again a subjective preference), the lead decision maker may prefer the alternative with better cost performance. In the dry port location problem, the best alternative, *A2*, is also superior in terms of cost (*CS*) which strengthens the ultimate choice.

Table 3.10 Normalisation of the aggregated matrix and weight of each alternative

TR					*NE*				
	A1	*A2*	*A3*	*Priority*		*A1*	*A2*	*A3*	*Priority*
A1	0.26	0.48	0.28	0.34	*A1*	0.22	0.33	0.22	0.26
A2	0.05	0.08	0.11	0.08	*A2*	0.07	0.10	0.12	0.09
A3	0.69	0.44	0.60	0.58	*A3*	0.71	0.57	0.66	0.65
IS					*TL*				
	A1	*A2*	*A3*	*Priority*		*A1*	*A2*	*A3*	*Priority*
A1	0.18	0.18	0.19	0.18	*A1*	0.64	0.71	0.51	0.62
A2	0.18	0.18	0.19	0.18	*A2*	0.24	0.24	0.40	0.29
A3	0.64	0.64	0.61	0.63	*A3*	0.11	0.06	0.09	0.09
PE					*CS*				
	A1	*A2*	*A3*	*Priority*		*A1*	*A2*	*A3*	*Priority*
A1	0.10	0.07	0.08	0.08	*A1*	0.18	0.20	0.25	0.21
A2	0.58	0.41	0.68	0.56	*A2*	0.70	0.68	0.64	0.67
A3	0.32	0.51	0.24	0.36	*A3*	0.12	0.12	0.11	0.12

Table 3.11 Priority of the criteria and weight of the alternatives

	TR	*NE*	*IS*	*TL*	*PE*	*CS*
	0.05	0.13	0.09	0.19	0.06	0.37
A1	0.34	0.26	0.18	0.62	0.08	0.21
A2	0.08	0.09	0.18	0.29	0.56	0.67
A3	0.58	0.65	0.63	0.09	0.36	0.12

Table 3.12 Final ranking scores for the alternatives

	TR	*NE*	*EL*	*IS*	*TL*	*PE*	*CS*	*Final priority weight*
A_1	0.02	0.04	0.01	0.02	0.12	0.00	0.08	0.28
A_2	0.00	0.01	0.03	0.02	0.06	0.03	0.25	**0.40**[a]
A_3	0.03	0.09	0.07	0.06	0.02	0.02	0.04	0.32

Example: A_1 *vs TR, 0.02 = 0.34*0.05 (inputs from* Table 3.11*)*
Example: A_2 *vs EL, 0.03 = 0.26*0.11 (inputs from* Table 3.11*)*
[a] *A_2 has the highest priority weight with 0.40 final score which makes it the best choice*

4 Fuzzy Analytic Hierarchy Process (FAHP)

In the last two decades, there is a growing interest and the use of fuzzy set theory in the AHP applications. Fuzzy set theory is first developed by Zadeh (1965), and it is already applied to thousands of different problems in engineering and computing. Fuzzy set theory is found a unique solution for rule-based systems to convert crisp sets to clusters (i.e. fuzzy sets) according to distribution and perception characteristics. Today, various home appliances employ fuzzy logic which a rule-based (IF-THEN) mechanism based on fuzzy set inputs and outputs.

The fundamental feature of fuzzy sets is the transformation of crisp, and in some cases, subjective inputs into a numerical cluster which can be available for arithmetic operations. Considering its major function, fuzzy sets are thought to be an outlet for decision makers' uncertainty and indecision under the complex problems. In the classical AHP, decision maker is asked to indicate a single sensitive index (1 ~ 9) which exposes the level of priority for given comparison problem. As it is mentioned in previous sections, task load is one of the critical challenge in applications. The extensive AHP scale amplifies the problem and reduces the reliability of results. The rationale behind the fuzzy extended AHP approach relies on the reduction of task load (less number of linguistic preferences) and pre-grouping of expert judgements in particular clusters of the crisp data space (experts preferring e.g. 1, 2 and 3 may be classified in a fuzzy set). Fuzzy set is also very connected with the practice. Experts usually cannot perceive indifference between a single number up or down. Decreasing the size of data space (Saaty's scale) helps experts to clarify their judgement much easily.

In the following sub-sections, the theory and algorithm of FAHP will be discussed, and each step of FAHP calculations will be reviewed through a case study of shipping asset selection.

4.1 *Fuzzy Sets and Triangular Fuzzy Numbers (TFNs)*

Fuzzy sets are the concept of the managing uncertainty, and its theory was first introduced by Zadeh (1965). The aim of the fuzzy sets is to generalize the concept of a set and resolution to accommodate the type of fuzziness or vagueness in many decision problems (Badiru and Cheung 2002). It has been widely applied to represent uncertain or flexible information in many different fields, such as investment management, engineering design, and production management.

The membership function denotes the degree of the relation in the fuzzy set, $\tilde{A}$, and is usually represented by $\mu_{\tilde{A}}(x)$.

The following basic definitions of a fuzzy set and its arithmetic operations are as follows:

Definition 1 Let X be a universe of discourse, $\tilde{A}$ is a fuzzy subset of X such that for all $x \in X$. $\mu_{\tilde{A}}(x) \in [0,1]$ which is assigned to stand for the membership of x to $\tilde{A}$, and $\mu_{\tilde{A}}(x)$ is called the membership function of fuzzy set $\tilde{A}$.

Definition 2 A fuzzy number is a fuzzy subset of the universe of discourse X, $X \subseteq R$, and it is both convex and normal (there exists at least one element $x \in X$ such that $\mu_{\tilde{A}}(x) = 1$).

Definition 3 A triangular fuzzy number is defined by its basic particulars which are;

$$\mu_{\tilde{A}}(x) = \begin{cases} 0, & x < l, \\ (x-l)/(m-l), & l \leq x < m, \\ 1, & x = m, \\ (u-x)/(u-m), & m < x \leq u, \\ 0, & u < x. \end{cases} \tag{3.8}$$

where l and u correspond to the lower and upper bounds of the fuzzy numbers $\tilde{A}$, respectively, and m is the midpoint.

$\tilde{A}$ is the triangular fuzzy numbers (TFNs) is a particular fuzzy set $\tilde{A}$, and its membership function $\mu_{\tilde{A}}(x)$ is a continuous linear function. Centre of gravity method is the most used for defining the crisp result of the fuzzy numbers, $\tilde{A} = (l, m, u)$.

Arithmetic operations of fuzzy numbers are defined in Zadeh (1965) by standard fuzzy arithmetic operations (Kaufmann and Gupta 1991).

4.2 FAHP

In order to overcome the deficiency of the fuzziness during decision making, the analytic hierarchy process was extended by using fuzzy logic to analyse and solve the decision making problems. First, van Laarhoven and Pedrycz (1983) use TFNs of the fuzzy set theory into the pairwise comparison matrix of the AHP to develop a fuzzy AHP (FAHP) approach.

Buckley (1985) determined trapezoidal fuzzy numbers to express the priorities of comparison ratios. Chang (1996) used triangular fuzzy membership value for pairwise comparison and introduced a new approach for handling FAHP named "*extent synthesis analysis*". Weck et al. (1997)present a method for evaluating different production cycle alternatives by adding the mathematics of fuzzy logic to the classical AHP. Lee et al. (1999) review the basic ideas behind the AHP and introduce the concept of comparison interval and propose a methodology based on stochastic optimisation to achieve global consistency and to accommodate the fuzzy nature of the comparison process. Duru et al. (2012) improve Chang's approach by considering coefficient based on CCI for each participant.

Since of Chang's extent synthesis method (Chang 1996) is widely preferred in the FAHP literature (the most cited version of the FAHP), it is applied in the following case study.

The mathematical algorithm of the extent synthesis method as follows:

Let $X = \{x_1, x_2, \ldots, x_n\}$ be an object set and $U = \{u_1, u_2, \ldots, u_m\}$ be a goal set. According to the method of extent analysis, each object is taken and extent analysis for each goal is performed, respectively (Chang 1996). Therefore, *m* extent analysis values for each object can be obtained, with the following signs:

$$M_{g_i}^1, M_{g_i}^2, \ldots, M_{g_i}^m, \quad i = 1, 2, \ldots, n, \tag{3.9}$$

where all the $M_g^j (j = 1,2,\ldots,m)$ are TFNs.

The steps of Chang's extent analysis can be given as in the following:

Step 1 The value of fuzzy synthetic extent with respect to the *i*th object is defined as;

$$S_i \sum_{j=1}^{m} M_{g_i}^j \otimes \left[\sum_{i=1}^{n} \sum_{j=1}^{m} M_{g_i}^j \right]^{-1} \tag{3.10}$$

To obtain $\sum_{j=1}^{m} M_{g_i}^j$, the fuzzy addition operation of *m* extent analysis values for a particular matrix is performed such as:

$$\sum_{j=1}^{m} M_{g_i}^j = \left(\sum_{j=1}^{m} l_j, \sum_{j=1}^{m} m_j, \sum_{j=1}^{m} u_j \right) \tag{3.11}$$

And to obtain $\left[\sum_{i=1}^{n} \sum_{j=1}^{m} M_{g_i}^j \right]^{-1}$, the fuzzy addition operation of $M_{g_i}^j$ $(j = 1, 2, \ldots, m)$ values is performed such as:

$$\sum_{i=1}^{n} \sum_{j=1}^{m} M_{g_i}^j = \left(\sum_{j=1}^{m} l_j, \sum_{j=1}^{m} m_j, \sum_{j=1}^{m} u_j \right) \tag{3.12}$$

and then the inverse of the vector in Eq. (3.12) is computed, such as:

$$\left[\sum_{i=1}^{n} \sum_{j=1}^{m} M_{g_i}^j \right]^{-1} = \left(\frac{1}{\sum_{i=1}^{n} u_i}, \frac{1}{\sum_{i=1}^{n} m_i}, \frac{1}{\sum_{i=1}^{n} l_i} \right). \tag{3.13}$$

Step 2 The degree of possibility of $M_2 = (l_2, m_2, u_2) \geq M_1 = (l_1, m_1, u_1)$ is defined as;

$$V(M_2 \geq M_1) = \sup_{y \geq x} \left\lfloor \min\left(\mu_{M_1}(x), \mu_{M_2}(y)\right) \right\rfloor \tag{3.14}$$

and can be expressed as follows:

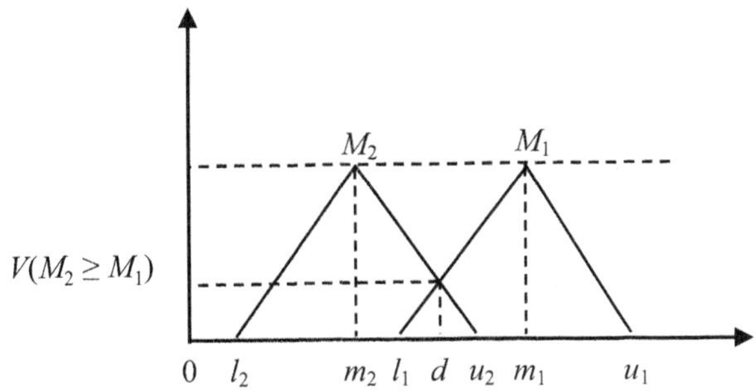

Fig. 3.9 The intersection between M_1 and M_2

$$V(M_2 \geq M_1) = hgt(M_1 \cap M_2)$$
$$= \mu_{M_2}(d) = \begin{cases} 1, & if \ m_2 \geq m_1, \\ 0, & if \ l_1 \geq u_2, \\ \dfrac{l_1 - u_2}{(m_2 - u_2) - (m_1 - l_1)}, & otherwise. \end{cases} \quad (3.15)$$

Figure 3.9 illustrates Eq. (3.15) where d is the ordinate of the highest intersection point D between μ_{M_1} andμ_{M_2}. To compare M_1 and M_2, we need both the values of $V(M_1 \geq M_2)$ and $V(M_2 \geq M_1)$.

Step 3 The degree possibility for a convex fuzzy number to be greater than k convex fuzzy M_i ($i = 1,2,\ldots, k$) numbers can be defined by

$$V(M \geq M_1, M_2, \ldots, M_k) = V[(M \geq M_1)and(M \geq M_2)and \ldots and(M \geq M_k)]$$
$$= minV(M \geq M_i), i = 1, 2, 3, \ldots, k. \quad (3.16)$$

Assume that $d'(A_i) = \min V(S_i \geq S_k)$ for $k = 1,2,\ldots, n; k \neq i$. Then the weight vector is given by

$$W' = \left(\left(d'(A_1)\right), d'(A_2), \ldots, d'(A_n)\right)^T \quad (3.17)$$

Where A_i ($i = 1, 2,\ldots, n$) are n elements.

Step 4 Via normalization, the normalized weight vectors are

$$W = (d(A_1), d(A_2), \ldots, d(A_n))^T, \quad (3.18)$$

where W is a non-fuzzy number.

4.3 Consistency Control for FAHP

In the existing literature, many studies applied the FAHP method to multi-attribute decision making problems. While Chang's approach (1996) is often used in FAHP studies, most of these studies ignore the consistency control including the study of Chang (1996) itself. Therefore, many of these FAHP applications are questionable in terms of reliability concern.

In the following example, the centric consistency index (CCI) is suggested to control the individual consistency of pairwise matrix (Bulut et al. 2012). The algorithm of the CCI is as follows:

Let $A = (a_{Lij}, a_{Mij}, a_{Uij})_{n\times n}$ be a fuzzy judgement matrix, and let $w = [(w_{L1}, w_{M1}, w_{U1}), (w_{L2}, w_{M2}, w_{U2}),\ldots,(w_{Ln}, w_{Mn}, w_{Un})]^T$ be the priority vector derived from Ausing the RGMM. The centric consistency index (CCI) is computed by Eq. (3.19).

$$CCI(A) = \frac{2}{(n-1)(n-2)}\sum_{i<j}\left(\log\left(\frac{a_{Lij}+a_{mij}+a_{Uij}}{3}\right) - \log\left(\frac{w_{Li}+w_{Mi}+w_{Ui}}{3}\right) + \log\left(\frac{w_{Lj}+w_{Mj}+w_{uj}}{3}\right)\right)^2 \quad (3.19)$$

When $CCI(A) = 0$, we consider A fully consistent. Aguarón and Moreno-Jiménez (2003) also provide the thresholds $(\overline{GCI})$ as $\overline{GCI}$=0.31 for $n = 3$; $\overline{GCI}$ =0.35 for $n = 4$ and $\overline{GCI}$=0.37 for $n > 4$. When $CCI(A) < \overline{GCI}$, it is considered that the matrix A is sufficiently consistent. Since the CCI is a fuzzy extended version of GCI, thresholds remain identical.

4.4 Case Study: Shipping Asset Selection

Transportation is of prime importance in world trading activities and world merchandise trade is broadly based on seaborne transportation. In recent figures, maritime transport is around 90% of import-export transportation (UNCTAD 2013). In the conventional economic analysis, shipping price plays a critical role and is used as a leading indicator (Israely 2009). Therefore, maritime transportation is a critical part of global economy.

Particularly in the rapid growth of the world economy, shipping capacity is more important and a certain size of the ship has advantages on employment and profit margins. An investor is basically asked to define characteristics of his ship investment. The ship size is one of the critical dimensions of ship investment. The ship investment can be performed by a new building contract or a second hand alternative. A new building option has a long delivery procedure and prices will be higher. Conversely, a second hand ship can be purchased in a reasonable term with cheaper prices. Differences between prices and delivery dates compose the main concern of the investment selection problem in the shipping industry.

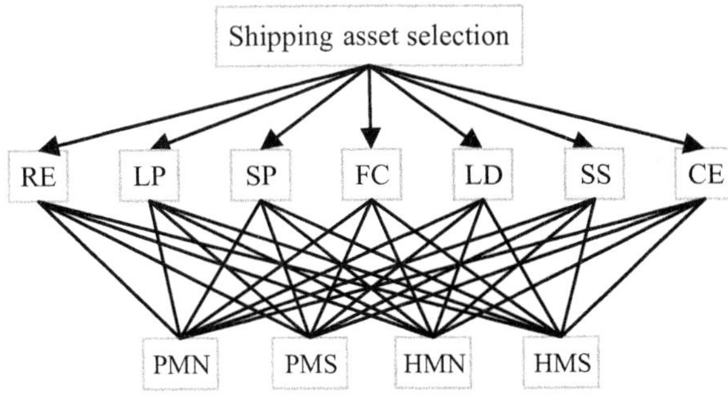

Fig. 3.10 Hierarchy for the ship investment selection by using FAHP

Shipping services are classified as either dry cargo shipments or wet cargo shipments. Dry cargoes are carried by dry bulk carriers; container ships, etc., while wet cargoes are transferred by tanker ships. In this case study for the FAHP application, the dry bulk shipping industry is concerned. According to potential cargo sizes, routes and service capabilities, dry bulk carriers are classified by their tonnage capacities. Tonnage classes are usually divided as follows: Capesize bulker (over 80,000 deadweight-DWT), Panamax bulker (around 80,000–60,000 DWT), Handymax bulker (60,000–45,000 DWT) and Handysize bulker (45,000–10,000 DWT) (Bulut et al. 2013). In the prepared scenario, four different projects based on two different type of ships Panamax new building (PMN), Panamax second hand (PMS), Handymax new building (HMN) and Handymax second hand (HMS) bulk carrier (Fig. 3.10) are performed for the FAHP application. The main reason for this selection is based on convenience of data collection, interactions of similar cargoes (competitiveness exists to some degree), and closeness of asset prices among others.

Under the FAHP framework, several criteria are defined by the industrial survey and literature review. Table 3.13 shows the major criteria which can be classified into two groups: Financial features and technical features, and all they have selective capabilities.

As mentioned in the example of the AHP method, each step of a solution is almost same in the FAHP application. In the first step, the questionnaire needs to be prepared and sent to participants to compare each criterion with each other. However, the evaluation scale used in the FAHP method is different from the numerical scale. The non-numerical expressions as fuzzy linguistic variables reflect the Saaty's nine-point fundamental scale (Saaty 1977) (Fig. 3.11). Assign the linguistic comparison terms and their equivalent fuzzy numbers considered in this chapter are presented in Table 3.14 (Bulut et al. 2014; Sahin and Senol 2015; Şen et al. 2010; Tadic et al. 2013).

In this case, there are five participants from the shipping industry, and their distribution is from two academicians, two shipbrokers and one shipowner. The selection of expert group is again a critical question. The context of the ship investment selection has both long-term and short-term aspects for the industry. Shipbrokers serve as intermediary in ship sale and purchase, and they usually have

Table 3.13 Criteria for the ship investment decision and their symbols.

Criterion of the ship investment selection	The symbols of each criterion
Return on equity	*RE*
Loss probability	*LP*
Second hand price	SP
Fuel consumption	*FC*
Loaded draught	*LD*
Ship's economic navigation speed	*SS*
Cargo crane existence	*CE*

Fig. 3.11 A fuzzy number of a linguistic variable set

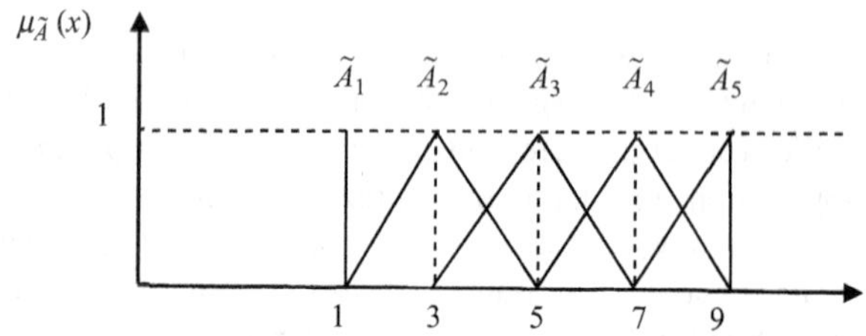

Table 3.14 Membership function of linguistic scale

Fuzzy number	*Linguistic scales*	*Membership*	*Inverse*
$\tilde{A}_1$	Equally important	(1,1,1)	(1,1,1)
$\tilde{A}_2$	Moderately important	(1,3,5)	(1/5,1/3,1)
$\tilde{A}_3$	More important	(3,5,7)	(1/7,1/5,1/3)
$\tilde{A}_4$	Strongly important	(5,7,9)	(1/9,1/7,1/5)
$\tilde{A}_5$	Extremely important	(7,9,9)	(1/9,1/9,1/7)

the latest market sentiment in terms right pricing of assets as well as right valuation of assets including various dimensions. On the other hand, shipbrokers have relatively short horizon due to recency bias (having very frequent data flow for the current market). Academician may have advantage of extensive horizon and future orientation. Therefore, academicians can reflect 'gross' trends and potential changes in the long-term. Shipowners definitely make the ultimate decision while their decision is composed of series of consultations with former two groups of experts (mostly shipbrokers in practice). So, vast majority of the problem is shared among shipbrokers and academicians while a portion of it is still asked to a practicing shipowner to reflect incentives behind the investor. The sample of decision making problem may be extended while maintaining the ratio (e.g. four shipbrokers, four academicians and two shipowners or their senior executives).

The second step in the process, each participant compares the criterion with each other to find the weight of each criterion by using the fuzzy linguistic terms and one of the individual pairwise matrix is as follows (Table 3.15):

We should start to calculate the consistency value for each participant's pairwise matrix whether accepting it or not. According to the Eq. 3.13, low, medium and upper value need to be calculated for each row. For the first row:

$$Low = (1 + 0.33 + 0.20 + 1 + 1 + 1 + 3)^{\frac{1}{7(number\ of\ criteria)}} = 0.79$$

$$Medium = (1 + 1 + 0.33 + 1 + 3 + 3 + 5)^{\frac{1}{7}} = 1.47$$

$$Upper = (1 + 1 + 1 + 3 + 5 + 5 + 7)^{\frac{1}{7}} = 2.45$$

After finding all results for each row, we need to make normalization;

$$Low = 0.79 + 0.79 + 1.60 + 0.58 + 0.38 + 0.32 + 0.26$$
$$= 4.74$$

$$Normalized\ value = \frac{0.79}{4.74} = 0.17$$

Value for low, medium and upper after normalisation is as follows, respectively:

$$Low = 0.17, 0.17, 0.34, 0.12, 0.08, 0.07, 0.06$$
$$Medium = 0.17, 0.15, 0.36, 0.15, 0.06, 0.06, 0.05$$
$$Upper = 0.17, 0.19, 0.31, 0.14, 0.08, 0.07, 0.05$$

After that, average value needs to be calculated for each row:

$$= \frac{0.17(low) + 0.17(medium) + 0.17(upper)}{3}$$
$$= 0.17$$

The average value set:

$$= 0.17, 0.17, 0.34, 0.14, 0.07, 0.06, 0.05$$

Also, we need to calculate the average value of the fuzzy linguistic terms to find *CCI*.

Calculation for some value;

$$RE\ vs.RE = \frac{1 + 1 + 1}{3}$$
$$= 1$$

Table 3.15 One of the individual fuzzy judgement matrix

	RE	*LP*	*SP*	*FC*	*LD*	*SS*	*CE*
RE	(1,1,1)	(0.33,1,1)	(0.20,0.33,1)	(1,1,3)	(1,3,5)	(1,3,5)	(3,5,7)
LP	(1,1,3)	(1,1,1)	(0.20,0.33,1)	(1,1,3)	(1.3,5)	(1,3,5)	(1,3,5)
SP	(1,3,5)	(1,3,5)	(1,1,1)	(1,3.5)	(3,5,7)	(3,5,7)	(3,5,7)
FC	(0.33,1,1)	(0.33,1,1)	(0.2,0.33,1)	(1,1,1)	(1,3,5)	(1,3,5)	(1,3,5)
LD	(0.20,0.33,1)	(0.20,0.33.1)	(0.14,0.20,0.33)	(0.20,0.33,1)	(1,1,1)	(1,1,3)	(1,1,3)
SS	(0.20,0.33,1)	(0.20,0.33.1)	(0.14,0.20,0.33)	(0.20,0.33,1)	(0.33,1,1)	(1,1,1)	(1,1,3)
CE	(0.14,0.20,0.33)	(0.20,0.33,1)	(0.14,0.20,0.33)	(0.20,0.33,1)	(0.33,1,1)	(0.33,1,1)	(1,1,1)

$$
\begin{aligned}
RE\ vs.LP &= \frac{0.33 + 1 + 1}{3} \\
&= 0.78
\end{aligned}
$$

After finding all results, the following matrix will be obtained:

0.17	0.17	0.34	0.14	0.07	0.06	0.05
1.00	0.78	0.51	1.67	3.00	3.00	5.00
	1.00	0.51	1.67	3.00	3.00	3.00
		1.00	3.00	5.00	5.00	5.00
			1.00	3.00	3.00	3.00
				1.00	1.67	1.67
					1.00	1.67
						1.00

The logarithmic operation need to be applied on this matrix as follows:
For the first value:

$$
\begin{aligned}
&= (\log(0.78) - \log(0.17) + \log(0.17))^2 \\
&= 0.01
\end{aligned}
$$

For the last value:

$$
\begin{aligned}
&= (\log(1.67) - \log(0.06) + \log(0.05))^2 \\
&= 0.02
\end{aligned}
$$

The following new matrix will be found after all logarithmic operations respectively:

0.01	0.00	0.02	0.01	0.00	0.04
	0.00	0.02	0.01	0.00	0.00
		0.01	0.00	0.00	0.01
			0.04	0.02	0.00
				0.03	0.01
					0.02

The total value of matrix = 0.26

$$
\begin{aligned}
CCI &= \frac{2}{(Number\ of\ criteria - 1)x(Number\ of\ criteria - 2)}^{0.26} \\
&= 0.02
\end{aligned}
$$

We can surely say that the pairwise matrix is consistent since the results are found less than 0.37. The CCI should be computed for each participant's pairwise matrix before proceeding to FAHP calculation. In this example, all CCI value for each individual pairwise matrix is found consistent.

The following calculation shows how to obtain the aggregated fuzzy judgement matrix (Table 3.17) based on individual fuzzy judgement matrix (Table 3.16).

(The first value of the TFNs represent low, the second value is medium and last one is upper value. Low value can be calculated according to the minimum value of the first column of TFNs)

$$\begin{aligned} l_{j1} &= \min(1, 1, 1, 1, 1) \\ &= 1 \\ l_{j2} &= \min(1, 1, 1, 0.14, 0.20) \\ &= 0.14 \\ l_{j3} &= \min(1, 0.20, 0.14, 0.33, 0.14) \\ &= 0.14 \\ l_{j4} &= \min(0.33, 0.14, 0.14, 0.14, 0.14) \\ &= 0.14 \\ l_{j5} &= \min(0.20, 0.11, 0.14, 0.11, 0.20) \\ &= 0.11 \\ l_{j6} &= \min(0.20, 0.11, 0.11, 0.11, 0.20) \\ &= 0.11 \\ l_{j7} &= \min(0.14, 0.11, 0.11, 0.11, 0.20) \\ &= 0.11 \end{aligned}$$

Medium value is the average value of the second column of TFNs:

$$\begin{aligned} l_{j2} &= \min(1, 1, 1, 0.33, 0.20) \\ &= 0.71 \end{aligned}$$

Upper value is the maximum value of the last column of TFNs:

$$\begin{aligned} l_{j2} &= \min(3, 1, 1, 3, 0.33) \\ &= 3 \end{aligned}$$

According to the Equations between 13 and 22, the extent synthesis analysis is performed for the shipping investment decision from Table 3.17 as follows:

Table 3.16 Individual fuzzy judgement matrix for the criterion

DM1	*RE*	*LP*	*SP*	*FC*	*LD*	*SS*	*CE*
RE	(1,1,1)	(0.33,1,1)	(0.20,0.33,1)	(1,1,3)	(1,3,5)	(1,3,5)	(3,5,7)
LP	(1,1,3)	(1,1,1)	(0.20,0.33,1)	(1,1,3)	(1,3,5)	(1,3,5)	(1,3,5)
SP	(1,3,5)	(1,3,5)	(1,1,1)	(1,3,5)	(3,5,7)	(3,5,7)	(3,5,7)
FC	(0.33,1,1)	(0.33,1,1)	(0.20,0.33,1)	(1,1,1)	(1,3,5)	(1,3,5)	(1,3,5)
LD	(0.20,0.33,1)	(0.20,0.33,1)	(0.14,0.20,0.33)	(0.20,0.33,1)	(1,1,1)	(1,1,3)	(1,1,3)
SS	(0.20,0,33,1)	(0.20,0.33,1)	(0.14,0.20,0.33)	(0.20,0.33,1)	(0.33,1,1)	(1,1,1)	(1,1,3)
CE	(0.14,0.20,0.33)	(0.20,0.33,1)	(0.14,0.20,0.33)	(0.20,0.33,1)	(0.33,1,1)	(0.33,1,1)	(1,1,1)
DM2	*RE*	*LP*	*SP*	*FC*	*LD*	*SS*	*CE*
RE	(1,1,1)	(1,1,1)	(1,3,5)	(3,5,7)	(5,7,9)	(5,7,9)	(5,7,9)
LP	(1,1,1)	(1,1,1)	(3,5,7)	(3,5,7)	(3,5,7)	(3,5,7)	(5,7,9)
SP	(0.20,0.33,1)	(0.14,0.20,0.33)	(1,1,1)	(1,3,5)	(1,3,5)	(1,3,5)	(1,1,1)
FC	(0.14,0,2,0,33)	(0.14,0.20,0.33)	(0.20,0.33,1)	(1,1,1)	(1,1,1)	(1,1,1)	(1,3,5)
LD	(0.11,0.14,0.20)	(0.14,0.20,0.33)	(0.20,0.33,1)	(1,1,1)	(1,1,1)	(1,1,1)	(1,3,5)
SS	(0.11,0.14,0.20)	(0.14,0.20,0.33)	(0.20,0.33,1)	(1,1,1)	(1,1,1)	(1,1,1)	(1,1,3)
CE	(0.11,0.14,0.20)	(0.11,0.14,0.20)	(1,1,1)	(0.20,0.33,1)	(0.20,0.33,1)	(0.33,1,1)	(1,1,1)
DM3	*RE*	*LP*	*SP*	*FC*	*LD*	*SS*	*CE*
RE	(1,1,1)	(1,1,1)	(3,5,7)	(3,5,7)	(3,5,7)	(5,7,9)	(5,7,9)
LP	(1,1,1)	(1,1,1)	(3,5,7)	(3,5,7)	(3,5,7)	(5,7,9)	(3,5,7)
SP	(0.14,0.20,0.33)	(0.14,0.20,0.33)	(1,1,1)	(1,1,3)	(1,3,5)	(3,5,7)	(1,3,5)
FC	(0.14,0.20,0.33)	(0.14,0.20,0.33)	(0.33,1,1)	(1,1,1)	(1,1,3)	(1,3,5)	(1,3,5)
LD	(0.14,0.20,0.33)	(0.14,0.20,0.33)	(0.20,0.33,1)	(0.33,1,1)	(1,1,1)	(1,3,5)	(1,3,5)
SS	(0.11,0.14,0.20)	(0.11,0.14,0.20)	(0.14, 0.20,0.33)	(0.20,0.33,1)	(0.20,0.33,1)	(1,1,1)	(1,1,1)
CE	(0.11,0.14,0.20)	(0.14,0.20,0.33)	(0.20,0.33,1)	(0.20,0.33,1)	(0.20,0.33,1)	(1,1,1)	(1,1,1)
DM4	*RE*	*LP*	*SP*	*FC*	*LD*	*SS*	*CE*
RE	(1,1,1)	(1,3,5)	(1,1,3)	(3,5,7)	(5,7,9)	(5,7,9)	(5,7,9)

(continued)

Table 3.16 (continued)

LP	(0.20,0.33,1)	(1,1,1)	(0.33,1,1)	(3,5,7)	(3,5,7)	(3,5,7)	(3,5,7)
SP	(0.33,1,1)	(1,1,3)	(1,1,1)	(3,5,7)	(5,7,9)	(5,7,9)	(5,7,9)
FC	(0.14, 0.20,0.33)	(0.14,0.20,0.33)	(0.14, 0.20,0.33)	(1,1,1)	(1,1,3)	(1,1,3)	(1,3,5)
LD	(0.11,0.14,0.20)	(0.14,0.20,0.33)	(0.11,0.14,0.20)	(0.33,1,1)	(1,1,1)	(1,1,1)	(1,1,1)
SS	(0.11,0.14,0.20)	(0.14,0.20,0.33)	(0.11,0.14,0.20)	(0.33,1,1)	(1,1,1)	(1,1,1)	(1,1,1)
CE	(0.11,0.14,0.20)	(0.14,0.20,0.33)	(0.11,0.14,0.20)	(0.20,0.33,1)	(1,1,1)	(1,1,1)	(1,1,1)
DM5	*RE*	*LP*	*SP*	*FC*	*LD*	*SS*	*CE*
RE	(1,1,1)	(3,5,7)	(3,5,7)	(3,5,7)	(1,3,5)	(1,3,5)	(1,3,5)
LP	(0.14,0.20,0.33)	(1,1,1)	(1,3,5)	(1,3,5)	(3,5,7)	(3,5,7)	(3,5,7)
SP	(0.14,0.20,0.33)	(0.20,0.33,1)	(1.1.1)	(1,3,5)	(1,3,5)	(1,3,5)	(1,3,5)
FC	(0.14,0.20,0.33)	(0.20,0.33,1)	(0.20,0.33,1)	(1,1,1)	(1,3,5)	(1,1,1)	(1,3,5)
LD	(0.20,0.33,1)	(0.14,0.20,0.33)	(0.20,0.33,1)	(0.20,0.33,1)	(1,1,1)	(1,3,5)	(1,1,3)
SS	(0.20,0.33,1)	(0.14,0.20,0.33)	(0.20,0.33.1)	(1,1,1)	(0.20,0.33,1)	(1,1,1)	(1,1,3)
CE	(0.20,0.33,1)	(0.14,0.20,0.33)	(0.20,0.33,1)	(0.20,0.33,1)	(0.33,1,1)	(0.33,1,1)	(1,1,1)

Table 3.17 Aggregated fuzzy judgement matrix for the criterion

	RE	*LP*	*SP*	*FC*	*LD*	*SS*	*CE*
RE	(1,1,1)	(0.33,2.20,7)	(0.20,2.87,7)	(1,4.20,7)	(1,5,9)	(1,5.40,9)	(1,5.80,9)
LP	(0.14,0.71,3)	(1,1,1)	(0.20,2.86,7)	(1,3.80,7)	(1.4.6,7)	(1,5,9)	(1,5,9)
SP	(0.14,0.94,5)	(0.14,0.94,5)	(1,1,1)	(1,3,7)	(1.4.2,9)	(1,4.60,9)	(1,3.80,9)
FC	(0.14,0.36,1)	(0.14,0.38,1)	(0.14,0.44,1)	(1,1,1)	(1,1.8,5)	(1,1.80,5)	(1,3,5)
LD	(0.11,0.23,1)	(0.14,0.22,1)	(0.11,0.26,1)	(0.20,0.73,1)	(1,1,1)	(1,1.80,5)	(1,1.80,5)
SS	(0.11,0.21,1)	(0.11,0.21,1)	(0.11,0.24,1)	(0.20,0.73,1)	(0.2,0.73,1)	(1,1,1)	(1,1,3)
CE	(0.11,0.19,1)	(0.11,0.21,1)	(0.11,0.40,1)	(0.20,0.33,1)	(0.2,0.73,1)	(0.33,1,1)	(1,1,1)

Each value of $\sum_{j=1}^{n} l_j$, $\sum_{j=1}^{n} m_j$ and $\sum_{j=1}^{n} u_j$ need to be calculated for each row and for the first row (Eq. (3.11)).

$$\sum_{j=1}^{n} j = (1 + 0.33 + 0.20 + 1 + 1 + 1 + 1)$$

$$= 5.53$$

$$\sum_{j=1}^{n} m_j = (1 + 2.20 + 2.87 + 4.20 + 5 + 5.40 + 5.80)$$

$$= 26.47$$

$$\sum_{j=1}^{n} l_j = (1 + 7 + 7 + 7 + 9 + 9 + 9)$$

$$= 49.00$$

After calculation sum of row for each value of TFNs, the following matrix will be found:

5.53	26.47	49.00
5.34	22.97	43.00
5.29	18.49	45.00
4.43	8.79	19.00
3.57	6.06	15.00
2.73	4.14	9.00
2.07	3.88	7.00

Then, the sum of each column needs to be computed according to the Eq. (3.13):

$$\frac{1}{\sum_{i=1}^{n} l_i} = \frac{1}{(5.53 + 5.34 + 5.29 + 4.43 + 3.57 + 2.73 + 2.07)}$$

$$= \frac{1}{28.96}$$

$$\frac{1}{\sum_{i=1}^{n} m_i} = \frac{1}{(26.47 + 22.97 + 18.49 + 8.79 + 6.06 + 4.14 + 3.88)}$$

$$= \frac{1}{90.80}$$

$$\frac{1}{\sum_{i=1}^{n} u_i} = \frac{1}{(49 + 43 + 45 + 19 + 15 + 9 + 7)}$$
$$= \frac{1}{187.00}$$

According to the Eq. (3.10):

$$S_{RE} = (5.53.26.47.49.00) \otimes \left(\frac{1}{187.00}.\frac{1}{90.80}.\frac{1}{28.96}\right)$$
$$= (0.03.0.29.1.69)$$

$S_{LP} = (0.03.\ 0.25.\ 1.49)$
$S_{SP} = (0.03.\ 0.20.\ 1.55)$
$S_{FC} = (0.02.\ 0.10.\ 0.66)$
$S_{LD} = (0.02.\ 0.07.\ 0.52)$
$S_{SS} = (0.01.\ 0.05.\ 0.31)$
$S_{CE} = (0.01.\ 0.04.\ 0.24)$

According to the Eq. (3.15):

$V(S_{RE} \geq S_{LP}) = 1$
$V(S_{RE} \geq S_{SP}) = 1$
$V(S_{RE} \geq S_{FC}) = 1$
$V(S_{RE} \geq S_{LD}) = 1$
$V(S_{RE} \geq S_{SS}) = 1$
$V(S_{RE} \geq S_{CE}) = 1$

The priority weights are calculated by using Eq. (3.16):

$d'(S_{RE}) = \min(1, 1, 1, 1, 1, 1) = 1$
$d'(S_{LP}) = \min(0.97, 1, 1, 1, 1) = 0.97$
$d'(S_{SP}) = \min(0.97, 0.95, 1, 1, 1) = 0.95$
$d'(S_{FC}) = \min(0.76, 0.80, 0.85, 1, 1) = 0.76$
$d'(S_{LD}) = \min(0.68, 0.72, 0.78, 0.94, 1) = 0.68$
$d'(S_{SS}) = \min(0.53, 0.58, 0.64, 0.85, 0.93, 1) = 0.53$
$d'(S_{CE}) = \min(0.46, 0.50, 0.57, 0.80, 0.90, 0.99) = 0.46$

Priority weights form $W' = (1.\ 0.97.\ 0.95.\ 0.76.\ 0.68.\ 0.53.\ 0.46)$ vector (Eq. (3.17)). After the normalization of these values, the priority weights for each criterion are as follows (Eq. (3.18)):

$$W_{RE} = 0.19, W_{LP} = 0.18, W_{SP} = 0.18, W_{FC} = 0.14, W_{LD} = 0.13, W_{SS} = 0.10, W_{CE} = 0.09$$

Table 3.18 Aggregated fuzzy judgement matrix for alternatives under each criterion

RE	*HMN*	*HMS*	*PMN*	*PMS*
HMN	(1.1.1)	(0.11.0.2.1)	(0.11.0.42.1)	(0.11.0.13.0.2)
HMS	(1.5.28.9)	(1.1.1)	(0.2.2.33.5)	(0.11.0.19.1)
PMN	(1.4.14.9)	(0.2.0.8.5)	(1.1.1)	(0.11.0.14.0.33)
PMS	(5.7.57.9)	(1.5.85.9)	(3.7.9)	(1.1.1)
LP	*HMN*	*HMS*	*PMN*	*PMS*
HMN	(1.1.1)	(0.14.0.79.5)	(0.2.1.5)	(0.11.0.88.7)
HMS	(0.2.2.61.7)	(1.1.1)	(1.2.42.5)	(0.11.0.55.3)
PMN	(0.2.1.76.5)	(0.2.0.52.1)	(1.1.1)	(0.11.0.7.5)
PMS	(0.14.4.6.9)	(0.33.3.85.9)	(0.2.4.33.9)	(1.1.1)
SP	*HMN*	*HMS*	*PMN*	*PMS*
HMN	(1.1.1)	(0.2.1.5)	(0.2.1.66.7)	(0.14.1.24.7)
HMS	(0.2.1.76.5)	(1.1.1)	(0.33.2.42.7)	(0.11.1.04.5)
PMN	(0.14.1.36.5)	(0.14.0.67.3)	(1.1.1)	(0.2.0.8.5)
PMS	(0.14.2.31.7)	(0.2.2.61.9)	(0.2.2.33.5)	(1.1.1)
FC	*HMN*	*HMS*	*PMN*	*PMS*
HMN	(1.1.1)	(1.1.85.5)	(0.33.2.14.7)	(0.33.3.9)
HMS	(0.2.0.71.1)	(1.1.1)	(0.33.1.85.7)	(0.33.1.85.7)
PMN	(0.14.0.77.3)	(0.14.0.79.3)	(1.1.1)	(1.1.85.7)
PMS	(0.11.0.57.3)	(0.14.0.79.3)	(0.14.0.79.1)	(1.1.1)
LD	*HMN*	*HMS*	*PMN*	*PMS*
HMN	(1.1.1)	(1.1.28.5)	(1.2.14.5)	(1.2.14.5)
HMS	(0.2.0.9.1)	(1.1.1)	(1.2.14.5)	(1.2.42.7)
PMN	(0.2.0.61.1)	(0.2.0.61.1)	(1.1.1)	(1.1.57.7)
PMS	(0.2.0.61.1)	(0.14.0.6.1)	(0.14.0.88.1)	(1.1.1)
SS	*HMN*	*HMS*	*PMN*	*PMS*
HMN	(1.1.1)	(1.2.42.7)	(1.2.14.7)	(1.3.28.9)
HMS	(0.14.0.6.1)	(1.1.1)	(1.1.57.7)	(1.2.14.7)
PMN	(0.14.0.69.1)	(0.14.0.88.1)	(1.1.1)	(1.2.71.7)
PMS	(0.11.0.47.1)	(0.14.0.69.1)	(0.14.0.5.1)	(1.1.1)
CE	*HMN*	*HMS*	*PMN*	*PMS*
HMN	(1.1.1)	(1.1.57.7)	(1.2.42.7)	(1.2.71.7)
HMS	(0.14.0.88.1)	(1.1.1)	(1.1.85.7)	(1.1.85.7)
PMN	(0.14.0.6.1)	(0.14.0.79.1)	(1.1.1)	(1.1.3)
PMS	(0.14.0.58.1)	(0.14.0.79.1)	(0.33.1.1)	(1.1.1)

As defined clearly above, all these steps are also applied to calculate weights of the alternative under each criterion. After the consistency control, the aggregated fuzzy judgement matrix based on individual fuzzy judgement matrix of participants will be found as in Table 3.18.

Table 3.19 Weights of alternatives under each criterion

	HMN	HMS	PMN	PMS
ROI	0.01	0.30	0.28	0.41
LP	0.24	0.25	0.23	0.28
SP	0.25	0.25	0.24	0.26
FC	0.27	0.26	0.25	0.22
LD	0.29	0.29	0.25	0.18
SS	0.31	0.28	0.26	0.16
CE	0.32	0.29	0.21	0.18

Table 3.20 Final assessment for the ship investment decision

	HMN	*HMS*	*PMN*	*PMS*	*Criteria weight*	*HMN*	*HMS*	*PMN*	*PMS*
ROI	0.01	0.30	0.28	0.41	0.27	0.00	0.08	0.07	0.11
LP	0.24	0.25	0.23	0.28	0.25	0.06	0.06	0.06	0.07
SP	0.25	0.25	0.24	0.26	0.19	0.05	0.05	0.05	0.05
FC	0.27	0.26	0.25	0.22	0.22	0.06	0.06	0.05	0.05
LD	0.29	0.29	0.25	0.18	0.04	0.01	0.01	0.01	0.01
SS	0.31	0.28	0.26	0.16	0.02	0.01	0.01	0.01	0.00
CE	0.32	0.29	0.21	0.18	0.01	0.00	0.00	0.00	0.00
						0.19	0.27	0.25	**0.29**

*Example: HMN, 0.00 = 0.01*0.27*
*Example: HMS, 0.08 = 0.30*0.27 (inputs from* Table 3.11*)*
[a]*PMS has the highest priority weight with 0.29 final score which makes it the best choice*

After calculation of the mathematical algorithm of the extent synthesis analysis same as defining weight of criterion process, the following weights for the alternatives will be found (Table 3.19):

The results by FAHP for the ship investment decision problem are summarised in Table 3.20. For the final assessment, each weight of alternatives need to be multiplied with related to the weight of criterion, and then the sum of the column of each alternative need to be computed to find the best alternative. In this case, Panamax second hand purchasing is found the most feasible investment than other alternatives by considering expert consultation. According to the results, Handymax second hand is found the second best alternative for the owner's investment.

5 Conclusions

Among the MCDM methods, AHP is obviously one of the most sophisticated and functional approach for various selection and ranking problems. In this chapter, the theory of conventional AHP approach and its fuzzy set extension are investigated with a number of case studies in the maritime and logistics industry. As it is frequently addressed in the relevant sections, theoretical background of the AHP is usually overlooked, and such shortcuts of scholars lead to impractical or invalid applications. Therefore, this chapter is particularly dedicated to fundamentals of AHP with an extensive discussion on potential errors.

Although AHP method has a formal procedure which is employed for several decades, it is also very flexible for improvements and justifications in particular cases. For example, Duru et al. (2012) once developed a novel version of FAHP for alternative (choice) oriented prioritisation problem. In their study, expert judgement is expected to change based on given set of alternatives. The rationale behind the approach is that some criteria may not relevant for a subset of alternatives, and experts' comparison among criteria (criteria vs. criteria) may have significant difference for a group of choices. By classification of choices and choice oriented prioritisation, experts are given a flexibility of clustering their judgements based on subsets.

Another improvement may be gained from classification of experts based on criteria. In several AHP applications, experts are not really expert in all given set of criteria. Some of them may be superior in a subset of criteria while another group of experts are better at a different subset of criteria. An AHP problem may be consists of financial, technical and political aspects. A group of experts may be selected from financial practice, and they may be asked to compare alternatives under the financial criteria. Another group may be preferred for technical assessment, and that group may compare alternatives based on technical criteria. Technical experts do not respond to financial assessment while financial experts do not compare in terms of technical superiority.

As seen on above examples, AHP should not be assumed as a static approach due to some methodological doctrine. Scholar and practitioners are well encouraged to develop and improve AHP approach based on their particular needs and features of their MCDM problems.

Pairwise comparison is not an essential stage of the AHP as it is frequently introduced in the literature. In *Sect. 2.2*, an alternative of pairwise comparison, *rank by rating*, is introduced which is also suggested by the founder of AHP, Thomas Saaty. Considering the task load induced by long and time-consuming surveys, 'rank by rating' may be utilised whenever it is suitable for the objective or a particular comparison matrix.

Using fuzzy set approach is also not an essential need for the AHP. Especially when there is only a few criteria and alternatives, conventional AHP may work well. However, when the number of criteria and alternatives rises, uncertainty of decision makers also increase accordingly. In such cases, representative linguistic clusters (fuzzy sets) would be very functional and time saving. In the classical form, validation of consistency with large number of criteria is a problem which in turn requires reapplication of surveys (DELPHI-like iterations). FAHP may be more complicated at the moderators' side (complexity of arithmetic) while experts would find it easier.

Finally, we would like to emphasise the gap between theory and practice and how those insufficient applications may cause loss of time and waste of efforts. An

AHP application is strongly required to be designed in such a way that considers its fundamentals and counter-effects as well as the characteristics of human judgement and its erroneous nature.

References

Aguarón J, Moreno-Jiménez JM (2003) The geometric consistency index: approximated thresholds. Eur J Oper Res 147(1):137–145
Akao Y, Mazur GH (2003) The leading edge in QFD: past, present and future. Int J Qual Reliab Manag 20(1):20–35
Badiru AB, Cheung J (2002) Fuzzy engineering expert systems with neural network applications. Wiley, New York
Buckley JJ (1985) Fuzzy hierarchical analysis. Fuzzy Sets Syst 17(3):233–247
Bulut E, Duru O, Kececi T, Yoshida S (2012) Use of consistency index, expert prioritization and direct numerical inputs for generic fuzzy-AHP modeling: a process model for shipping asset management. Expert Syst Appl 39:1911–1923
Bulut E, Duru O, Koçak G (2014) Rotational priority investigation in fuzzy analytic hierarchy process design: an empirical study on the marine engine selection problem. Appl Math Model 39(2):913–923
Bulut E, Duru O, Yoshida S (2010) Multi-attribute decision making for crew nationality pattern selection in the shipping business: an empirical study for Turkish shipping case. Asian J Shipp Logist 26(1):139–152
Bulut E, Duru O, Yoshida S (2013) Market entry, asset returns, and irrational exuberance: asset management anomalies in dry cargo shipping. Int J Shipp Transp Logist 5(6):652–667
Callaway MR, Esser JK (1984) Groupthink: effects of cohesiveness and problem-solving procedures on group decision making. Soc Behav Personal Int J 12(2):157–164
Chang D-Y (1996) Applications of the extent analysis method on fuzzy AHP. Eur J Oper Res 95 (3):649–655
Cochran WG (1977) Sampling techniques, 3rd edn. Wiley, New York
Davey L (2017) If your team agrees on everything, working together is pointless. Harv Bus Rev, January 31, 2017 (online article)
Dhar R, Simonson I (2003) The effect of forced choice on choice. J Mark Res 40(2):146–160
Duru O, Bulut E, Yoshida S (2012) Regime switching fuzzy AHP model for choice-varying priorities problem and expert consistency prioritization: a cubic fuzzy-priority matrix design. Expert Syst Appl 39(5):4954–4964
Esser JK (1998) Alive and well after 25 years: a review of groupthink research. Organ Behav Hum Decis Process 73(2):116–141
Gilboa I (2010) Rational choice. MIT Press, Cambridge, USA
Hanbin B, Nuanchen W (2010) Research on the selection of scale in AHP. In: Advanced computer theory and engineering (ICACTE), 2010 3rd international conference on, vol 6, pp 1–108
Israely J (2009) The least known key economic indicator. TIME Magazine
Ji P, Jiang R (2003) Scale transitivity in the AHP. J Oper Res Soc 54(8):896–905
Joshi R, Banwet DK, Shankar R (2011) A Delphi-AHP-TOPSIS based benchmarking framework for performance improvement of a cold chain. Expert Syst Appl 38(8):10170–10182
Ka B (2011) Application of fuzzy AHP and ELECTRE to China dry port location selection. Asian J Shipp Logist 27(2):331–353
Kahneman D, Tversky A (1979) Prospect theory: an analysis of decision under risk. Econometrica 47:263–291
Kaufmann A, Gupta MM (1991) Introduction to fuzzy arithmetic: theory and applications. Van Nostrand Reinhold, New York

Lee M, Pham H, Zhang X (1999) A methodology for priority setting with application to software development process. Eur J Oper Res 118:375–389
Levin IP, Schnittjer SK, Thee SL (1988) Information framing effects in social and personal decisions. J Exp Soc Psychol 24(6):520–529
Lootsma FA (1993) Scale sensitivity in the multiplicative AHP and SMART. J Multi-Criteria Decis Anal 2(2):87–110
Ma D, Zheng X (1991) 9/9-9/1 scale method of AHP. In: Proceedings of the second international symposium on the AHP, vol 1. University of Pittsburgh, Pittsburgh, pp 197–202
Marks G, Miller N (1987) Ten years of research on the false-consensus effect: an empirical and theoretical review. Psychol Bull 102(1):72
McCauley C (1989) The nature of social influence in groupthink: compliance and internalization. J Pers Soc Psychol 57(2):250
Millet I, Saaty TL (2000) On the relativity of relative measures–accommodating both rank preservation and rank reversals in the AHP. Eur J Oper Res 121(1):205–212
Muchnik L, Aral S, Taylor SJ (2013) Social influence bias: a randomized experiment. Science 341 (6146):647–651
Prasad AVS, Somasekhara N (1990) The analytic hierarchy process for choice of technologies: an application. Technol Forecast Soc Chang 38(2):151–158
Robbins L (1932) The nature and significance of economic science. Philos Econ Anthol 1:73–99
Saaty TL (1977) A scaling method for priorities in hierarchical structures. J Math Psychol 15 (3):234–281
Saaty TL (1980) The analytic hierarchy process. McGraw-Hill, New York
Saaty TL (1987) Rank generation, preservation, and reversal in the analytic hierarchy decision process. Decis Sci 18(2):157–177
Saaty TL (2000) Fundamentals of decision making and priority theory with the analytic hierarchy process, vol 6. Rws Publications, Pittsburgh
Saaty TL (2006) Rank from comparisons and from ratings in the analytic hierarchy/network processes. Eur J Oper Res 168(2):557–570
Sahin B, Senol YE (2015) A novel process model for marine accident analysis by using generic fuzzy-AHP algorithm. J Navig 68(1):162–183
Salo AA, Hämäläinen RP (1997) On the measurement of preferences in the analytic hierarchy process. J Multi-Criteria Decis Anal 6(6):309–319
Schenkerman S (1994) Avoiding rank reversal in AHP decision-support models. Eur J Oper Res 74 (3):407–419
Scheufele DA, Tewksbury D (2007) Framing, agenda setting, and priming: the evolution of three media effects models. J Commun 57(1):9–20
Şen CG, Şen S, Başlıgila H (2010) Pre-selection of suppliers through an integrated fuzzy analytic hierarchy process and max-min methodology. Int J Prod Res 48(6):1603–1625. Retrieved from
Simon HA (1982) Models of bounded rationality: empirically grounded economic reason, vol 3. MIT Press, Cambridge, USA
Tadic D, Gumus AT, Arsovski S, Aleksic A, Stefanovic M (2013) An evaluation of quality goals by using fuzzy AHP and fuzzy TOPSIS methodology. J Intell Fuzzy Syst 25(3):547–556
Tavana M, Kennedy DT, Rappaport J, Ugras YJ (1993) An AHP-Delphi group decision support system applied to conflict resolution in hiring decisions. J Manag Syst 5(1):49–74
Thaler R (1985) Mental accounting and consumer choice. Mark Sci 4(3):199–214
UNCTAD (2013) Review of maritime transport. United Nations, New York
van Laarhoven PJM, Pedrycz W (1983) A fuzzy extension of Saaty's priority theory. Fuzzy Sets Syst 11(1–3):199–227
Wang Y-M, Elhag TMS (2006) An approach to avoiding rank reversal in AHP. Decis Support Syst 42(3):1474–1480
Weck M, Klocke F, Schell H, Rüenauver E (1997) Evaluating alternative production cycles using the extended fuzzy AHP method. Eur J Oper Res 100:351–366
Zadeh LA (1965) Fuzzy sets. Inf Control 8(3):338–353

Chapter 4
Identification of Success Factors for Green Shipping with Measurement of Greenness Based on ANP and ISM

Jingzheng Ren, Marie Lützen, and Hanna Barbara Rasmussen

Abstract Green shipping as an emerging concept which aims to mitigate the negative environmental impacts caused by shipping activities has received more and more attentions recently. However, there is a gap in knowledge how to take the efficacious measures, which makes it difficult for the stakeholders of shipping activities to promote green shipping. In order to fill this gap, this chapter proposed a generic methodology for establishing a criterion evaluation system for greenness assessment of shipping, including the identification of the success factors, the development of some strategic measures, and the analysis of the measures for enhancing the greenness of shipping. A criterion evaluation system which consists of multiple criteria in five aspects including: technological aspect, economic aspect, environmental aspect, social aspect, and managerial aspect has been firstly established. Subsequently, Analytic Network Process (ANP) has been employed to determine the relative importance of these factors in green shipping with the consideration of the interdependences and interactions among these criteria for evaluating the greenness of shipping, and they have been ranked from the most important to the least. Accordingly, the key success factors for green shipping have been obtained. Then, some strategic measures for helping the stakeholders enhance the greenness of shipping have been proposed. Finally, Interpretative Structuring Modeling (ISM) has been employed to analyze the cause-effect relationships among these measures and the features of these measures.

J. Ren (✉)
Department of Industrial and Systems Engineering, The Hong Kong Polytechnic University, Hong Kong, SAR, China
e-mail: jire@iti.sdu.dk

M. Lützen
Mechanical Engineering, Department of Technology and Innovation, University of Southern Denmark, Odense, Denmark

H.B. Rasmussen
Centre of Maritime Health and Society, University of Southern Denmark, Odense, Denmark

P.T.-W. Lee, Z. Yang (eds.), *Multi-Criteria Decision Making in Maritime Studies and Logistics*, International Series in Operations Research & Management Science 260,
DOI 10.1007/978-3-319-62338-2_4

Keywords Green shipping • Success factors • Analytic network process • Interpretative structuring modeling • Structural self-interaction matrix

1 Introduction

Shipping has been playing a dominant role for the world trade among all the transportation modes for the advantages of the large quantities of cargo with high efficiency and low cost for transportation. It is no doubt that shipping is significantly important for the world's economy development. However, it also caused many environmental problems and negative social impacts. The dominate emissions from ships consist of SO_x, NO_x, PM, and green-house gas (GHG) emissions, which further lead to serious environmental problems, i.e. climate change, acid rain, and damage on the ecosystem, and human health problems, i.e. asthma, cardiovascular disease, and lung cancer (Lam and van de Voorde 2012; Bailey and Solomon 2004).

The awareness of environment protection and human health has dramatically increased in the past few years. Accordingly, the concept of green shipping has attracted attentions from academics, industries and policy-makers (Prpić-Oršić et al. 2014). One of the most typical examples is the implementation of MARPOL Annex VI drafted by the International Maritime Organization (IMO) (IMO 2006), and it was put into effect for emissions control from shipping as an administrative measure. Accordingly, the stakeholders in environmental issues took their own effort to achieve the compliance to this kind of regulations. Many scholars started to study the ways for emissions reduction and energy-saving from shipping, i.e. speed reduction and ship hull design (Prpić-Oršić et al. 2014). The stakeholders (e.g. ship owners) in the maritime industry took various measures for green shipping, for instance they reuse waste heat and the utilization of wind power (Nuttall et al. 2014; Lun 2013). All the effort aims at promoting the development of green shipping.

There are various factors (e.g., fuel efficiency, GHG emissions, and social acceptability) affecting green shipping and also measures (e.g., innovations in ship design, R&D on green shipping, and education and training) that can be used for promoting the development of green shipping, while the stakeholders do not understand well the relative effects of these success factors and the effectiveness of the measures for promoting the development of green shipping. In order to fill this gap, this chapter aims at establishing the criterion evaluation system for green shipping, prioritizing the success factors of green shipping in terms of their relative importance, proposing measures for promoting green shipping, and investigating the interrelationships among the measures for ensuring the effectiveness of the selected measures.

2 Methodology

The objective of this chapter is to identify the key success factors for green shipping, to propose some strategic measures for enhancing the greenness, and to analyze these measures. This chapter firstly established an evaluation criterion system, which consists of multiple success factors for assessing the greenness of shipping. The first method used was the Analytic Network Process (ANP) method to prioritize the success factors in terms of their relative importance in green shipping. Based on the identified key factors that have significant effect on green shipping, strategic measures were developed and recommended, as presented in Fig. 4.1. The second method was Interpretative Structuring Modeling (ISM) to analyze the strategic measures for enhancing the greenness of shipping, which is illustrated in Fig. 4.2.

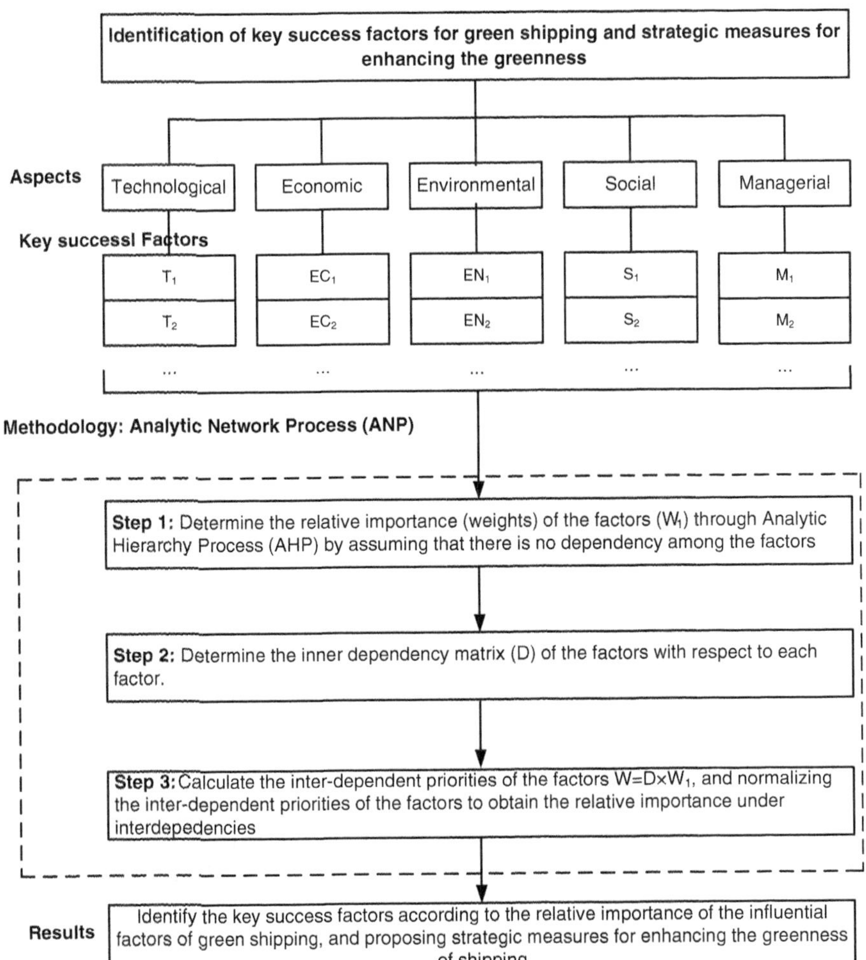

Fig. 4.1 Framework of the methodology

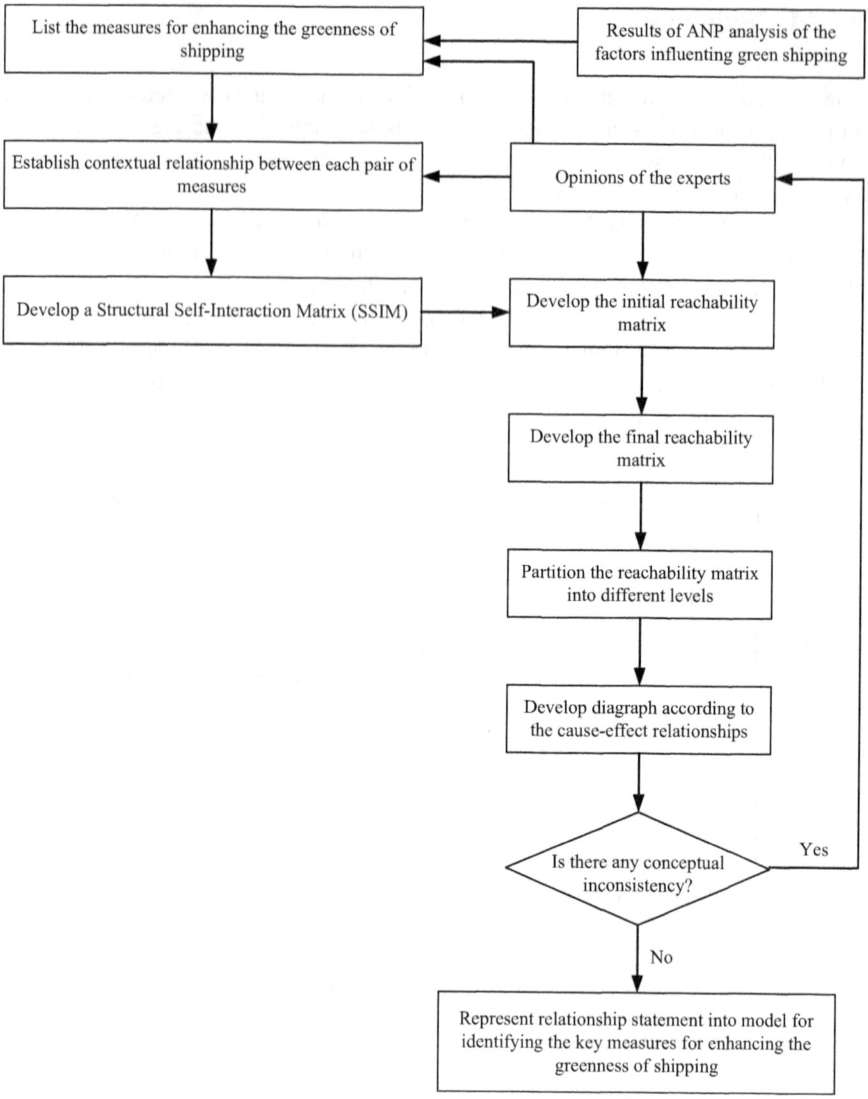

Fig. 4.2 Procedures of ISM methodology for identifying the influencing factors of green shipping (Modified from Kannan et al. (2009))

2.1 Evaluation Criterion System

The evaluation criterion system for green shipping is established based on literature review, and a total of 17 factors under the technological, economic, environmental, social, and managerial aspects have been obtained and presented in Table 4.1.

Table 4.1 Kay success factors for green shipping

Aspect	Factors	Code
Technological (T)	Fuel efficiency	T_1
	Advancement of ship design, engines and machinery	T_2
	Reuse of waste heat	T_3
	Proportion of clean energy use	T_4
Economic(EC)	Fuel cost	EC_1
	Investments on emissions reduction from shipping	EC_2
Environmental (EN)	NO_x emission	EN_1
	SO_x emission	EN_2
	GHG emissions	EN_3
	Particulate matters	EN_4
	Ballast water	EN_5
Social (S)	Social acceptability	S_1
	Corporate Social Responsibility	S_2
Managerial (M)	Inspection and maintenances on ship	M_1
	Compliance with regulations and standards	M_2
	Advance logistics and performance systems	M_3
	Perception of shipping stakeholders on environment protection	M_4

2.1.1 Technological Aspect

- Fuel efficiency

  Fuel efficiency, also usually called "energy efficiency", is defined by Eq. 4.1 in this chapter, and it is used to measure the level of the efficiency in terms of use of fuels (Lam and Lai 2015).

$$\text{Fuel efficiency} = \text{Energy consumption}/\text{Transport work} \tag{4.1}$$

where the total energy consumption including propulsion and auxiliary engines, and the transport work is calculated by multiplying the ship's capacity (dwt), as designed, with the ship's design speed measured at the maximum design load condition and at 75% of the rated installed shaft power (AMSA 2012).

- Advancement of ship design, engines and machinery

  This factor is to measure the advancement of ship design, engines and machinery as better ship design, engines and machinery can increase fuel efficiency, reduce energy consumption, mitigate the possibility of oil spill, and lower emissions (Lam and Lai 2015).
- Reuse of waste heat

  The waste heat can be reused through waste heat recovery systems to reduce fuel consumption (Lun 2013).
- Proportion of clean energy use

The proportion of clean energy use is defined as the percentage of the clean energy (e.g., low sulfur fuels, wind power, solar power, and electricity from hydropower) used in the total consumed energy, and it is a measure of the effort for reducing the dependency on traditional fossil fuels (Nuttall et al. 2014).

2.1.2 Economic Aspect

- Fuel cost

  Fuel cost refers to the cost associated with fuel consumption in the propulsion and auxiliary engines, and boilers (Doudnikoff and Lacoste 2014).
- Investments on emissions reduction from shipping

  Investments refer to the capital cost, installation cost and annual operating cost for taking techniques (e.g. seawater scrubbing technology, selective catalytic reduction technology, and exhaust gas scrubber) for emissions reduction from shipping (Yang et al. 2012; Ogbonnaya et al. 2013; Zhang 2009), and it is a measure of the effort made by the stakeholders for reducing environmental impacts of shipping. However, the high investment is also an important obstacle that hinders the ship owners to take techniques for emission reductions from shipping.

2.1.3 Environmental Aspect

- NOx emissions

  NOx emissions refer to nitrogen monoxide (NO) and nitrogen dioxide (NO_2) emissions that can cause acid rain phenomenon, and this phenomenon will further lead to over-fertilization of smog formation (Turliainen 2005).
- SOx emissions

  SOx emissions consist of sulfur dioxide (SO_2) and sulfur trioxide (SO_3) that also cause to generate acid rain and have negative impact on vegetation, human health and building (Turliainen 2005).
- GHG emissions

  GHG emissions represent the gases that cause global warming in association with carbon dioxide (CO_2), methane (CH_4), nitrous oxide (N_2O), and hydrofluorocarbons (HFCs), and etc. (Helfre and Boot 2013).
- Particulate matters

  The particulate matters (PM) is related to the poor quality of marine fuel and is produced from the combustion process and mainly refer to $PM_{2.5}$ and PM_{10}, and mainly in the form of soot and ash (Cullinane and Cullinane 2013). The composition and properties of PM vary greatly, and the estimated amount of global PM released from shipping in 2007 was 1.8 million tones (Helfre and Boot 2013; EMCC 2009–2010).

- Ballast water
  Ballast water management is quite important because the inappropriate treatment of ballast would lead to serious water pollutions and further influence the biodiversity of the sea (Voorham 2013).

2.1.4 Social

- Social acceptability
  Social acceptability is used to measure the attitudes of the public to environmental impacts caused by shipping activities as well as the satisfaction level to the effort taken by shipping owners and operators for environment protection (Liu 2012).
- Corporate Social Responsibility
  The concept of Corporate Social Responsibility (CSR) is defined as "a concept whereby companies integrate social and environmental concerns in their business operations and in their interaction with the stakeholders on a voluntary basis" by the European Commission (EC 2010; Walmsley 2012).

2.1.5 Managerial

- Compliance with regulations and standards
  This factor is to measure how shipping companies can fulfill the regulation and standards for green and clean shipping (Lam and Lai 2015).
- Inspection and maintenance on ship
  Inspection and maintenance on ship refer to the replacement of the parts in the propulsion and auxiliary engines of ship, painting, and fouling control and the like. (Lam and Lai 2015; Fan 2012).
- Advanced logistics and performance systems
  Advanced logistics and performance systems aim at using the measures such as weather forecast, speed control, planning routes, optimizing sailing schedules and changing ports for reducing voyage distance, delivering on time, lowering energy consumption, ad mitigating emission (Lam and Lai 2015; Lun 2013).
- Perception of shipping stakeholders on environment protection

2.2 *Analytic Network Process*

Analytic Network Process (ANP) is derived from Analytic Hierarchy Process (AHP) which was developed by Saaty (1980). AHP can decompose a complex multi-criteria decision-making problem into different levels which may consist of several sub-factors by establishing a hierarchal structure, but it cannot consider the interdependences among the sub-factors in each level. ANP is an extension of AHP

Table 4.2 Scale of relative importance used in the comparison matrix (Saaty 1980)

Scales	Definition	Note
1	Equal importance	i is equally important to j.
3	Moderate importance	i is moderately important to j.
5	Essential importance	i is essentially important to j.
7	Very Strong importance	i is very strongly important to j.
9	Absolute importance	i is very absolutely important to j.
2, 4, 6, 8	Intermediate value	The relative importance of i to j is between to adjacent judgment.
Reciprocal	The value had been assigned to i when compared to j, then j has the reciprocal value compared to i.	

to address this problem. The process of ANP consists of three steps as follows (Shahabi et al. 2014; Lam and Lai 2015; Dağdeviren and Yüksel 2010):

Step 1: Determine the relative importance (weights) of the factors through Analytic Hierarchy Process (AHP) by assuming that there is no dependency among the factors.

Conducting pairwise comparisons by assuming that all the studies factors are independent, and a ratio scale of 1 to 9 is used to compare two factors (as presented in Table 4.2). Assuming that there are n factors, then, the comparison matrix can be determined, as shown in Eq. 4.2.

$$A = \begin{vmatrix} 1 & a_{12} & \cdots & a_{n1} \\ a_{21} & 1 & \cdots & a_{n2} \\ \vdots & \cdots & \ddots & \vdots \\ a_{n1} & a_{n2} & \cdots & 1 \end{vmatrix} \tag{4.2}$$

where A is the comparison matrix, and a_{ij} denotes the relative importance of criteria i comparing with j.

According to the comparison matrix, the weight of each criterion can be obtained by calculating the principal eigenvector of the comparison matrix, as shown in Eq. 4.3.

$$\begin{vmatrix} 1 & a_{12} & \cdots & a_{n1} \\ a_{21} & 1 & \cdots & a_{n2} \\ \vdots & \cdots & \ddots & \vdots \\ a_{n1} & a_{n2} & \cdots & 1 \end{vmatrix} \begin{vmatrix} w_1 \\ w_2 \\ \vdots \\ w_n \end{vmatrix} = \lambda_{\max} \begin{vmatrix} w_1 \\ w_2 \\ \vdots \\ w_n \end{vmatrix} \tag{4.3}$$

where $(w_1, w_2, \cdots, w_n)^T$ is the maximal eigenvector of matrix A, $\lambda_{\max}$ is the maximal eigenvalue of matrix A.

Accordingly, the weight vector of the n factors can be determined, denoted by $W_1 = (w_1, w_2, \cdots, w_n)^T$.

Table 4.3 The value of the average random consistency index RI

n	1	2	3	4	5	6	7	8	9	10	11	12
RI	0	0	0.58	0.90	1.12	1.24	1.32	1.41	1.45	1.49	1.52	1.54

The consistency check can be carried out to judge whether a comparison matrix is consistent or not by calculating the consistency ratio according to Eqs. 4.4–4.5.

$$CI = \frac{\lambda_{\max} - n}{n - 1} \tag{4.4}$$

$$CR = \frac{CI}{RI} \tag{4.5}$$

where CI is consistency index, $\lambda_{\max}$ represents the maximal eigenvalue of the comparison matrix A, n represents the dimension of the matrix, CR is consistency ratio, and RI is the average random index with the same dimension with A. RI can be determined according to Table 4.3.

Accordingly, when CR <0.1, the matrix could be regarded as an acceptable matrix, otherwise, the comparison matrix should be modified until to an acceptable one.

Step 2: Determine the inner dependency matrix (*D*) of the factors with respect to each factor. The elements of the j-th column vector in matrix D represent the relative effects of the factors on the j-th factor, and this vector can be obtained through establishing the comparison matrix with respect to the j-th factor. Similarly, all the column vectors in matrix D can be obtained.

$$D = \begin{vmatrix} 1 & d_{12} & \cdots & d_{n1} \\ d_{21} & 1 & \cdots & d_{n2} \\ \vdots & \cdots & \ddots & \vdots \\ d_{n1} & d_{n2} & \cdots & 1 \end{vmatrix} \tag{4.5}$$

where D is the inner dependency matrix.

It is worth pointing out that the i-th row vector represents the relative effect of the i-th factor on the other factors, and all the diagonal elements in matrix D equal 1 according to Ref. (Dağdeviren et al. 2008; Zamani et al. 2014).where d_{ij} represents the relative effect of the i-th factor on the j-th factor.

Step 3: Calculating the inter-dependent priorities of the n factors by Eq. 4.7, then normalizing the inter-dependent priorities of the n factors by Eq. 4.8.

$$\mathrm{W}' = D \times \mathrm{W}_1 = [\omega_1, \omega_2, \cdots, \omega_n] \tag{4.7}$$

$$W = \left[\omega_1 / \sum_{i=1}^{n} \omega_i, \omega_2 / \sum_{i=1}^{n} \omega_i, \cdots, \omega_n / \sum_{i=1}^{n} \omega_i\right] \quad (4.8)$$

where W' represents weight vector of the inter-dependent priorities of the investigated factors, and W represents the normalized weight vector of the inter-dependent priorities of the factors.

2.3 *Interpretative Structuring Modeling*

ISM developed by Warfield (1974a and 1974b) is a computer-assistant methodology for investigating the complex interrelationships among multiple elements involved in a complex system (Kannan et al. 2009). This method has been employed to investigate the complex cause-effect relationships among the strategic measures for enhancing the greenness of shipping that will be proposed according to the results of ANP analysis of the success factors of green shipping. The procedures of ISM have been illustrated in Fig. 4.2, for more specified procedures can refer the works of Kannan et al. (2009) and Luthra et al. (2014).

Step 1: Determine the structural self-interaction matrix (SSIM). 'V', 'A', 'X', and 'O' are used to represent the relationship between each pair of strategic measures.

V: The i-th measure will exert the j-th measure;
A: The j-th measure will be exerted by the i-th measure;
X: The i-th measure and the j-th measure will exert each other; and.
O: No direct relationship between the i-th measure and the j-th measure.

Step 2: Determine the initial reachability matrix. The SSIM will be transformed into the initial reachability matrix according to the following rules:

- If the element in the cell (i, j) of the SSIM is V, the elements in the cell (i, j) and cell (j, i) of the initial reachability matrix should be 1 and 0, respectively.
- If the element in the cell (i, j) of the SSIM is A, the elements in the cell (i, j) and cell (j, i) of the initial reachability matrix should be 0 and 1, respectively.
- If the element in the cell (i, j) of the SSIM is X, both the elements in the cell (i, j) and cell (j, i) of the initial reachability matrix should be 1.
- If the element in the cell (i, j) of the SSIM is O, both the elements in the cell (i, j) and cell (j, i) of the initial reachability matrix should be 0.

Step 3: Determine the final reachability matrix (FRM), and the driving power and dependence power of each measure. The final reachability matrix (FRM) can be obtained through transitivity. The transitivity of the contextual relation is a basic assumption in ISM, and it represents that if the i-th measure exerts the j-th measure, and the j-th measure exerts the k-th measure, then the i-th measure is regarded to exert the k-th measure. According to the final reachability matrix, the driving power and the dependence power of each measure can be determined by all ones in the rows and all ones on the columns with respect that measure, respectively.

Step 4: Level partitions. The reachability set with respect to each strategic measure consists of all the measures which were exerted by that measure and the measure itself. The antecedent set with respect to each strategic measure consists of all the measure that exerted that measure and the measure itself. The intersection set with respect to each measure represents the intersection between the reachability set and the antecedent set. After determining the reachability set, antecedent set, and interaction set with respect to each measure from the final reachability matrix, the measures belonging to level I in ISM hierarchy can be determined through identifying the measures with respect to which the reachability set and intersection set are the same. The measures belonging to level I should be discarded to identify the measures belonging to level II in ISM hierarchy through repeating the process of finding the measures belonging to level I. Similarly, the measures in the other levels of ISM hierarchy can be identified with iterations.

Step 5: Determine the ISM hierarchy model. According to the final reachability matrix, and various ISM hierarchy levels determined in Step 5, the measures in the same level and that in the adjacent levels are connected in the format of diagraph by vertices and edges. Then, the ISM hierarchy model can be determined after removing the transitivities among the measures.

3 Results

3.1 Identification of Success Factors of Green Shipping by ANP

ANP has been employed to determine the relative weights of the success factors in terms of their relative importance, the comparison matrices were determined based on a focus group meeting in which three experts in Denmark about green shipping and maritime technology were invited to participate, and it is worth pointing out that the judgments of the three experts were based on the inquiries of many other scholars and engineers and literature reviews. The procedures are specified as follows:

1. AHP was firstly used to determine the weights of the five aspects and that of the factors in each aspect with the assumption that there is no interdependence among the five aspects and also no interdependence among the factors in each aspect, and the results are presented in Tables 4.4, 4.5, 4.6, 4.7, 4.8 and 4.9. It is worth pointing out that the comparison matrices are determined by comparing

Table 4.4 Comparison matrix for calculating the weights of the five aspects (W_1)

Aspects	T	EC	EN	S	MO	Weights
Technological (T)	1	**3**	1/2	5	1	0.2241
Economic (EC)	1/3	1	1/5	3	1/3	0.0881
Environmental (EN)	2	5	1	7	2	0.4066
Social (S)	1/5	1/3	1/7	1	1/7	0.0409
Managerial (M)	1	3	1/2	7	1	0.2402
$\lambda_{max} = 5.0719$, CI = 0.0180, CR = 0.0160 < 0.1						

Note: λ_{max} is the maximal eigenvalue of the comparison, CI represents the consistency index, and CR represents the consistency ratio, it is less than 10% means that the comparison matrix is acceptable for consistency check

Table 4.5 Comparison matrix for calculating the weights of the factors in technological

Technological	T_1	T_2	T_3	T_4	Weights
Fuel efficiency (T_1)	1	1/3	2	2	0.2090
Advancement of ship design, engines and machinery (T_2)	3	1	5	5	0.5725
Reuse of waste heat (T_3)	1/2	1/5	1	1	0.1093
Proportion of clean energy use (T_4)	1/2	1/5	1	1	0.1093
$\lambda_{max} = 4.0042$, CI = 0.0014, CR = 0.0015 < 0.1					

Table 4.6 Comparison matrix for calculating the weights of the factors in economic aspect

Economic	EC_1	EC_2	Weights
Fuel cost (EC_1)	1	1	0.50
Investments on emissions reduction from shipping (EC_2)	1	1	0.50
NA for consistency check			

Table 4.7 Comparison matrix for calculating the weights of the factors in environmental aspect

Environmental	EN_1	EN_2	EN_3	EN_4	EN_5	Weights
NO_x emission (EN_1)	1	1	2	3	3	0.3133
SO_x emission (EN_2)	1	1	2	3	3	0.3133
GHG emission (EN_3)	1/2	1/2	1	2	2	0.1763
Particulate matters (EN_4)	1/3	1/3	1/2	1	1	0.0986
Ballast water treatment (EN_5)	1/3	1/3	1/2	1	1	0.0986
$\lambda_{max} = 5.0133$, CI = 0.0033, CR = 0.0030 < 0.1						

Table 4.8 Comparison matrix for calculating the weights of the factors in social aspect

Social	S_1	S_2	Weights
Social acceptability (S_1)	1	1/4	0.2000
Corporate social responsibility (S_2)	4	1	0.8000
NA for consistency check			

Table 4.9 Comparison matrix for calculating the weights of the factors in managerial aspect

Managerial (M)	M_1	M_2	M_3	M_4	Weights
Inspection and maintenances on ship (M_1)	1	1/2	1/4	1/5	0.0812
Compliance with regulations and standards (M_2)	2	1	1/3	1/3	0.1399
Advanced logistics and performance systems (M_3)	4	3	1	1	0.3794
Perception of shipping stakeholders on environment protection (M_4)	5	3	1	1	0.3996

$\lambda_{max} = 4.0155$, $CI = 0.0052$, $CR = 0.0057 < 0.1$

each pair of factors by using 1, 3, 5, 7, 9 representing the phrases ‘equal importance’, ‘moderate importance’, ‘essential importance’, ‘very strong importance’, and ‘absolute importance’, respectively, and the intermediate values to answer the question:” how do you think the importance of the factor i compared with the factor j?”. Taking the comparison matrix for calculating the weights of the five aspects as an example (see Table 4.4), the element in cell (1,2) of Table 4.4 represents the relative importance of the technological aspect compared with the economic aspect. The decision-maker can find this value of this element by answering “how do you think the importance of the technological aspect compared with the economic aspect?”. Based on knowledge and experience on shipping, and the understanding from literature reviews the authors determined the importance of the technological aspect compared with the economic aspect as ‘moderate importance’; accordingly, the corresponding value of the phrase ‘moderate importance’ is 3, so 3 was put in cell (1,2) of Table 4.4. Similarly, the other elements of Table 4.4 can also be determined.

2. The inner dependency matrix of the five aspects with respect to each aspect can also be determined. Taking the relative priorities of economic, environmental, social, managerial aspects with respect to the technological aspect as an example, the comparison matrix for determining the relative importance of the four aspects including economic, environmental, social, and managerial aspects with respect to the technological can be firstly determined by comparing each pair of the two aspects within the four aspects. For example, the cell (1,2) of Table 4.10a represents the relative importance of the economic aspect compared with the environmental aspect with respect to technological aspect. The authors held the view that the booming of economy (economic) could significantly stimulate the development of new technologies for green shipping and promote the shipping stakeholders to adopt clean technologies for green shipping. The environmental deterioration could awake stakeholders to aware environmental protection (environmental aspect), and could further promote the development of science and technology for green shipping. Based on the above-mentioned analysis, the authors thought that the relative importance of the economic aspect compared with the environmental aspect with respect to technological aspect should be

Table 4.10 Comparison matrices for determining the inner dependency matrix of the five aspects with respect to each aspect (a–e)

a

Technological (T)	EC	EN	S	M	Weights
Economic (EC)	1	2	5	1	0.3587
Environmental (EN)	1/2	1	3	1/3	0.1713
Social (S)	1/5	1/3	1	1/5	0.0689
Managerial (M)	1	3	5	1	0.4011

$\lambda_{max} = 4.0341$, CI = 0.0114, CR = 0.0126 < 0.1

b

Economic (EC)	T	EN	S	M	Weights
Technological (T)	1	3	5	1/2	0.3005
Environmental (EN)	1/3	1	2	1/5	0.1098
Social (S)	1/5	1/2	1	1/7	0.0630
Managerial (M)	2	5	7	1	0.5267

$\lambda_{max} = 4.0201$, CI = 0.0067, CR = 0.0074 < 0.1

c

Environmental (EN)	T	EC	S	M	Weights
Technological (T)	1	3	7	1	0.3962
Economic (EC)	1/3	1	3	1/4	0.1327
Social (S)	1/7	1/3	1	1/6	0.0559
Managerial (M)	1	4	6	1	0.4152

$\lambda_{max} = 4.0409$, CI = 0.0136, CR = 0.0152 < 0.1

d

Social (S)	T	EC	EN	M	Weights
Technological (T)	1	1/3	1/5	1/2	0.0851
Economic (EC)	3	1	1/3	2	0.2398
Environmental (EN)	5	3	1	3	0.5232
Managerial (M)	2	1/2	1/3	1	0.1519

$\lambda_{max} = 4.0593$, CI = 0.0198, CR = 0.0220 < 0.1

e

Managerial (M)	T	EC	EN	S	Weights
Technological (T)	1	1/2	1/4	1/3	0.0954
Economic (EC)	2	1	1/3	1/2	0.1601
Environmental (EN)	4	3	1	2	0.4673
Social (S)	3	2	1/2	1	0.2772

$\lambda_{max} = 4.0310$, CI = 0.0103, CR = 0.0115 < 0.1

Table 4.11 Inner dependency matrix with respect to the five aspects (D)

	T	EC	EN	S	M
Technological (T)	1	0.3005	0.3962	0.0851	0.0954
Economic (EC)	0.3587	1	0.1327	0.2398	0.1601
Environmental (EN)	0.1713	0.1098	1	0.5232	0.4673
Social (S)	0.0689	0.0630	0.0559	1	0.2772
Managerial (M)	0.4011	0.5267	0.4152	0.1519	1

Table 4.12 Global weights of the factors determined by ANP

Aspect	Factors	Local weights	Global weights
Technological (0.2191)	T_1	0.2090	0.0458
	T_2	0.5725	0.1254
	T_3	0.1093	0.0239
	T_4	0.1093	0.0239
Economic(0.1354)	EC_1	0.50	0.0677
	EC_2	0.50	0.0677
Environmental (0.2942)	EN_1	0.3133	0.0922
	EN_2	0.3133	0.0922
	EN_3	0.1763	0.0519
	EN_4	0.0986	0.0290
	EN_5	0.0986	0.0290
Social (0.0756)	S_1	0.2000	0.0151
	S_2	0.8000	0.0605
Managerial (0.2758)	M_1	0.0812	0.0224
	M_2	0.1399	0.0386
	M_3	0.3794	0.1046
	M_4	0.3996	0.1102

between 'equal importance' (corresponding to the value 1) and 'moderate importance' (corresponding to the value 3), so that the element in cell (1,2) of Table 4.10a should be 2. Similarly, all the other elements in the comparison matrices for determining the inner dependency matrix of the five aspects with respect to each aspect can also be determined, and the results are presented in Table 4.10a–e. Then, the inner dependency matrix with respect to the five aspects can be determined, as presented in Table 4.11.

3. The inter-dependent priorities of the five aspects can be determined according to Eq. 4.7, and $W' = [0.4381, 0.2707, 0.5883, 0.1512, 0.5515]$.
4. Then, the normalized inter-dependent priorities of the five aspects can be obtained according to Eq. 4.8, and $W = [0.2191, 0.1354, 0.2942, 0.0756, 0.2758]$.

Finally, the global weight of each success factor can be determined by combining the local weight of each criterion and that of the aspect to which it belongs to, The global weights of the factors determined by ANP are presented in Table 4.12.

These success factors can be divided into three types according to their relative importance in green shipping according to Table 4.12. The first category is significantly important type, which consists following elements:

- advancement of ship design, engines and machinery,
- perception of shipping stakeholders on environment protection,
- advanced logistics and performance systems, and
- NO_x emission, SO_x emission, fuel cost, and investments on emissions reduction from shipping.

The second category is moderately important type, including:

- corporate social responsibility,
- GHG emission,
- fuel efficiency, and
- compliance with regulations and standards.

The third category is less important type, which includes the other success factors. Therefore, advancement of ship design, engines and machinery, perception of shipping stakeholders on environment protection, advanced logistics and performance systems, NO_x emission, SO_x emission, fuel cost, and investments on emissions reduction from shipping are the key factors for successfully achieving green shipping.

3.2 *Strategic Measures for Enhancing the Greenness*

In order to promote the development of green shipping, eight strategic measures are obtained by considering how to address the factors in significantly important type, as presented in Table 4.13. For instance, advancement of ship design, engines and machinery is the most important success for green shipping through the ANP analysis, and two measures including innovations in ship design and R&D on green shipping are recommended to achieve the advancement of ship design, engines and machinery. In order to increase perception of shipping stakeholders on environment protection, the education and training is prerequisite. Advanced logistics and performance systems can be achieved by smart management and operations, and R&D on green shipping. Policy measures for emission reduction from shipping, and adoption of clean techniques are beneficial for mitigating NO_x emission and SO_x emission. Recycling and reuse of energy and natural resource on board can save the consumption of energy and further lower fuel cost. Subsidies and tax credits can help the shipping stakeholders overcome the financial difficulties, and encourage them to invest on technologies and measures for green shipping.

Table 4.13 Key success factors for green shipping and corresponding measures for address these measures

No.	Factors	Measures
1	Advancement of ship design, engines and machinery	Innovations in ship design; R&D on green shipping
2	Perception of shipping stakeholders on environment protection	Education and training
3	Advanced logistics and performance systems	Advanced logistics and performance systems; R&D on green shipping
4	NO_x emission	Policy measures for emission reduction from shipping; adoption of clean techniques
5	SO_x emission	Policy measures for emission reduction from shipping; adoption of clean techniques
6	Fuel cost	Recycling and reuse of energy and natural resource on board
7	Investments on emissions reduction from shipping	Subsidies and tax credits

3.2.1 Innovations in Ship Design (M_1)

Innovations in ship design refer to the innovative efforts on equipment of ships for energy saving and emissions reduction, i.e. optimization of hull design, turbo charger cut off for improving fuel efficiency of propulsion, air lubrication for reducing energy consumption of vessel, hull coating for reducing roughness and drag, and propeller optimization to increase propeller efficiency (Voorham 2013; Cullinane and Cullinane 2013).

3.2.2 R&D on Green Shipping (M_2)

R&D on sustainable shipping means Research, Development and Demonstration of green shipping, and it aims at setting special funding for encouraging the research on ship innovation for energy-saving and emissions reduction, promoting the development of green shipping through demonstrations.

3.2.3 Education and Training (M_3)

Education and training aiming at educating and training on the crew to improve their environmental awareness and the operating skills can increase the consciousness of corporate social responsibility, reduce fuel consumption, mitigate emissions, correctly treat ballast water (McConnell 2002).

3.2.4 Policy Measures for Emission Reduction from Shipping (M_4)

The policy measures consist of three types: command and control measures, economic/market based measures, information strategies/self-regulations. Command and control measures refer the mandatory regulations and standards draft by national governments or international organizations, i.e. mandatory Energy Efficiency Design Index (EEDI) and MARPOL Annex VI. Economic/market based measures represent the measures for emissions reduction from shipping based on economic-oriented or market-oriented incentive strategies, i.e. emissions trade and emissions tax. Information strategies/self-regulations are the voluntary measures drafted or actions taken by the shipping stakeholders (i.e. ship owners and ship operators), i.e. voluntary EEDI and Corporate Social Responsibility (CSR) (Gunningham et al. 1998; Rehmatulla 2011).

3.2.5 Adoption of Clean Techniques (M_5)

Adoption of clean techniques refers to adopting the technologies for emissions (i.e. CO_2, SO_x and NO_x) reduction and energy-saving by refitting or installing new equipment, e.g. scrubbers, selective catalytic reduction, the installation of ballast water treatment system, and using alternative clean fuels (Voorham 2013).

3.2.6 Subsidies and Tax Credits (M_6)

Implementing the techniques for energy-saving and emission reduction is usually high-cost and low-return in a short term, thus, subsidies and tax credits can effectively encourage the ship owners to invest on the technologies for energy-saving and emission reduction. For instance, some governments have implements subsidies to give refund on investments that can reduce the initial investment and effectively influence the investment decisions (Voorham 2013).

3.2.7 Smart Management and Operations (M_7)

Smart management and operations refer to the management strategies and operational measures adopted by the operators (i.e., the crew on ships and the staff on ports) for emissions (i.e., CO_2, SO_x and NO_x) reduction and energy-saving, i.e. speed reduction, route and schedule optimization, and enhanced fleet management and others (Lam and Lai 2015; Voorham 2013).

3.2.8 Recycling and Reuse of Energy and Natural Resource on Board (M_8)

Recycling and reuse of energy and natural resource on board aims at saving energy and natural resources through recycling and reuse, i.e. waste heat recovery and lubricant recycling (Lam and Lai 2015; Voorham 2013).

3.3 Analysis of the Strategic Measures

ISM has been employed to analyze the interrelationships among the eight strategic measures for promoting green shipping. The SSIM was firstly established by using 'V', 'A', 'X', and 'O', as presented in Table 4.14. For instance, the first measure 'innovations in ship design (M_1)' is regarded to exert the eighth measure 'recycling and reuse of energy and natural resource on board (M_8)', so 'V' is put in cell (1, 8) of the SSIM.

Subsequently, the initial reachability matrix can be obtained by transforming V′, 'A', 'X', and 'O used in the SSIM into 0 and 1, as shown in Table 4.15.

Accordingly, the final reachability matrix can be determined according to the transitivity rule. For instance, the first measure 'innovations in ship design (M_1)' exerts the fifth strategic 'adoption of clean techniques (M_5)', and M_5 exerts the

Table 4.14 Structural self-interaction matrix (SSIM)

Measures	M_8	M_7	M_6	M_5	M_4	M_3	M_2	M_1
M_1	V	O	A	V	A	A	A	–
M_2	X	X	A	X	A	A	–	
M_3	V	V	O	V	A	–		
M_4	V	V	A	V	–			
M_5	O	O	A	–				
M_6	V	V	–					
M_7	O	–						
M_8	–							

Table 4.15 The initial reachability matrix

Measures	M_1	M_2	M_3	M_4	M_5	M_6	M_7	M_8
M_1	1	0	0	0	1	0	0	1
M_2	1	1	0	0	1	0	1	1
M_3	1	1	1	0	1	0	1	1
M_4	1	1	1	1	1	0	1	1
M_5	0	1	0	0	1	0	0	0
M_6	1	1	0	1	1	1	1	1
M_7	0	1	0	0	0	0	1	0
M_8	0	1	0	0	0	0	0	1

Table 4.16 The final reachability matrix

Measures	M_1	M_2	M_3	M_4	M_5	M_6	M_7	M_8	Driving
M_1	1	1[a]	0	0	1	0	1[a]	1	5
M_2	1	1	0	0	1	0	1	1	5
M_3	1	1	1	0	1	0	1	1	6
M_4	1	1	1	1	1	0	1	1	7
M_5	1[a]	1	0	0	1	0	1[a]	1[a]	5
M_6	1	1	1[a]	1	1	1	1	1	8
M_7	1[a]	1	0	0	1[a]	0	1	1[a]	5
M_8	1[a]	1	0	0	1[a]	0	1[a]	1	5
Dependence	8	8	3	2	8	1	8	8	

[a]Incorporating transitivity

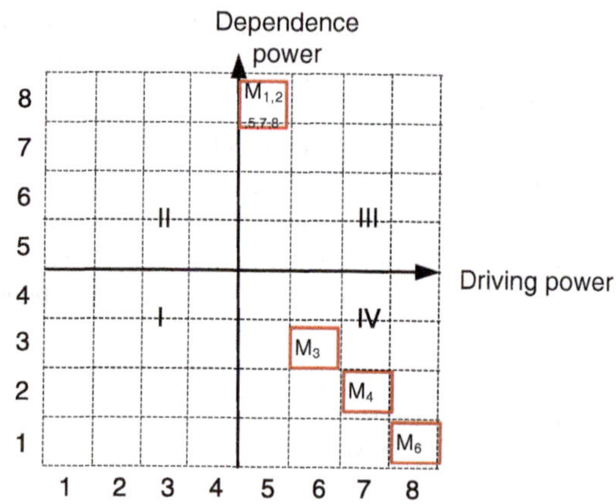

Fig. 4.3 Driving power and dependence power diagram

second measure 'R&D on green shipping (M_2)' so M_1 exerts M_2. In a similar way, the final reachability matrix was obtained, as presented in Table 4.16.

Then, the driving power and dependence of each measure can be obtained by calculating the sum of the corresponding row and column in the final reachability matrix, respectively (see Table 4.16). The driving power and dependence power diagram has been presented in Fig. 4.3. The driving power and dependence power diagram can be divided into four quadrants including autonomous quadrant (Quadrant-I), dependent quadrant (Quadrant-II), linkage quadrant (Quadrant-III), and independent quadrant (Quadrant-V). According to Ref. (Kannan et al. 2009), the measures in Quadrant-I have weak driving power and dependence power, and all the measures are highly connected. The measures in Quadrant-II have weak driving power but strong dependence power, and these three measures were highly affected by others. The measures in Quadrant-III have both strong driving power and strong dependence power, but they are unstable, they would affect other measures with a disturbance, and they would also be affected by feedbacks. The measures in Quadrant-IV have weak dependence power but strong driving power. It

Table 4.17 Level partition of the measures-Iteration 1

	Reachability set	Antecedent set	Intersection set	Level
S_1	M_1, M_2, M_5, M_7, M_8	$M_1, M_2, M_3, M_4, M_5, M_6, M_7, M_8$	M_1, M_2, M_5, M_7, M_8	I
S_2	M_1, M_2, M_5, M_7, M_8	$M_1, M_2, M_3, M_4, M_5, M_6, M_7, M_8$	M_1, M_2, M_5, M_7, M_8	I
S_3	$M_1, M_2, M_3, M_5, M_7, M_8$	M_3, M_4, M_6	M_3	
S_4	$M_1, M_2, M_3, M_4, M_5, M_7, M_8$	M_4, M_6	M_4	
S_5	M_1, M_2, M_5, M_8	$M_1, M_2, M_3, M_4, M_5, M_6, M_7, M_8$	M_1, M_2, M_5, M_8	I
S_6	$M_1, M_2, M_3, M_4, M_5, M_6, M_7, M_8$	M_6	M_6	
S_7	M_1, M_2, M_5, M_7, M_8	$M_1, M_2, M_3, M_4, M_6, M_7, M_8$	M_1, M_2, M_7, M_8	I
S_8	M_1, M_2, M_5, M_7, M_8	$M_1, M_2, M_3, M_4, M_5, M_6, M_7, M_8$	M_1, M_2, M_5, M_7, M_8	I

is apparent that education and training (M_3), policy measures for emission reduction from shipping (M_4), and subsidies and tax credits (M_6) belong to Quadrant-IV, thus, they have weak dependence power but strong driving power, in other words, these measures are the key driving force for promoting the development of green shipping. While all the other measures belong to Quadrant-III, thus, they also play important roles in the green shipping but highly depend on the other measures. In other words, these measures are also important for promoting the development of green shipping, but the successful implementation of each of these measures highly relies on some other measures, especially those belonging to Quadrant-IV.

Finally, these measures have been partitioned. Taking the level partition of the measures in iteration 1 as an example (see Table 4.17), the reachability set and the antecedent set with respect to each measure can firstly be determined. Subsequently, the intersection set with respect to each measure can also be determined; then, the measures with respect to which the reachability set and intersection set are the same belong to level I. Similarly, the measures belonging to the other levels can also be determined.

According the result of level partition, the ISM based hierarchical model can also be determined, as presented in Fig. 4.4. It is worth pointing out that only the relationships between two measures in the same level or that between a measure and another measure in the adjacent level are presented in the ISM based hierarchical model. It is apparent that the three measures, namely subsidies and tax credits, policy measures for emission reduction from shipping, and education and training are the foundation of green shipping. The results are reasonable as these three measures are the foundation of the specific measures for emission reduction from shipping. Subsidies and tax credits can effectively encourage the shipping stakeholders to take measures for obeying the policies, standards and regulations for emission reduction from shipping, and these policies measures make the shipping stakeholders participate in the education and training spontaneously to

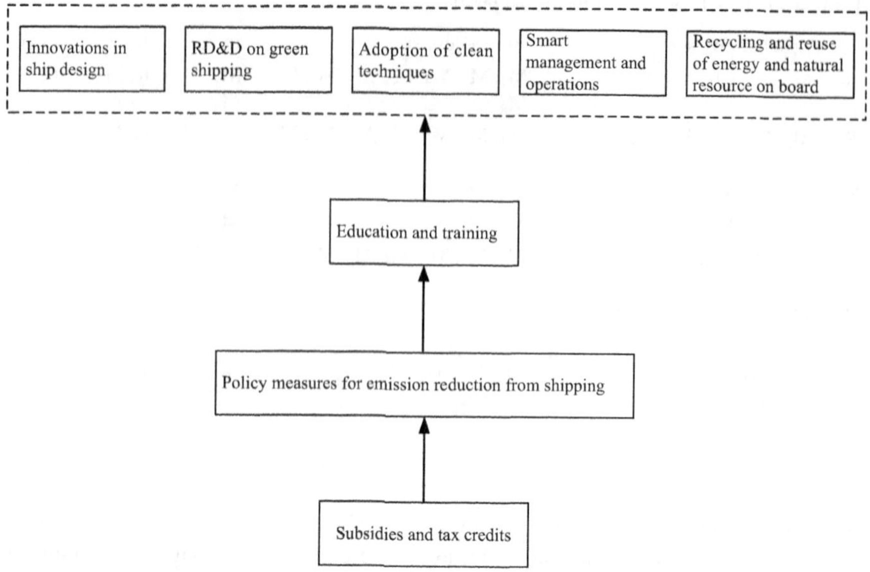

Fig. 4.4 ISM based hierarchical model for strategic measures for green shipping

improve the awareness of environment protection and the skills for green shipping. After overcoming the obstacle of lacking funds, establishing the policy, standard and regulatory system for emissions reduction from shipping, improving the awareness of environment protection, and mastering the skills for green shipping, the stakeholders can implement some specific measures such as innovations on ship design and adoption of clean techniques for promoting the development of green shipping.

4 Discussion

Advancement of ship design, engines and machinery, perception of shipping stakeholders on environment protection, NO_x emission, SO_x emission are the four most success factors that influence green shipping, followed by fuel cost, investments on emissions reduction from shipping, Corporate Social Responsibility, compliance with regulations and standards. Fuel efficiency, GHG emission in descending order, and the other factors have relatively low influence on green shipping.

Eight strategic measures have been proposed to address the most important factors for enhancing the greenness of shipping. Interpretative Structural Modeling has been used to study the characteristics of these measures and investigate the inner relationships among them. Subsidies and tax credits, policy measures for emission reduction from shipping, and education and training are the foundation of green shipping have been identified as the foundation of green shipping.

Therefore, the corresponding administrators should focus on three tasks for promoting the development of green shipping in the near future: (1) drafting the measures for subsidies and tax credits in national and international levels for green shipping, especially for emissions mitigation and energy-saving; (2) drafting more complete the policy, standard and regulatory system for green shipping; (3) organizing more courses for the education and training of shipping stakeholders, especially the crew, the ship owners and the staff in the ports, to improve their awareness of environment protection and the skills for green shipping.

5 Conclusion

This chapter aims at analyzing the success factors of green shipping and studying the strategic measures for enhancing the greenness of shipping. Seventeen success factors that have significant effects on green shipping in five aspects including technological, economic, environmental, social, and management and operational aspects have been obtained based on literature review. ANP which can consider the interdependences among the success factors has been employed to prioritize the success factors in terms of their relative importance, and the prior sequence of these factors has been obtained.

Some implications for promoting the development of green shipping have also been presented for the shipping stakeholders according to the relative importance of the success factors and the effects of the strategic measures. The cause-effect relationships among the measures for promoting green shipping were investigated by ISM analysis.

It could be summarized that this chapter has the following advantages:

1. Multi-dimensional success factors including technological, economic, environmental, social, and management and operational aspects of green shipping have been incorporated;
2. The interdependences among the success factors of green shipping are considered by using ANP to rank these factors;
3. The effects of the strategic measures for green shipping and the inner relationships among them are obtained by using ISM for telling the stakeholders what the most urgent needed measures.

Besides the advantages, there are also some drawbacks in this chapter:

1. Lack of group decision-making

   The determination of the comparison matrices used in ANP and the SSIM used in ISM analysis is only based on the experience and preferences of the stakeholders from academics. More shipping stakeholders should be invited to participate in the decision-making to assure the accuracy of the results.

2. Inaccuracy in establishing comparison matrices by using crisp numbers
   The comparison matrices used in ANP are based on intuitional judgments, while it is usually difficult to use crisp numbers to establish comparison matrices as human's judgments usually involve vagueness, ambiguity and subjectivity.

Accordingly, the future work of the authors will focus on developing a method by combining group decision-making method and fuzzy theory that allows multiple shipping stakeholders to use linguistic terms to express their opinions for analyzing the success factors of green shipping.

References

AMSA (Australian Maritime Safety Authority), Canberra ACT Australia (2012) Adoption of an energy efficiency design index for international shipping

Bailey D, Solomon G (2004) Pollution prevention at ports: clearing the air. Environ Impact Assess Rev 24(7):749–774

Cullinane K, Cullinane S (2013) Atmospheric emissions from shipping: the need for regulation and approaches to compliance. Transp Rev 33(4):377–401

Dağdeviren M, Yüksel İ, Kurt M (2008) A fuzzy analytic network process (ANP) model to identify faulty behavior risk (FBR) in work system. Saf Sci 46(5):771–783

Dağdeviren M, Yüksel İ (2010) A fuzzy analytic network process (ANP) model for measurement of the sectoral competititon level (SCL). Expert Syst Appl 37(2):1005–1014

Doudnikoff M, Lacoste R (2014) Effect of a speed reduction of containerships in response to higher energy costs in sulphur emission control areas. Transp Res Part D: Transp Environ 28:51–61

EC (European Commission) (2010) Corporate Social Responsibility. This file is available at: http://ec.europa.eu/enterprise/policies/sustainable-business/corporate-social-responsibility/index_en.htm

EMCC (European Marine Equipment Council) (2009–2010) Green ship technology book. This file is available at: http://www.oecd.org/sti/ind/48365856.pdf

Fan N (2012) The study on evaluation system of energy consumption in shipping enterprise. Master Dissertation of Wuhan University of Technology. (in Chinese)

Gunningham N, Grabosky P, Sinclair D (1998) Smart regulation: designing environmental policy. Oxford University Press, Oxford

Helfre JF, Boot PAC (2013) Emission reduction in the shipping industry: regulations, exposure and solutions. The file is available at: http://www.sustainalytics.com/sites/default/files/shippingemissions_july2013.pdf

IMO (2006) MARPOL Annex VI: regulations for the prevention of air pollution from ships, Consolidated Ed., MARPOL, IMO, London

Kannan G, Pokharel S, Kumar PS (2009) A hybrid approach using ISM and fuzzy TOPSIS for the selection of reverse logistics provider. Resour Conserv Recycl 54(1):28–36

Lam JSL, Lai K (2015) Developing environmental sustainability by ANP-QFD approach: the case of shipping operations. J Cleanr Prod 105:275–284

Lam JSL, van de Voorde E (2012) Green port strategy for sustainable growth and development. In: Transport logistics for sustainable growth at a new level, International Forum on Shipping, Ports and Airports (IFSPA), pp 27–30. https://www.polyu.edu.hk/lms/icms/ifspa2012/Papers/M25.pdf

Liu X (2012) Research on low-carbon green development in Chinese shipping. Master dissertation of Dalian Maritime University. (in Chinese)

Lun YHV (2013) Development of green shipping network to enhance environmental and economic performance. Polish Marit Res 20:13–19

Luthra S, Kumar S, Kharb R, Ansari MF, Shimmi SL (2014) Adoption of smart grid technologies: an analysis of interactions among barriers. Renew Sust Energ Rev 33:554–565

McConnell M (2002) Capacity building for a sustainable shipping industry: a key ingredient in improving coastal and ocean and management. Ocean Coast Manag 45(9):617–632

Nuttall P, Newell A, Prasad B, Veitayaki J, Holland E (2014) A review of sustainable sea-transport for Oceania: providing context for renewable energy shipping for the Pacific. Mar Policy 43:283–287

Ogbonnaya EA, Poku R, Adigio E (2013) Effects of green shipping to the maritime industry. Eng Technol Appl Sci Res 3(2):402–407

Prpić-Oršić J, Faltinsen OM, Valčić M (2014) Development strategies for greener shipping. In: 56th International (IEEE) Symposium ELMAR-2014

Rehmatulla N (2011) Applying systems thinking approach for qualitative analysis of GHG emissions regulations in shipping. http://discovery.ucl.ac.uk/1450778/2/Rehmatulla%20(2011)%20Applying%20systems%20thinking%20approach %20for%20qualitative%20analysis%20of%20GHG%20emissions%20regulations%20in%20shipping.pdf

Saaty TL (1980) The analytic hierarchy process. McGraw-Hill, New York

Shahabi RS, Basiri MH, Kahag MR, Zonouzi SA (2014) An ANP–SWOT approach for interdependency analysis and prioritizing the Iran's steel scrap industry strategies. Resources Policy 42:18–26

Turliainen M (2005) Green shipping. ABB Rev 3:54–57. This file is available at: http://www05.abb.com/global/scot/scot271.nsf/veritydisplay/417b651cf0dd8298c125707b004ba25c/$file/54-57%203M554_ENG72dpi.pdf (Access date 18th February, 2016)

Voorham CHM (2013) Assessment of sustainable investment in shipping. Master Dissertation of Delft University of Technology

Walmsley S (2012) Shipping and sustainability. This file is available at: http://www.wwf.at/de/view/files/download/showDownload/?tool=12&feld=download&sprach_connect=2395 (Access date 16th March, 2016)

Warfield JN (1974a) Toward interpretation of complex structural modeling. IEEE Trans Syst Man Cybern 4(5):405–417

Warfield JN (1974b) Developing interconnection matrices in structural modeling. IEEE Trans Syst Man Cybern 4(1):81–87

Yang ZL, Zhang D, Caglayan O, Jenkinson ID, Bonsall S, Wang J, Huang M, Yan XP (2012) Selection of techniques for reducing shipping NOx and SOx emissions. Transp Res Part D Transp Environ 17(6):478–486

Zamani M, Rabbani A, Yazdani-Chamzini A, Turskis Z (2014) An integrated model for extending brand based on fuzzy ARAS and ANP methods. J Bus Econ Manag 15(3):403–423

Zhang Z (2009) Research on the evaluation method of energy-saving and emission reduction for the port and shipping system. Master dissertation of Wuhan University of Technology. (in Chinese)

Chapter 5
Use of Fuzzy Evidential Reasoning for Vessel Selection Under Uncertainty

Zaili Yang, Lefteris Maistralis, Stephen Bonsall, and Jin Wang

Abstract The selection of appropriate vessels to carry out shipping activities is crucial for many maritime stakeholders including charterers, shipowners, brokers, surveyors and safety engineers. The task is essentially a process of multiple criterion decision making under uncertainty requiring analysts to derive rational decisions from uncertain and incomplete data contained in different quantitative and qualitative forms. The difficulty and complexity of such a task is obvious and thus stimulates the study of developing a novel decision making technique under uncertainty with multiple criteria. This chapter aims to use the new function of fuzzy Evidential Reasoning to improve the vessel selection process of dealing with multiple criteria with insufficient and ambiguous information. A numerical case study of selecting an oil tanker based on a voyage charter party is presented to demonstrate the proposed method.

Keywords Vessel selection • MCDM • Evidential reasoning • Uncertainty modelling

1 Introduction

Selecting vessels, which are often treated as complicated marine engineering systems, is affected by a great number of factors associated with their design, manufacturing, installation, operation, commissioning and maintenance. A recent literature survey has shown that most previous studies in vessel selection focused on ship design and structure analysis (Brown and Mierzwicki 2004; Busch 1999;

Z. Yang (✉)
Liverpool Logistics, Offshore and Marine Research Institute (LOOM), Liverpool John Moores University, Liverpool, UK
e-mail: Z.Yang@ljmu.ac.uk

L. Maistralis
Risk & Reliability Services, ABS Maritime Services (Hellas) LLC, Piraeus 185 36, Greece

S. Bonsall • J. Wang
School of Engineering, Liverpool John Moores University, Liverpool L3 3AF, UK

P.T.-W. Lee, Z. Yang (eds.), *Multi-Criteria Decision Making in Maritime Studies and Logistics*, International Series in Operations Research & Management Science 260,
DOI 10.1007/978-3-319-62338-2_5

Xie et al. 2008) and overlooked its operation and economic and social impacts which are of particular significance in chartering ships. Selecting an appropriate vessel to perform a specific shipping activity does also depend on many.

considerations including its capacity, type, pollution and running cost, and etc. This means that in many cases, there are always some imprecisely or inaccurately known parameters resulting in that decision makers lacking of confidence in establishing a sound complete mathematical model to present the scenario with high level uncertainty are inevitably present therefore.

In the process of analysing vessel selection, the main uncertainties that decision makers may encounter include (a) different types of assessments (numbers, linguistic terms or stochastic values) depending on the characteristics of decision criteria; (b) imprecise assessment due to insufficient data, shortcomings in expertise, small time intervals for evaluation or inability of experts to provide a fully detailed assessment; and (c) proper and robust aggregation of subjective and objective assessments made on multiple decision criteria (Wang and Yang 2001).

One realistic and efficient way to compensate the unavailability or incompleteness of data is to incorporate expert judgements based on linguistic assessments. Consequently decision criteria have a dual nature of being both qualitative and quantitative depending on their data sources. In order to aggregate all input information, it is necessary to convert different types of assessments into a unique plane either by transforming the quantitative assessment into qualitative forms based on the pre-defined linguistic terms or by assigning the linguistic terms quantitative values. This will be determined by the nature of the decision scenario and the multiple criteria decision-making (MCDM) method selected to analyse it. One of the most typical MCDM techniques, Evidential Reasoning (ER) (Yang and Singh 1994) requiring the transformation from quantitative to qualitative assessments is appropriate for carrying out vessel selection analysis.

The vessel selection analysis requires the construction of a hierarchical structure accommodating many criteria and sub-criteria with the appropriate presentation of their relationships. In such a hierarchical structure, it is usually the case that the selection analysis at a higher level makes use of the information produced at lower levels. It is therefore important to synthesise the evaluations of the lowest level criteria in a rational way so as to enable the selection of the best vessel at the highest level. However, when the qualitative assessment using linguistic terms is involved in the analysis, it is difficult to use normal mathematically logical operations to conduct the synthesis. An ER method is well suited to modelling subjective credibility induced by partial evidence. The kernel of this approach is an ER algorithm developed on the basis of the Dempster-Shafer (D-S) theory, which requires modelling the narrowing of the hypothesis set with the requirements of the accumulation of evidence (Yang and Xu 2002).

The current study therefore aims at developing a conceptual methodology for optimal vessel selection when multiple criteria under uncertainty need to be considered in decision analysis. In order to achieve this purpose, the ER approach is revisited in Sect. 2 with the introduction of the rest of the techniques needed in this chapter. The proposed methodology is presented in Sect. 3 and its feasibility is validated in Sect. 4 by using a numerical case study of oil tanker selection. This chapter is concluded in Sect. 5 with the discussion of results obtained.

2 Background

2.1 *Dempster-Shafer Theory and ER Approach*

ER is a process of drawing plausible conclusions from uncertain or incomplete information. The theory of evidence was first introduced by Dempster (1967) and it was further developed by his student Shafer (1976). Therefore, it is common to encounter ER as the D-S theory (Wang and Yang 2001; Yang and Xu 2002; Liu et al. 2004).

The D-S theory is essentially based on probability theory, yet it is more flexible in a manner that it allows probabilistic judgements to capture the inaccurate nature of the examined factor. This results in degrees of likelihood being measured by probability intervals, as opposed to point probabilities in a Bayesian approach. The D-S theory uses a number between 0 to 1 to set the degree of belief for a proposition, which could be deduced from multiple grades (linguistic terms such as good, average, fair and poor). For example, the reliability of a ship can be evaluated as 60% good and 20% average. Such an example clearly indicates that the evaluation can be assigned to more than one grade according to the supporting evidence and the subjective experience of assessors. Another advantage of this approach is the fact that the grades of belief do not have to sum up to 1. In the above example, the unassigned belief, the remaining 20% could be the result of uncertain data, lack of information or evidence or even insufficient expertise.

When dealing with a decision making problem, analysts are asked to use their knowledge in terms of preference and evaluation to make the best possible decision. The ER approach was developed by Yang and Singh (1994) specifically for problems incorporating both qualitative and quantitative criteria under uncertainties. The strongest point of ER is its ability to deal with incomplete and vague as well as complete and precise data. It is also useful as it enables the experts involved in a decision-making problem to reach their decisions either in a subjective or a quantitative way. This inherently means that judgements can be made in terms of both verbal descriptors and specific numbers. It has therefore been applied in areas where uncertainty in data is high such as maritime security analysis.

2.2 *ER Algorithm*

The ER approach can be used to effectively synthesise pieces of evaluation from different assessors to various criteria in MCDM. In continuously researching and practicing processes, the kernel of this approach, the ER algorithm has been developed, improved and modified to achieve greater rationality (Yang and Xu 2002). The latest algorithm has also been analysed and explained in Chap. 2.

Using the ER approach, multiple sets from the evaluations of more sub-criteria or the judgements from multiple persons can also be combined. However, the application of the approach requires the assumption that all evaluations are assessed or obtained on the basis of the same linguistic expressions (one common utility space), which is often not the case in decision making. Therefore, the evaluations of both upper-level criteria and lower-level sub-criteria need to be transformed before being aggregated using a belief distribution based utility mapping technique.

2.3 Fuzzy Mapping Technique

There are different types of fuzzy mapping techniques available in the literature. They use various mechanisms (i.e. distance, overlapping area or interaction point) to investigate the similarity between two fuzzy sets representing linguistic expressions. In order to unify the linguistic terms associated with different criteria, a knowledge-based fuzzy mapping technique is presented here using belief distribution based utility theory (Yang et al. 2007a). For example, assume the criterion "Cost" has its upper level criterion "Selection Control Option (SCO)" and lower level sub-criteria "Investment" and "Maintenance" in a decision making hierarchy. The top level event "SCO" can be expressed using such linguistic terms as "Slightly preferred", "Moderately preferred", "Average", "Preferred" and "Greatly preferred". The attribute "Cost" is described linguistically in terms of "Very High", "High", "Average", "Low" and "Very Low". The linguistic terms used to assess the parameters "Investment" and "Maintenance" are individually the sets of ("Substantive", "Large", "Moderate", "Little") and ("Excessive", "Reasonable", "Marginal", "Negligible"). Then, a belief structure link between the linguistic variables expressing different criteria at the three levels can be generated for the transformation from fuzzy input to output and shown in Fig. 5.1.

In Fig. 5.1, w represents the relative (normalised) weights of each criterion (same-level factors) under the same upper level criterion. The values attached to the arrows are the degrees of belief β distributed by experts for indicating the relationships between linguistic variables of different-level decision factors. Note that the sum of the belief values from one linguistic variable is equal to one. For example, the parameter "Investment" with "Large" expression indicates that the level of the attribute "Cost" can be believed as 0.8 ($\beta_{i=2}{}^{c=2}$) "High" and 0.2 ($\beta_{i=2}{}^{c=3}$) "Average" without the presence of other evidence. As far as selecting the best "*SCO*" is concerned, the "High" cost evaluation can support "*SCO*" to 1 ($\beta_{c=2}{}^{r=2}$) "Moderately preferred" and the "Average" cost evaluation can be transformed into 1 ($\beta_{c=3}{}^{r=3}$) "Average" on the universe expressing "*SCO*". Such a linked belief structure can be used as a channel to transform the fuzzy input to fuzzy output by aggregating all values of fuzzy input, decision factor weights and degrees of belief. The detailed transformation process and aggregating calculations can be described as follows.

Suppose each I^i ($i = 1, 2, 3, 4$) represents the fuzzy input (subjective assessment) associated with the criterion "Investment"; each I^c ($c = 1, 2, \ldots, 5$) stands for the corresponding fuzzy input of the criterion "Cost" transformed from the

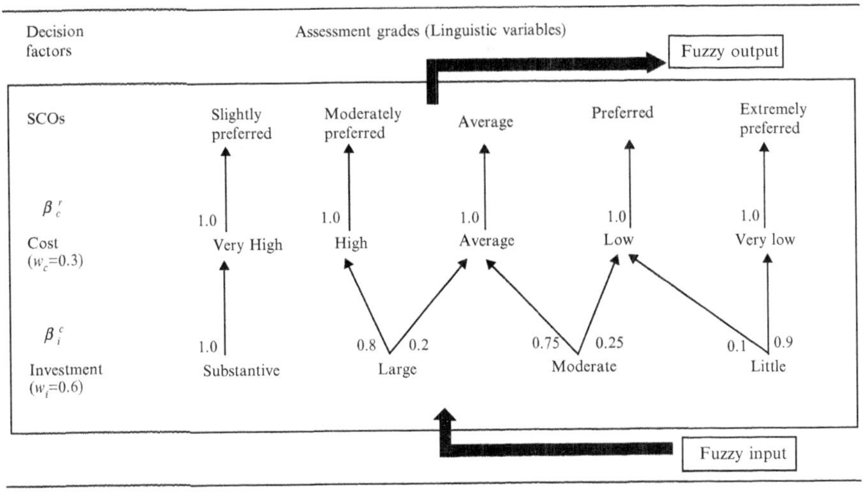

Fig. 5.1 An example of transforming fuzzy input to output

"Investment" related fuzzy input I^i; and each O^r ($r = 1, 2, ..., 5$) indicates the fuzzy output transformed from I^c. Then,

$$O^r = \sum_{c=1}^{5} I^c \beta_c^r = \sum_{c=1}^{5} \left(\sum_{i=1}^{4} (I^i \beta_i^c) \beta_c^r \right) \quad (r = 1,\ 2, \ldots,\ 5) \tag{5.1}$$

Where $\sum_{r=1}^{5} O^r = 1$.

Assume that w^r indicates the relative weight associated with the fuzzy output transformed by the fuzzy input associated with the criterion "Investment". Then,

$$w^r = w_{investement} \cdot w_{cost} \tag{5.2}$$

Where $w_{investement}$ and w_{cost} represent the weights of criteria "Investment" and "Cost" respectively. Note that the sum of the relative weights of the fuzzy output transformed by the input associated with all the lowest level factors equals one.

3 A Novel Fuzzy ER Approach in Vessel Selection

A vessel selection scenario is better visualized through the application of a hierarchical decision tree. In the first level, the main goal of the problem is discussed. In the second level, there are several criteria, each of which has a different contribution to measuring the overall goal. Then it is common that many of the second level criteria could be broken further down to sub-criteria in order to facilitate the assessment as completely as possible. The decomposition of these criteria reaches a point where decision makers are confident that they have adequate information to start the decision process. Once the sub-division of criteria is complete, the decision makers will evaluate each alternative based on the lowest level criteria. The results will be transformed from the lowest level criteria to their respective upper levels and eventually towards the main goal. This can be achieved through the application of the ER approach, which is described as a hierarchical evaluation into which all criteria are aggregated into the top goal of the problem. An MCDM framework incorporating fuzzy mapping and ER to deal with uncertainties is presented within the area of vessel selection. The proposed framework consists of the following steps.

1. *Define the problem and construct an analytical hierarchy.* The first step is to describe the specific decision problem in detail using an effective method. Using decision tables or decision trees enables the simplification of the complex decision analysis. These two formats can be interchangeable once a decision situation has been established in any of the two forms. Decision trees are chosen in this chapter given that they are much easier to use compared to decision tables. In real life problems, decision makers do not always know the true nature of the problem, but are aware of the states that exist. A decision tree gives an illustration of a system along with all the criteria (used to evaluate the system) involved which enable the decision makers to observe the relation of all conditions applicable. It also provides the flexibility to add any new data found during the process of decision making (Sen and Yang 1995). This means that not only addition but also modification can be instantly made to the decision tree according to the updated information.
2. *Set the criterion grades.* From the above, after the initial goal is set, all criteria and sub-criteria are defined. In this step, all criteria are required to be given appropriate grades for decision makers' assessment using either objective data or subjective judgments. For those criteria with qualitative data input, various sets of linguistic terms are defined to reflect the nature of the criteria. For example, a set of linguistic terms {Very high, High, Average, Low, Very low, Minimum} can be used to categorise the evaluation of the criterion "Running cost". For those criteria with quantitative data, numerical grades will be developed. For example, a set of numerical grades {15, 12, 9, 6, 3 (years)} can be used to describe the evaluation of "Vessel age".
3. *Evaluate each alternative based on the lowest level criteria.* The lowest level criteria are defined as the criteria without any further sub-criteria presented in the hierarchy. In order to find out how well a candidate vessel can perform given a specific activity, the lowest criteria need to be evaluated qualitatively or quantitatively depending on the data sources. From Step 1, it is believed that in any case, the evaluation can be conducted successfully since the assessors have sufficient data or feel confident in using subjective judgements. The result can be expressed by degrees of belief belonging to either linguistic terms or numerical grades. While it is straightforward to obtain degrees of belief associated with linguistic terms using expert judgement (i.e. the running cost of a vessel is 60%

high, 20% average), some location measurement techniques such as linear, bi-linear, non-linear and judgemental (Yang et al. 2007b) may be employed to produce the degrees of belief distributed to numerical grades. Taking "Vessel age" as an example, a vessel with 10.5 years old can be linearly evaluated to have 50% degrees of belief belonging to 9 years and 50% belonging to 12 years.

4. *Transform the evaluation from the lowest level to top level criteria.* The creation of different/ specific evaluation grades to each criterion facilitates raw data collection. The grades defined will need to be transformed into the same form for further analysis and assessment. The transformation takes place with the aid of the decision maker's expertise and knowledge. The process has been demonstrated in Sect. 2.3 with the assistance of Fig. 5.1. It can be further explained using the concept of "equivalent rules", which means that different grades in different criteria may be used to describe equivalent standards based on utility theory.
5. *Use the ER algorithm for the synthesis.* Having made the transformation from the lowest level to the top level criteria, the information is fed into the ER algorithm and its calculation software package for the analysis of a multilevel decision problem. The software which will assist in the decision making process is called Intelligent Decision System via ER, IDS (Yang and Xu 2000). It is a Window-based tool, which can be used to build up a model, define alternatives and criteria and perform different assessments according to the decision makers' requirements. For example, different stakeholders in this scenario may have different foci such as cost or reliability and require the model to deliver different assessment results when criteria are given changeable weights.
6. *Rank alternatives and make decisions in a dynamic environment.* As soon as the aggregated values are derived for each candidate vessel in question the ranking takes place according to their priority values in terms of preference. This can be obtained by assigning utility values to the (set of) linguistic terms at the highest level in the hierarchy. However, it is particularly noteworthy that the aggregated values will be changeable due to the different weights of the criteria given when the analysis is used by different stakeholders or under different circumstances. Based on the combination of the steps above, the decision makers can come to a certain conclusion concerning the decision problem that they analyse. The results from IDS as well as the criteria and alternatives selected will be able to provide the prime factors that will set the boundaries for further discussion, if necessary.

This procedure will be illustrated through a case study described in the next section. The requirements for a verification experiment are essential to assess the validity of the results obtained. The contribution of industrial expert's judgement in the form of a structured interview is significant. The successful application of the proposed framework in this case has partially validated the reliability of the method, while more verification experiments conducted through its real application in the future should further increase the confidence of using it in the area of selecting the best vessel for a designed task in general and assessing a vessel's performance in (sub-criteria) quality, reliability and economic advantages, etc. in particular.

4 Application of the ER Approach to a Vessel Selection Process

A voyage chartering example is chosen to demonstrate that the proposed ER approach is able to facilitate the development of maritime business through modelling the vessel selection process for a particular transfer of cargo. The example illustrates how ER can be used to assess multiple criteria containing both qualitative and quantitative data with uncertainties in information.

Within the marine industry's boundaries, the selection of a proper vessel for the transfer of a liquid oil cargo is a process consisting of three different stages. The first stage is to request/invite a vessel from a shipbroker for a particular cargo by a charterer who is normally a refinery with discharging facilities or an independent customer having the cargo transferred by other means away from the initial storage tanks. Then it is the stage of the shipbroker trying to find a list of proper vessels that match the criteria set by the charterer. Finally, it is the stage of vessel selection among the ones pre-selected by the charterer or its representatives.

Looking at port safety during cargo handling operations, there are two main factors involved in this process. One is the vessel that is required to fulfil certain characteristics and the other is the port of loading/unloading cargo that has different regulations to follow. This case study aims at selecting a suitable vessel to approach a port on the west coast of the United States. When a request is made from a charterer to his/her broker, concerning a particular cargo and a port of destination, the broker will normally find several vessels matching the criteria set by the charterer at the first instance. Selecting the best vessel is a complex decision making process, requiring the criteria to be simultaneously measured and evaluated. Due to the nature of the criteria, they may conflict with each other leading to one criterion being influenced at the expense of another.

4.1 *Step 1: Define the Problem*

The case examined is to select an appropriate oil tanker with a capability of delivering 80,000 tonnes of cargo from a European port to the west coast of the United States. The ship broker employed in this case has identified that there are five candidate vessels "open" in the area of the original port. Therefore, the main or top goal of this decision making process is to select the best vessel from the five vessels based on the information required from the charterer side, shipowner side and port specifications. Next, a hierarchical decision tree requires to be established to describe the general main goal to a more specific degree, to which the above information can be appropriately used for the corresponding evaluations. In order to derive the criteria and their assessment grades, a structured interview has been conducted and presented to the senior managers (i.e. Director, Head of Risk Service

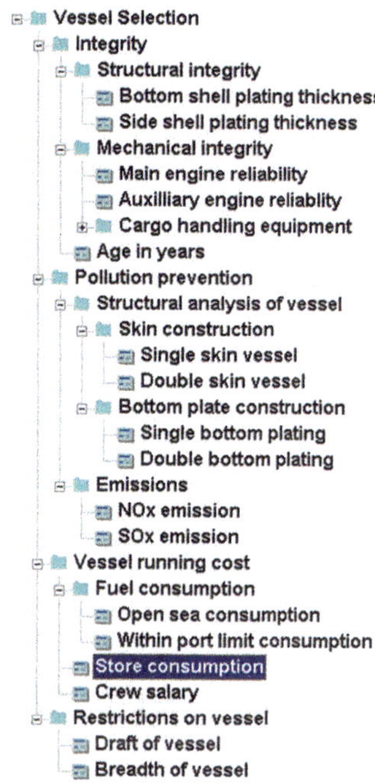

Fig. 5.2 The hierarchy of vessel selection analysis (Mastralis 2007)

and Senior Surveyor) from a ship management company, (Interunity Management Corporation), two leading ship broker companies (Clarksons and Himatiki Marine Ltd) and two classification societies (Bureau Veritas and American Bureau of Shipping). More details about the structured interview and the relevant analysis have been presented in the work by Mastralis (2007). Based on the interview result of identifying criteria, the hierarchy is constructed in Fig. 5.2 with five levels of criteria and sub-criteria. The criteria chosen are the most significant ones taken under consideration from the stakeholders in this analysis. Their relevant descriptions are given as follows:

- Integrity. This criterion is concerned with the structural and mechanical integrity as well as the age of the vessels investigated. The structural integrity is determined by the thickness of the bottom plating, side shell, cargo tanks as well as brackets and frames around the hull. The mechanical integrity can be measured using the reliability data gathered for the main and auxiliary engines as well as the condition of cargo handling equipment. The actual age of the vessel is of great importance as the conditions of both mechanical and structural components are directly related to age.
- Pollution prevention. Pollution control is playing an increasingly important role in maritime transportation. Stakeholders are therefore paying more attention to the relevant factors in vessel selection, especially when the vessels' sailing passages are through waters in Europe and the United States. Structural

characteristics like double bottoms and double side skins are required in Europe and United States as they prevent a great percentage of possible leakages of cargo from being spilt into sea. Finally, the emission values for both NOx and SOx are important in order to call at specific ports. The port of call is based at the west coast of the United States, where the permitted emission levels are very low and pollution regulations are extremely strict.

- Vessel running costs. During the operation of the vessel there are certain factors like fuel (including diesel, lubricating oil and cylinder oil), store consumption and crew salary that need to be investigated. A vessel that has the capability to be run with less crew members is more desirable in terms of daily expenditure during time at sea.
- Restrictions on vessel. They are mainly imposed by geographical factors. Since vessels will sail from Europe to the west coast of the United States through the Panama Canal there will be limitations as far as the draft and breadth of the vessels are concerned in order to fit into the locks.

4.2 *Step 2: Set the Criterion Grades*

The criterion assessment grades (linguistic terms or numerical grades) have been obtained in the interview process. They have been set in a way that the experts can feel confidence in using their domain knowledge. For example, Table 5.1 contains the main criteria used to assess each vessel at the second level in Fig. 5.2. Each criterion is characterised by a set of assessment grades. Assessment grades are not the same in each criterion as each condition can be better assessed into grades based on the personal intuition and preferences of decision makers. For example, based on the interviews, the assessment grades of Criterion "Integrity" have been explained in Table 5.2. The assessment grades of the other criteria have been investigated in this chapter. This provides a guideline to set the criterion grades in the relevant research areas. In a similar way, the grades related to the third, fourth and fifth level sub-criteria are defined and detailed in Mastralis (2007).

Table 5.1 Assessment grades defined for the second level criteria

Main criteria	Assessment grades					
Integrity	Very bad	Bad	Average	Good	Very good	
Pollution prevention	Worst	Poor	Average	Good	Very good	Excellent
Vessel running costs	Very high	High	Average	Low	Very low	Minimum
Restrictions on vessel	Bad		Average		Good	

Table 5.2 The explanations of the assessment grades defined for the second level criteria

Assessment grades of "integrity"	Explanations and definitions
Very bad	The vessel's integrity at both the mechanical and structural level is approaching an unacceptable level. They are a lot of outstanding remarks of class and a probable detention between the last two special surveys. The majority of the vessel's certificates have expired
Bad	The vessel's integrity condition can be at a critical state at either the mechanical or the structural side. Outstanding remarks that have not been resolved yet will be noted in the vessel's record. Some certificates have expired
Average	The vessel's integrity condition is at such a state that it can barely pass the margin between being acceptable or unacceptable. The majority of its certificates are still valid but more work is required to bring it to the acceptable region
Good	The vessel's integrity is above the average condition and within the acceptable region. The vessel's certificates are updated and in the vessel's class records some recommendations may appear
Very good	The vessel is newly built within the last five years. It is classed at a reputable classification society, with no remarks in its class records

4.3 *Step 3: Evaluate Five Candidate Vessels Using the Lowest Level Criteria*

From Figs. 5.2, 19 criteria have been identified to have no sub-criteria and therefore defined as the lowest level criteria in the hierarchy. They include Bottom shell plate thickness, Side shell plate thickness, Main engine reliability, Auxiliary engine reliability, Loading pumps and valves, discharging pumps and valves, Ages in years, etc. Based on such criteria and their assessment grades presented in Step 2, each vessel can be appropriately evaluated using both expert judgement and observed data/fact in the form of degrees of belief attached to the grades. For example, the general information concerning the overall condition of Vessel 1 is given in the following discussion and its evaluation in terms of the 19 lowest level criteria is provided in Table 5.3. The other four vessels have also been evaluated in a similar way in this chapter.

Vessel 1 has a good record of overall structure and engineering systems such as the main and auxiliary engines. It has just had a major servicing after a special dry dock survey which resulted in 300 tones of steel being changed where required and a full overhaul of the main and auxiliary engines bringing it above the average operational standards. It is due to the special survey amendments that the pollution control systems have been checked and updated accordingly making it a strong candidate for the USA port of call. Due to the fact that a complete overhaul was given to its engines it is expected to maintain reasonably low daily running costs. This vessel complies with all the geographical requirements in terms of draft and breadth in order to pass through the Panama Canal.

Table 5.3 The evaluation of vessel 1 in terms of the lowest level criteria

Lowest level criteria	The evaluation of vessel 1
Bottom shell plating thickness	0% very thin, 15% thin, 15% average, 20% thick, 40% very thick
Side shell plating thickness	0% very thin, 10% thin, 20% average, 20% thick, 50% very thick
Main engine reliability	0% very bad, 0% bad, 0% average, 10% good, 80% very good (10% unknown)
Auxiliary engine reliability	0% very bad, 0% bad, 0% average, 0% good, 85% very good, (15% unknown)
Loading pumps, valves	0% malfunction, 0% very unreliable, 0% unreliable, 0% average, 0% reliable, 0% very reliable, 95% fully operational, (5% unknown)
Discharging pumps, valves	0% malfunction, 0% very unreliable, 0% unreliable, 0% average, 0% reliable, 0% very reliable, 90% fully operational, (10% unknown)
Age in years	0% 15 years, 33% 12 years, 67% 9 years, 0% 6 years, 0% 3 years
Single skin vessel	0% very weak, 20% weak, 20% average, 40% strong, 20% very strong
Double skin vessel	0% very weak, 20% weak, 20% average, 40% strong, 20% very strong
Single bottom plating	0% very thin, 0% thin, 20% average, 30% thick, 40% very thick, (10% unknown)
Double bottom plating	0% very thin, 0% thin, 20% average, 30% thick, 40% very thick, (10% unknown)
NOx emission	0% very high, 20% high, 20% low, 60% very low
SOx emission	0% very high, 20% high, 20% low, 60% very low
Open sea consumption	0% very high, 0% high, 0% average, 0% low, 80% very low, 0% minimum, (20% unknown)
Within port limit consumption	0% very high, 0% high, 0% average, 50%low, 35% very low, 0% minimum, (15% unknown)
Store consumption	0% very high, 0% high, 10% average, 40% low, 50% very low
Crew salary	0% very high, 0% high, 70% good, 0% average, 0% bad, (30% unknown)
Draft of vessel	0% more than 12 meters, 100% less than (or equal to) 12 meters
Breadth of vessel	0% more than 32.3 meters, 100% less than (or equal to) 32.3 meters

4.4 *Step 4: Transform the Evaluation from the Lowest Level to Top Level Criteria*

Equivalent rules can be used in this section to establish links between the grades of different level criteria. For instance, a "15 year" old vessel means that its integrity is "very bad" and furthermore, the "very bad" integrity of the vessel is said to be equivalent to a grade "slightly preferred" in vessel selection. Similarly, if "15 years" is equivalent to "slightly preferred", "12 years" to "moderately preferred", "9 years" to "Average", "6 years" to "preferred" and "3 years" to "greatly preferred", then it can be said that the set of grades {15, 12, 9, 6, 3 (years)} in ages is equivalent to the set {slightly preferred, moderately preferred, average, preferred, greatly preferred} in vessel selection. However, it is highly possible that the grades of different level criteria are not equivalent to a 100% degree of belief, especially when different numbers of grades exist in different criteria. To deal with this problem, degrees of belief can be incorporated to keep the link/equivalence between the grades of different criteria to a reasonable extent. For example, to link two criteria, store consumption and vessel's running cost, "very high" in store consumption is said to be equivalent to "very high" in vessel's running cost to a 100% degree of belief. Such an equivalent rule can be kept between the grades of store consumption and vessel's running cost to a 100% degree of belief until the relation between "very low" in store consumption and "very low" and "minimum" in vessel's running cost is investigated. Such relation is further evaluated as that "very low" in store consumption is equivalent to "very low" to a 80% degree of belief and "minimum" to a 20% degree of belief in vessel's running cost. Therefore, the grade set of {very high, high, average, low, very low} in store consumption is said to be equivalent to the set of {very high, high, average, low, (80% very low and 20% minimum)} in vessel's running cost. Similarly, the grade set of {very high, high, average, low, very low, minimum} in vessel's running cost can be said to be equivalent to the set of {slightly preferred, (20% slightly preferred and 80% moderately preferred), (50% moderately preferred and 50% average), (50% average and 50% preferred), (80% preferred and 20% greatly preferred), greatly preferred} in vessel selection. The equivalent relations between grade sets can be virtually expressed in the form of a hierarchy similar to the one in Fig. 5.1. Consequently, the evaluation from the lowest level can be transformed to the top level using Eqs. (5.1) and (5.2). For example, the evaluation of Vessel 1 in terms of age in years and store consumption (in Table 5.3) can be separately transformed to the top level as the set of {0% slightly preferred, 33% moderately preferred, 67% average, 0% preferred, 0% greatly preferred} and the set of {0% slightly preferred, 5% moderately preferred, 25% average, 40% preferred, 30% greatly preferred}. In this process, it is noteworthy that weights of each criterion or sub-criterion can be calculated using an AHP method. However, it has not been described in detail due to the fact that the weights, which have been considered as a variable in a dynamic decision making model to reflect the different importance of criteria in various analyses, differ with respect to intuition and the needs of decision makers.

4.5 Step 5: Synthesise All Evaluations Using the ER Algorithm and Its Calculation Software IDS

Having transformed all the original evaluations on the grade set of vessel selection at the top level, the ER algorithm described in Sect. 2.2 can be used to synthesise them and obtain the overall estimation of Vessel 1 as {1.38% slightly preferred, 7.03% moderately preferred, 13.97% average, 24.51% preferred, 39.94% greatly preferred, 13.17% unknown}. This result is obtained on the condition that the weights of criteria at the same level are distributed evenly. In a similar way, the overall estimations of the other vessels can be obtained and shown in Fig. 5.3.

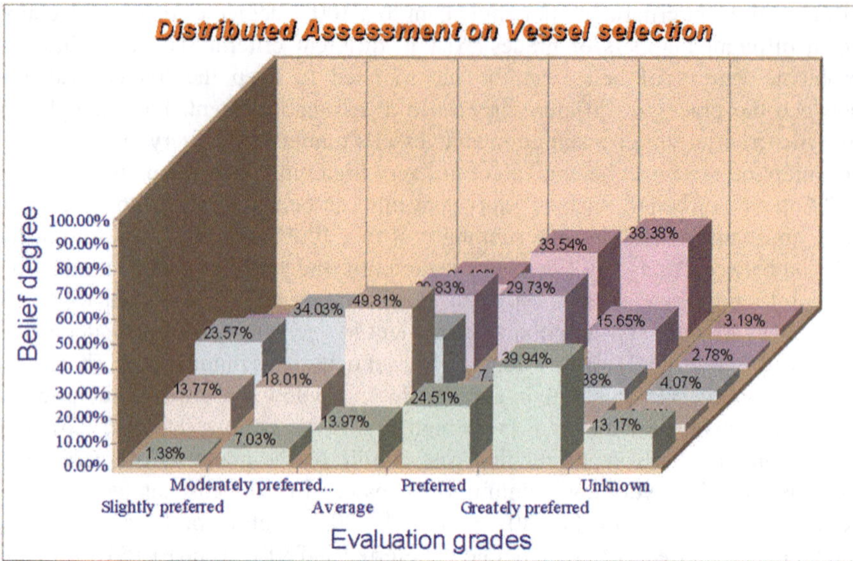

	Slightly preferred	Moderately preferred	Average	Preferred	Greatly preferred	Unknown
Vessel 1	1.38%	7.03%	13.97%	24.51%	39.94%	13.17%
Vessel 2	13.77%	18.01%	49.81%	11.34%	3.64%	3.44%
Vessel 3	23.57%	34.03%	25.72%	7.23%	5.38%	4.07%
Vessel 4	12.7%	9.3%	29.83%	29.73%	15.65%	2.78%
Vessel 5	0.26%	3.47%	21.16%	33.54%	38.38%	3.19%

Fig. 5.3 The overall estimations of all vessels

4.6 Step 6: Choose the Best Vessel for the Voyage Planned

It is difficult for decision makers to choose the best vessel based on the results expressed by linguistic terms with degrees of belief. The best way to rank the vessels would be through their respective utility values generated by quantifying the assessment grades in vessel selection. For example, the assessment grades are given their corresponding values as the set of {0.2, 0.4, 0.6, 0.8, 1}. IDS uses the concept of a utility interval to characterize the unassigned degree of belief (unknown percentage). The ER algorithm produces a utility interval enclosed by the two extreme cases where the unassigned belief goes either to slightly preferred with a minimum utility value or to greatly preferred with a maximum utility value.

A graphical representation of utility intervals is illustrated in Fig. 5.4. The vessels can be ranked based on the average utility. However, it is particular noteworthy that in order to have a vessel being absolutely better than another, the preferred vessel's minimum utility must be greater than the compared vessel's maximum utility. Therefore, when all criteria at the same level are distributed even weights, the result of this vessel selection scenario is that Vessels 5 and

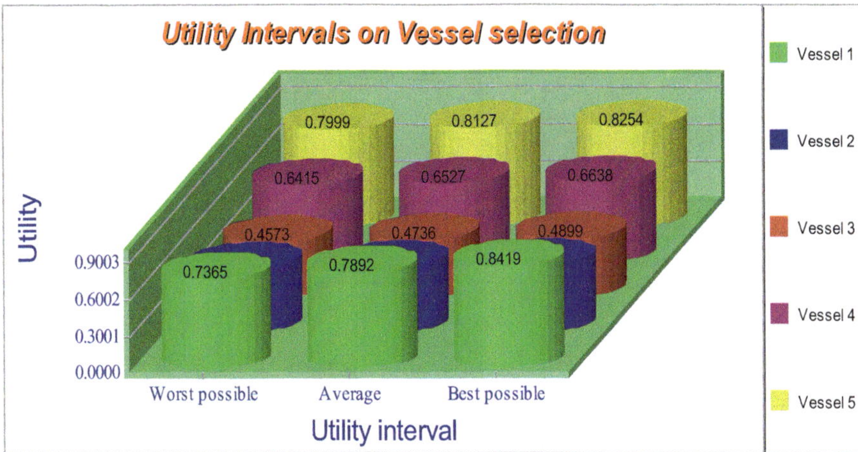

	Worst possible	Average	Best possible
Vessel 1	0.7365	0.7892	0.8419
Vessel 2	0.5324	0.5462	0.5599
Vessel 3	0.4573	0.4736	0.4899
Vessel 4	0.6415	0.6527	0.6638
Vessel 5	0.7999	0.8127	0.8254

Fig. 5.4 Ranking of the five candidate vessels

1 are absolutely better than Vessels 2, 3 and 4, while Vessel 5 with a higher average utility value is more preferred than Vessel 1.

Selecting vessels is a dynamic process given that for different voyages with delivery of various cargoes, stakeholders may consider the criteria playing different roles with different levels of importance involved. For example, if the vessel is not chosen for sailing through the waters in Europe and the United States, then the requirement related to emissions and pollution may be relatively loose and therefore, the weights of such criteria will decrease accordingly. This change will result in the new ranking of vessels. Such a dynamic analysis has been virtually modelled in Fig. 5.5, where the importance ratio between pollution prevention and the other criteria (i.e. integrity, vessel's running cost and restrictions on vessels) is changed in the interval [1:10–10:1]. Obviously, when the pollution issue becomes more important (its importance being twice more than the others') in decision making, Vessel 1 is more preferred than Vessel 5. This result also well reflects the fact that the estimate of Vessel 1 in terms of pollution prevention is better than that of Vessel 5, which can be observed from the original relevant evaluations associated with these two vessels.

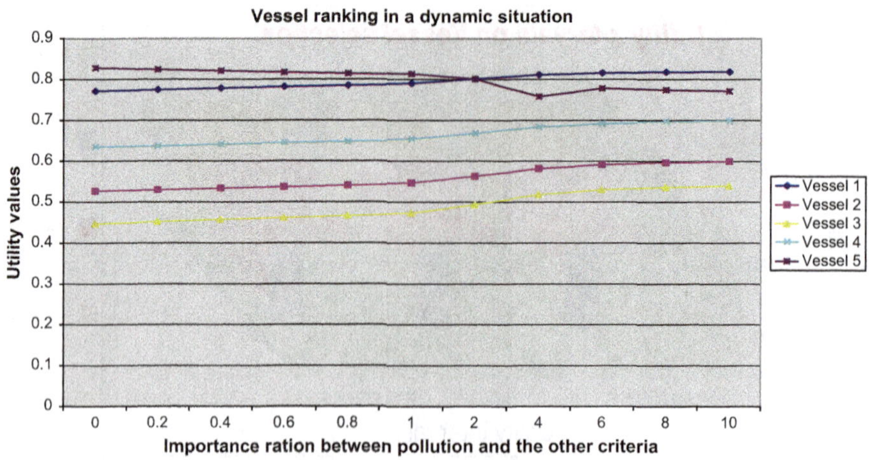

Fig. 5.5 Vessel ranking with different weight distribution of criteria

5 Conclusion

An ER approach is used in this chapter to tackle the problem of using uncertain and multiple criteria to select an appropriate vessel carrying out a specific shipping activity. The information available either quantitative or qualitative may contain uncertainties due to many factors such as an incomplete report on the specified vessel by its independent surveyor. The steps followed within the framework are, with the aid of IDS, capable of producing adequate results for decision makers to choose the best vessel in a dynamic environment. A supporting tool in the case of uncertainty treatment is that the data can be presented in the form of a degree of belief with respect to each linguistic variable and the assessment can be conducted at different levels if necessary.

The vessel selection for a shipping activity is an important but difficult analysis as it involves a large capital sum to be invested both for the transport of the cargo as well as for the operation and maintenance of the vessel chartered. If the stakeholders who are going to make the selection only depend on the data available and ignore or inappropriately deal with the information of having a qualitative nature, it will be highly possible to make wrong and costly decisions. The ER approach and IDS software enable the decision makers to make use of both tabular and graphical data and make decisions based on any necessary comparison. The marriage of ER and fuzzy mapping can provide the appropriate foundation to model any type of vessel selection scenarios under uncertainties and propose a reasonable solution. It can also prove to be useful in the areas of asset management, risk assessment and cost benefit analysis, etc. in the maritime industry. The results produced from the case study match to a great extent the initial evaluation and data input used for the vessels in question, thus partially validating the feasibility of the ranking procedure.

It can be reasonably expected that the continuous applications of this vessel selection framework will facilitate the development of a cost-effective and safer shipping industry.

Acknowledgements An earlier version was presented in Proceedings of the Institution of Mechanical Engineers, Part M, Journal of Engineering for the Maritime Environment by the SAGE publication (Yang et al. 2009).

References

Brown A, Mierzwicki T (2004) Risk metric for multi-objective design of naval ships. Nav Eng J 116:55–71

Busch I (1999) Buying the boat – choosing a ship for the sealift conversion program. Mar Technol SNAME News 36:1–10

Dempster AP (1967) Upper and lower probabilities induced by a multivalued mapping. Ann Math Stat 38:325–339

Liu J, Yang JB, Wang J, Sii HS, Wang YM (2004) Fuzzy rule-based evidential reasoning approach for safety analysis. Int J Gen Syst 23(2–3):183–204

Maistralis E (2007) Formal safety assessment of marine applications, PhD Thesis. Liverpool John Moores University, Liverpool

Sen P, Yang JB (1995) Multiple criteria decision making in design selection and synthesis. J Eng Des 6:207–230

Shafer G (1976) A mathematical theory of evidence. Princeton University Press, Princeton

Wang J, Yang JB (2001) A subjective safety based decision making approach for evaluation of safety requirements specifications in software development. Int J Reliab Saf Eng 8:35–57

Xie X, Xu X, Yang JB, Wang J, Ren J, Yu SJ (2008) Ship selection using a multiple-criteria synthesis approach. J Mar Sci Technol 13:50–62

Yang JB, Singh MG (1994) An evidential reasoning approach for multiple attribute decision making with uncertainty. IEEE Trans Syst Man Cybern 24:1–18

Yang JB, Xu DL (2000) An introduction to an intelligent decision system: IDS. In: Proceedings of the 42nd annual conference of the UK Operational Research Society, Sep. 12–14, Swansea

Yang JB, Xu DL (2002) On the evidential reasoning algorithm for multiple attribute decision analysis under uncertainty. IEEE Trans Syst Man Cybern Part A Syst Hum 32:289–304

Yang ZL, Bonsall S, Fang QG, Wang J (2007a) Maritime security assessment and management. J Int Assoc Marit Univ 5:56–72

Yang ZL, Bonsall S, Wang J (2007b) Use of hybrid multiple uncertain attribute decision making techniques in safety management. Expert Syst Appl. doi:10.1016/j.eswa.2007.11.054

Yang ZL, Maistralis L, Bonsall S, Wang J (2009) Incorporating uncertainty and multiple criteria in vessel selection. Proc Inst Mech Eng Part M J Eng Marit Environ 223:177–188

Yang ZL, Ng AKY, Wang J (2015) Towards robust management of maritime risk and security. In: Lee P, Cullinane K (eds) Dynamic shipping and port development in the globalized economy, Palgrave

Chapter 6
Probabilistic Assessment of Vessel Collision Risk: An Evidential Reasoning and Artificial Potential Field-Based Method

Feng Ma and Yu-Wang Chen

Abstract This chapter proposes a novel method to estimate the collision probabilities of monitoring targets for coastal radar surveillance. Initially, the probability of a monitoring target being a real moving vessel is estimated using the records of manual operations and the Evidential Reasoning (ER) rule. Subsequently, the bridges, piers and other encountering vessels in a waterway are characterized as collision potential fields using an Artificial Potential Field (APF) model, and the corresponding coefficients can be trained in terms of the historical vessel distributions. As a result, the positional collision potential of any monitoring vessel can be obtained through overlapping all the collision potential fields together. The probabilities of authenticity and the collision potential are further formulated as two pieces of evidence on which the Dempster's rule of combination is used to reason the collision probability of a monitoring target. The vessels associated with high collision probabilities can be highlighted for supervisors' attention, as they potentially pose high risks to safety. A preliminary field test was conducted to validate the proposed method.

Keywords Dempster's rule • Vessel collision risk • Artificial Potential Field • Nonlinear optimization • Evidential Reasoning-based Method

---

F. Ma
Intelligent Transport System Research Center, Wuhan University of Technology, Wuhan, People's Republic of China

National Engineering Research Center of Water Transportation Safety (WTS), Wuhan, People's Republic of China
e-mail: martin7wind@whut.edu.cn

Y.-W. Chen (✉)
Decision and Cognitive Sciences Research Centre, The University of Manchester, Manchester M15 6PB, UK
e-mail: yu-wang.chen@manchester.ac.uk

P.T.-W. Lee, Z. Yang (eds.), *Multi-Criteria Decision Making in Maritime Studies and Logistics*, International Series in Operations Research & Management Science 260,
DOI 10.1007/978-3-319-62338-2_6

1 Introduction

The safety concerns for coastal surveillance have been imposing high requirements to sensors and supervisors. To obtain detailed and real-time information of vessels and the associated navigational environments, costal surveillance sensors have been improved rapidly, leading to an explosion of information. For example, the International Maritime Organization (IMO) requires all ports and vessels to be equipped with radar, Automatic Identification System (AIS) and satellite facilities, so that three monitoring systems can complement each other to ensure the validation of tracking (Guerriero et al., 2008; IEC 62288, 2014). These active detection systems are kernels in ship safety management. However, a radar system usually provides a plethora of objects indiscriminately. The useful objects, e.g., moving vessels, only take a small proportion of all the monitoring targets. Even worse, radar systems sometimes misinterpret the behaviours of moving targets, leading to management difficulties. To address such issues, radar manufactures have paid enormous efforts to improve the sensitivity and capability of noise suppression. In the past decades, radar sensors have been improved significantly. For example, the latest marine radar is capable of tracking a 0.5-square meter target in a distance of 5 miles. However, a radar system is not capable of differentiating such a 0.5-square meter target as a drowning person, a clump of sea-grass, or a canoe. In fact, high sensitivity will bring many false and useless targets, which might confuse radar operators. Therefore, radar systems need human assistance to identify the targets on the screens. Nevertheless, too many non-distinctive false or un-important targets cluttering on the screens will distract the attention of vessel operators, with the potential of threatening ship safety. In the increasingly crowded harbours and inland rivers, manual identification becomes impractical. For instance, on the downstream of the Yangtze River, there might be 4000 large vessels passing by each day. It is impossible to ensure that any single radar target is inspected manually.

Furthermore, even if a radar target has been confirmed to be a real moving vessel, it might not always need much attention. For instance, a vessel anchoring in a berth is generally safe, and no much attention is needed in surveillance. In fact, only a real moving vessel posing a threat to safety that means a potential collision needs a close inspection (Li et al. 2007). Therefore, the avoidance of collision is one of the core objectives of coastal surveillance. In other words, it is important to develop an intelligent method to estimate the collision probabilities of radar targets, to lower the burden of radar operators observably and to improve safety at sea.

Vessel collision risk is widely described as the product of the corresponding collision probability and the collision consequence (Fujii et al. 1974). Since the collision consequence is difficult to quantify, much work has been conducted for estimating the collision probability. However, the main challenge in the assessment of collision probability is that it cannot be inferred from frequency analysis, as the collision accidents might not happen frequently. Thus, the classical theory of probability might not be applicable. In relevant research findings, the estimation of the collision probability generally takes into account macro perspectives or ship handling (Eleye-Datubo et al. 2008). The relevant studies are not capable of describing the successive variation of collision probabilities in microscopic adjacent positions (e.g. Dong and Frangopol 2015). Thus, these studies can be used to estimate the overall collision probability of "black spot", for setting a speed limit. However, they are not capable of describing the collision probability differences between two points which are 50 meters far from each other in such a "black spot". The Artificial Potential Field (APF) model formulates the collision probabilities or potential as a continuous function (Volpe and Khosla 1990; Kim and Khosla, 1992) which has been proved to be fairly effective in the path planning of robots. In addition, some intelligent methods have been introduced to distinguish moving vessels from false or stationary objects to lower the burden of operators. For marine radar, Ma et al. (2015) proposed a fuzzy k-means (FCM) based classification method to identify the false targets among ARPA targets, and reported the accuracy of 91.0%. Zhou et al. (2013) invented a radar target-recognition method based on fuzzy optimal transformation using high-resolution range profiles. The ER rule originated from Dempster's and Bayes' rules provides an inference process that takes into account evidence weight and reliability as coefficients, meanwhile keeping the consistency with Bayes' rule, and it can be very practical in the probabilistic inference of radar blips being moving vessels.

On the basis of existing literature review, this chapter aims to propose an intelligent method to estimate the collision probabilities by combining the probabilistic ER inference and the APF model. The fundamentals of the APF model and obstacle avoidance modelling are briefly reviewed in Sect. 2. A novel method as the key contribution of this chapter is proposed to estimate the collision probabilities in Sect. 3. A case study is conducted to validate the proposed method in Sect. 4. Section 5 concludes this chapter.

2 Literature Review

2.1 Artificial Potential Field (APF)

The collision potential determined by the environment and the encountered vessel is a core factor of estimating the collision probability of a vessel, which, in fact, is affected by many factors, including ship handling, ship condition, and encountered vessels. Hence, it has been modelled from different perspectives (Hänninen and Kujala 2012). The static collision probability model proposed by Fujii et al. (1974) has been widely used in the research of ship handling. In the model, a collision probability is equal to the product of the geometrical probability of a collision course and the causation probability. Montewka et al. (2010) proposed a new approach for quantifying the geometrical probability to estimate collision probabilities on the basis of maritime and aviation experience. Pedersen (2010) reviewed the procedures of reducing the high economic environmental and human costs associated with ship collisions and grounding. Particularly, many researchers began to find characteristics of vessels using AIS data records, since such records are widely believed to be reliable and objective (Montewka et al. 2010). This research also uses AIS records as a fundamental data source. In summary, the research of collision probability generally starts with a multi-factor qualitative analysis involving ship handling, human factors, and geometrical collision model which are originated from ship domains or minimum distance modelling (Montewka et al. 2012). However, this information is mostly unavailable from the view of coastal surveillance; it can only be confirmed with very high frequency (VHF) radio. In its daily management, the verification of VHF radio is usually very brief; hence, the location of a blip seems to be the only direct and credible evidence of estimating the corresponding collision probability or potential, which is closely related to the dynamic navigation environment of waterways. The change of berths, piers, buoys and depths might have significant impacts on the distribution of collision probability or potential. Many other researchers investigated the collision probability of vessels using a ship safety domain model (Fujii et al. 1978). Although many researchers have proposed various methods for modelling collision risk based on these factors individually (Kujala et al. 2009; Qu et al. 2011), a widely acknowledged and comprehensive modelling method has not been invented yet.

As discussed above, it is also worth noting that the collision probability here cannot be estimated from frequency analysis, since collision accidents should be prevented from frequent occurrence. Hence, the research of collision probability estimation can only be started from a qualitative analysis of incident causation, including formal safety assessment (Wang 2001; Zhang et al. 2013; Dong and Frangopol 2015). Referring to the research of path planning for mobile robots, it might be appropriate to investigate the collision probability in coastal surveillance in the format of a potential field (Dellacherie and Meyer 2011).

2.2 Obstacle Avoidance Modelling with the APF Model

It was believed that the fundamental forces of nature can be modelled using potentials which satisfy Laplace's equation (Dellacherie and Meyer 2011). It is usual that objects might attract or repulse each other, and the so-called repulsions or attractions among them are actually very difficult to be measured or quantified, whilst the corresponding distances are the core factor in the attenuation of these forces. By this moment, the potential theory is considered to be attractive for use (Statheros et al. 2008).

Inspired by this, a collision probability or a collision potential can also be considered as a special "repulsion", which objectively repulses away the corresponding objects, generally vessels, to avoid possible collisions. In other words, the closer to obstacles the target is, the higher collision potential there should be. Under the framework of potential theory, the strength of "repulsion" is exactly consistent with the collision potential. As mentioned previously, there are in-sufficient records of collision accidents; with the help of potential theory, a collision potential might be quantified by the "repulsions". For instance, it is widely acknowledged that channels between the piers of a bridge are dangerous for passing vessels, or the corresponding collision potentials are high although the accidents that vessels collide with piers are rare. In fact, there are very strict regulations for the operators of vessels when crossing piers, including speed limit, no overtaking. These regulations reduce collision accidents objectively. Hence, a collision probability or a collision potential cannot be estimated with a frequency analysis. However, the high collision probabilities or potentials are objective existence, which are changing the behaviours of vessels, making them as far as possible away from the piers. It is might be logical to take the collision potential as "repulsions" that repulse these vessels away from the piers. In the potential theory, those "repulsions" are caused by the corresponding so-called "repulsive potential fields", which are exactly produced by the piers (Volpe and Khosla 1990).

The phenomenon of traffic flow between piers is illustrated in Fig. 6.1, where there are several piers in a waterway. Hundreds of vessels crossed these piers, and vessels' tracks fetched from AIS database are represented with blue circles and lines. Particularly, these tracks indicate that vessels were obviously willing to take routes which were far away from these piers to lower their collision potentials. On the other hand, such a phenomenon can be regarded as that these vessels were pushed into a narrow channel by some undetectable "repulsions". As shown in Fig. 6.1, these "repulsions" are represented as red arrows. Apparently, the closer to the piers, the greater of the repulsions there should be; the distance is the core factor in the attenuation of the repulsions. The strength of the "repulsions" is consistent with the corresponding collision potential. By analysing the distribution of passing vessels, the corresponding repulsions or repulsive potentials can be measured.

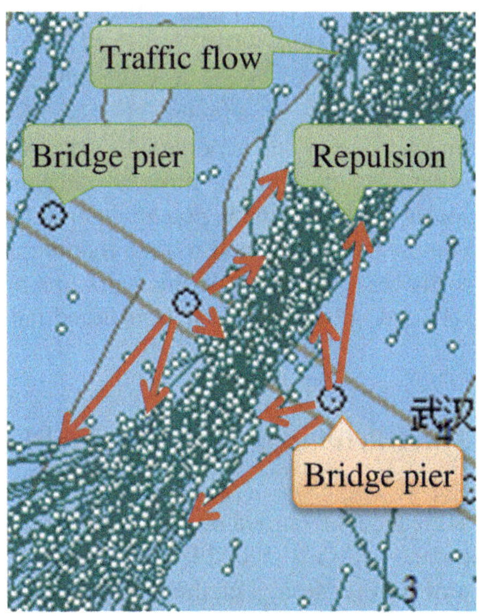

Fig. 6.1 Traffic flow between piers

In other words, the collision potential or probability of a position can be obtained from the distribution of passing vessels.

Statheros et al. (2008) used a Virtual Field Force (VFF) to describe the collision potential for collision avoidance in the unmanned surface vessel (USV) research. In fact, similar approaches are common in robot research, and the most frequently used methodology of field theory is the APF, which was invented by Khatib (1986). With the APF model, movements of a robot are governed by artificial potential fields, which are usually composed of two components, attractive potential and repulsive potential fields (Park et al. 2001). An attractive potential field is generally a bowl shape to draw the robot towards the goal. A repulsive potential filed is generally built at the location of an obstacle to repulse the robot away. With the use of the APF model, the collision potentials are modelled as continuous functions. Therefore, the differences of collision potential among adjacent positions can be described as the change of the values of these functions.

There is no unified formula for the APF model. In general, several potential functions are frequently used, which are mostly in quadratic and conical forms (Park et al. 2001). The following issue is to determine which potential function is appropriate for modelling collision potential in a waterway. In practice, the shape of the potential field should be compatible with the influences of corresponding obstacles. In addition, the corresponding influence range should conform to reality. Hence, the values of coefficients of the potential function should be assigned very carefully.

Presently, more researchers work towards addressing the problems of local minima and the modelling for arbitrarily shaped obstacles. There is very limited research on obtaining appropriate coefficients of potential field. Zhang et al. (2012) developed an evolved APF method by genetic algorithm, which uses a grid method to generate an obstacle avoidance path to address the local minimum problem. Montiel et al. (2015) used a bacterial evolutionary algorithm to address the same issue. Pêtrès et al. (2012) proposed an APF-based reactive navigation approach for vessels. In their approach, environment and local constraints are represented as potential fields around the vessels. Moreover, potential fields caused by wind directions and surrounding obstacles will be updated periodically, ensuring an optimal heading for the navigation. Overall, the APF model is an efficient method for modelling collision potentials in waterway transportation. However, there is no comprehensive method to obtain appropriate coefficients of potential fields, especially for a waterway. In this work, the distribution of passing vessels indicated by AIS and radar records might be good indications.

3 A Probabilistic ER and APF-Based Method

To reduce the burden of operators of coastal surveillance, this research proposes a probabilistic ER and APF-based method to identify targets that have high collision probabilities from a plethora of radar blips.

3.1 The Probability Estimation of a Radar Blip Being a Real Moving Vessel

In fact, experienced operators are able to achieve high accuracy in the identification of moving vessels under uncertainties, because they know the regularities of moving vessels after a long term observation. In other words, they can identify the patterns of blips or maritime targets using imprecise descriptions after sufficient experience or data has been accumulated. For instance, the speed of a moving vessel in a specific waterway is generally stable. Therefore, a velocity indicated by a blip is a piece of direct evidence for authenticity identification. Hence, it is possible to estimate the authenticity probability of a blip based on its velocity in adjacent frames. Moreover, operators can take several other attributes of a blip into consideration, in order to make comprehensive and accurate identification. In this research, the ER rule is adopted to simulate such a manual inference process.

In the ER rule, it is needed to constitute likelihoods of patterns or states from verified samples. The likelihoods stand for the probabilistic relationships between the attribute values of targets and the states. However, the attribute values of blips are indeed hidden in the sequential radar images. Therefore, the first step is to quantify these attributes of blips. As discussed previously, many attributes can be taken as evidence for identification, such as velocity, course, size, colour, width, and length. However, there is a precondition in the ER rule that the pieces of evidence for use should be independent of each other. Under this requirement, the evidence has to be chosen carefully. In this research, three types of evidence or attributes are selected, namely, velocity, motion direction (i.e., course), and blip shape. These attributes are generally considered to be independent of each other in terms of their contributions to moving vessel identification. The possible dependencies among them will be discussed in the future research.

The velocity and motion direction can be easily understood. Real moving vessels are more likely to move with a steady velocity and a steerable course, and noise objects are more likely to drift around a small area. The velocity and motion direction can be quantified as illustrated in Fig. 6.2a, b. In real life, operators are generally able to identify a blip in 10 consecutive frames. Therefore, this research extracts the velocity and direction attributes from the analysis of 10 frames. Less frames will be discussed in the future research.

Different from the velocity and direction, the blip shape is more related to the imaging principle of marine radar. Visually, a moving vessel's graph is generally more slender than others, and the principle is illustrated in Fig. 6.2d. In this sub-figure, a moving vessel blip possesses an afterglow, which is caused by an image delay function. This function is supported by most radar systems. The slenderness of a blip shape can be computed as the quotient of the blip's size (S_2) to the blip's circumcircle area (S_1), or S_2/S_1 in Fig. 6.2c.

After the quantification of a blip's attributes, the next step is to find out their probabilistic relationships to the authenticity.

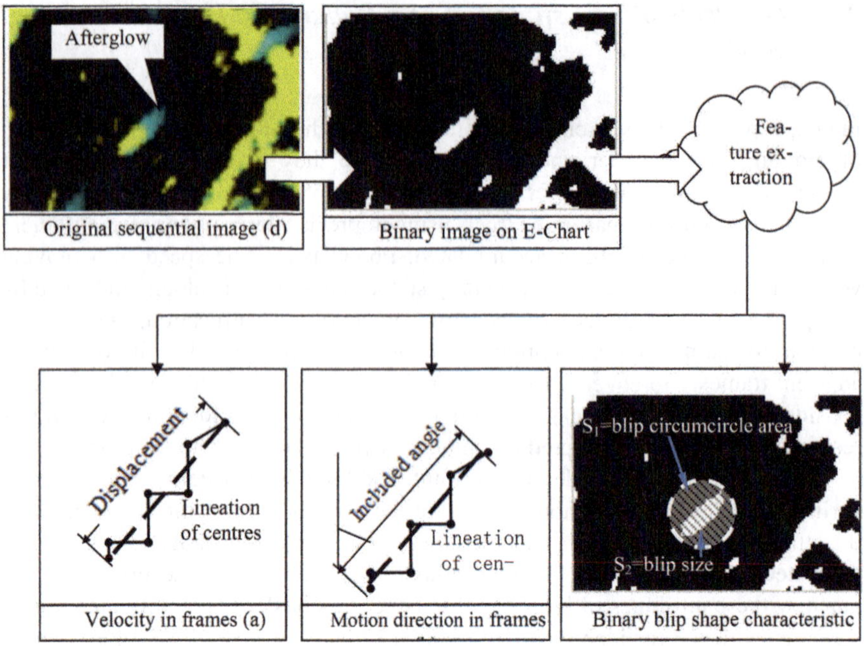

Fig. 6.2 The quantification of a blip's attributes

Suppose $\Theta = \{\theta_{True}, \theta_{False}\}$ is a set of mutually exclusive and collectively exhaustive propositions for the identification of blips, where θ_{True} is a True state and θ_{False} is a False state. Let $\emptyset$ represent the empty set. In practice, the Unknown state $\theta_{Unknown}$can be represented by the frame of discernment Θ itself, and it means the state that is neither True nor False. Thus, the power set of Θ consists of 4 subsets of Θ, and is denoted by 2^{Θ} or $P(\Theta)$, as follows:

$$P(\Theta) = \{\emptyset, \theta_{True}, \theta_{False}, \theta_{Unknown}\} \tag{6.1}$$

Different from the conventional probabilistic inference methods, a belief degree or a probability might also be assigned to the power set $P(\Theta)$ in the ER rule when there is a reliability problem in evidence.

A Basic Probability Assignment (*bpa*) is a function p: $2^{\Theta} \to [0, 1]$ that satisfies,

$$p(\emptyset) = 0, \sum_{\theta \subseteq \Theta} p(\theta) = 1 \tag{6.2}$$

where *the* basic probability $p(\theta)$ is assigned exactly to a proposition θ and not to any smaller subset of θ. $p(\theta)$ is generated from the values of attributes, including the velocity, direction or slenderness of a blip. Referring to the research conducted by Yang and Xu (2014), the likelihoods of authenticity states based on attribute values can be presented as follows.

Table 6.1 Verified samples

States	Observation attribute value of verified samples					Total
	Value 1	...	Value i	...	Value L	
False (0)	y_1^0	...	y_i^0	...	y_L^0	$Q^0 = \sum_{i=1}^{L} y_i^0$
True (1)	y_1^1	...	y_i^1	...	y_L^1	$Q^1 = \sum_{i=1}^{L} y_i^1$
Unknown (2)	y_1^2	...	y_i^2	...	y_L^2	$Q^2 = \sum_{i=1}^{L} y_i^2$

The core of likelihood modelling is to find the probabilistic relationships between the values of an attribute and the states (i.e. True, False and Unknown) of a blip. For example, when a blip is moving too fast on screen, the probability of a blip being a moving vessel is low. It is easy to know that there were rare vessels moving at such a high speed in this area previously. Therefore, the authenticity level and the velocity can be linked together by the prior data.

In any verified samples shown in Table 6.1, y_i^j denotes the frequency or the number of times that an attribute is equal to Value i for state j, with $i = 1, 2, \ldots, L$, and $j = 0$ for False, 1 for True, 2 for Unknown; Q^j denotes the total number of datasets for state j.

The likelihood transformation approach established by Yang and Xu (2014) underpins a new likelihood modelling method for moving target identification, which is described as follows. Based on the samples given in Table 6.1, the likelihood that an attribute is equal to Value i for a state of an object is calculated in Eq. (3) and presented in Table 6.2.

$$c_i^j = y_i^j / Q^j \quad \text{for i} = 1, 2, \ldots, L, j = 0, 1, 2. \tag{6.3}$$

where c_i^j denotes the likelihood to which the attribute is expected to be equal to Value i given that state j is true.

Let p_i^j denote the probability or belief degree that an attribute with Value i points to state j, which is independent of the prior distribution about the states. p_i^j is then acquired as normalised likelihood as follows.

$$p_i^j = c_i^j / \sum_{k=0}^{2} c_i^k \quad \text{for i} = 1, 2, \ldots, L, j = 0, 1, 2. \tag{6.4}$$

Belief distributions, given by $\{(False, p_i^0), (True, p_i^1), (Unknown, p_i^2)\}$ for $i = 1, \ldots, L$, represent the probabilistic relationships between the attribute of a blip and its states. Note that a belief distribution reduces to a conventional probability distribution when p_i^2 is equal to zero, or there is no ambiguity about the states of an object. Following the above procedure, an attribute value can be mapped to a belief distribution which is regarded as a piece of evidence.

It is worth mentioning that p_i^j for $j = 0$, 1 or 2 represents the inherent relationship between the attribute value of a blip and its states and it is not dependent on the prior distribution about the states from specific samples. For example, if a blip is

Table 6.2 Likelihoods without classification prior distribution

	Verified sample observation attribute value likelihood				
Classifications	Value 1	...	Value i	...	Value L
False (0)	c_1^0	...	c_i^0	...	c_L^0
True (1)	c_1^1	...	c_i^1	...	c_L^1
Unknown (2)	c_1^2	...	c_i^2	...	c_L^2

moving too fast, the probability of this blip being a normal moving vessel is very low. Such a low probability or belief degree should be reflected in any reliable historical records because it is unlikely that a normal vessel could move at such an abnormal velocity.

Subsequently, the ER rule is used to combine these pieces of evidence, and it also takes the reliability and weight of evidence into consideration. Evidence e_i is profiled by a belief distribution (BD) as follows:

$$e_i = \left\{ \left(\theta, p_i^\theta\right), \forall\theta \subseteq \Theta, \sum\nolimits_{\theta\subseteq\Theta} p_i^\theta = 1 \right\} \tag{6.5}$$

where $\left(\theta, p_i^\theta\right)$ represents that evidence e_i points to proposition (state) θ, which can be any subset of Θ or any element of $P(\Theta)$ except for the empty set. Particularly, $p_i^\theta > 0$, which denotes the probability or degree of belief of proposition (state) θ. In this occasion, p_i^θ is exactly obtained from the quantified attributes of a blip, given by Eqs. (3) and (4), where $\theta = 0(False), 1(True) \quad or \ 2(Unknown)$.

In addition, two coefficients, r_i and w_i, are introduced to measure the reliability and weight of evidence e_i in the ER rule. The reliability r_i is an inherent and objective property of the evidence e_i, and it represents the degree of support for or against a proposition given that the evidence points to the proposition. The weight w_i denotes the relative importance of evidence e_i in an evidence combination or a decision. In general, the weight w_i is assigned in accordance with who is making such a decision.

In the ER rule, it is required to take into account the three elements of the evidence in an evidence combination: its belief distribution (or probability distribution when there is no ambiguity in the evidence), reliability and weight. As a result, a so-called weighted belief distribution with reliability is defined as follows27:

$$m_i = \left\{ \left(\theta, \tilde{m}_{\theta,i}\right), \forall\theta \subseteq \Theta; \left(P(\Theta), \tilde{m}_{P(\Theta),i}\right) \right\} \tag{6.6}$$

where $\tilde{m}_{\theta,i}$ represents the degree of support for θ from e_i with both the weight and reliability of e_i taken into account, defined as follows:

$$\tilde{m}_{\theta,i} = \begin{cases} 0 & \theta = \phi \\ c_{rw,i} m_{\theta,i} & \theta \subset \Theta, \theta \neq \phi \\ c_{rw,i}(1 - r_i) & \theta = P(\Theta) \end{cases} \tag{6.7}$$

$$c_{rw,i} = 1/(1 + w_i - r_i) \tag{6.8}$$

where $c_{rw,i}$ denotes a normalisation factor, w_i denotes for weight, and r_i denotes reliability. $m_{\theta,i}$ is the degree of support for proposition (state) θ from evidence i, which is given by $m_{\theta,i} = w_i p_i^\theta$, with p_i^θ being the degree of belief that evidence *i* points to θ. As described previously, p_i^θ can be obtained using Table 6.1, Table 6.2, Eq. (3) and Eq. (4). $P(\Theta)$ is the power set of the frame of discernment Θ that contains all mutually exclusive hypotheses in question, and it has been defined in Eq. (6.1). It is worth mentioning that $P(\Theta)$ is treated as an independent element in the ER rule.

If each piece of evidence is completely reliable, e.g. $r_i = 1$ for any i, the ER rule reduces to Dempster's rule. In this research, each piece of evidence is apparently not fully reliable, or $r_i < 1$. The combination of two pieces of evidence e_1 and e_2 (defined in Eq. (6.5)) will be conducted as follows:

$$p_{\theta,e(2)} = \begin{cases} 0 & \theta \subseteq \phi \\ \dfrac{\widehat{m}_{\theta,e(2)}}{\sum_{D\subseteq\Theta}\widehat{m}_{D,e(2)}} & \theta \subseteq \Theta \end{cases} \tag{6.9}$$

$$\widehat{m}_{\theta,e(2)} = [(1 - r_2)m_{\theta,1} + (1 - r_1)m_{\theta,2}] + \sum_{B\cap C=\theta} m_{B,1}m_{C,2} \tag{6.10}$$

where $m_{\theta,1}$, $m_{\theta,2}$ $m_{B,1}$ and $m_{C,2}$ are given by Eqs. (6.6), (6.7) and (6.8); B, C and D denote any elements in the power set $P(\Theta)$ except for empty set, as Eq. (6.1); the $p_{\theta,e(2)}$ is the synthetic belief degree to proposition (state) θ when taking the two pieces of evidence e_1 and e_2 into consideration. Yang and Xu (2014) proved that the belief degree here is equivalent to the probability in Bayes' rule if each belief degree is assigned to a single state only and p_i^θ is calculated by Eq. (6.5). In Eqs. (6.7) and (6.8), the reliability and weight of evidence are parameters used to measure evidence quality. In Sect. 3.2, evidence was formulated from normalised likelihoods, which were generated independently of the prior distribution.

3.2 *The Modelling of Collision Potential Field Using the APF Model*

As described in Sect. 2.2, the APF model is adopted to model the collision potentials of static and encountered obstacles. In fact, there are a variety of APF formulations, among which the Yukawa function is widely used in collision avoidance potential modelling (Volpe and Khosla 1990),

$$U_{obs,m}(K) = A\frac{e^{-\alpha K}}{K} \tag{6.11}$$

where $U_{obs,m}$ denotes the avoidance or collision potential value to the mth obstacle. The constant A denotes a maximum value of collision or avoidance potential. α is also a constant, and it denotes the rate of decay, which is determined by the boundaries of APF. The variable K denotes the pseudo-distance to the mth obstacle, which may be different from the actual distance. It is required to take the characteristics of obstacles and the environmental factors into consideration to propose an appropriate formulation of the variable K (Volpe and Khosla 1990), especially in a waterway. Hence, the formulations of calculating variable K for different obstacles vary, including buoys, piers, rocks, shoals and encountered vessels. Two typical static obstacles (i.e. buoys and piers) have been discussed in the reference (Ma et al. 2016b).

By contrast, the collision potential distribution of an encountered vessel is more complicated, as the moving direction should be taken into consideration. Moreover, the dimension of this vessel is also very important. Hence, in this research, the rectangle model is also used to describe an encountered vessel after some improvement. Its pseudo-distance K is presented as follows,

$$K_v = \tau_{speed} \cdot \sigma_v \cdot \min\left(\sqrt{(x - x_v')^2 + (y - y_v')^2}\right) \tag{6.12}$$

where $(x_v', y_v') = \begin{cases} |x_v' - x_v| < l, y_v' = y_v \pm w_v \\ |y_v' - y_v| < w, x_v' = x_v \pm l_v \end{cases}$, and

$$\tau_{speed} = \begin{cases} 1/_{v \cdot \sigma_{speed}}, x - x_v > {}^{l_v}/_2 \\ 1, x - x_v \leq {}^{l_v}/_2 \end{cases}$$

(x_v, y_v) denotes the centre of the encountered vessel, l_v the length of the vessel, and w_v the length of the vessel. σ_p is an adjustment coefficient of the bridge pier pseudo-distance. τ_{speed} denotes the APF stretching caused by the moving of this vessel, v the speed of this vessel, and σ_{speed} the adjusting coefficient of the speed. Substituting Eq. (6.12) into Eq. (6.11), $K = K_p$, the rectangle equipotential lines are presented in Fig. 6.3. In particular, similar to the model of a pier, the potential edge rectangle that represents the maximum value of collision potentials is also larger than the actual geometrical dimensions of the corresponding vessel. The X axis here is set to be parallel to the moving direction.

Subsequently, all the buoys, piers and encountered vessels can be modelled as sources of collision (or repulsive) potential fields, which pose threats to the corresponding vessel. Moreover, in any place of the waterway, the corresponding collision potential can be considered as the combination of the different collision potential fields, which can be obtained with Eq. (6.11) and the equations of the reference (Ma et al. 2016b).

Different from a static obstacle, it is very difficult to get the distribution of encountered vessels, since the vessel under consideration is moving in the waterway. Moreover, AIS records are actually not synchronized (Ma et al. 2015).

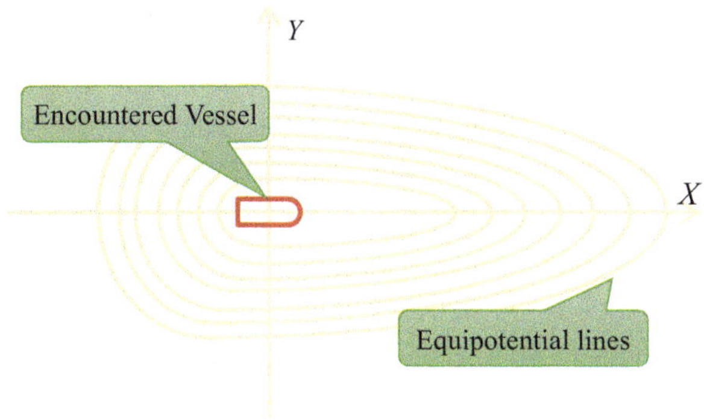

Fig. 6.3 A vessel repulsive potential field with equipotential lines

Therefore, the cues can only be conveyed by a radar system, which detects targets actively and synchronously. Therefore, the mutual distance distribution of moving vessel targets can be obtained from a radar system. Then, the coefficients of this APF model can be trained in accordance with this distribution. In particular, all the vessels may share the same coefficients of APF for simplicity. Moreover, such coefficient training should be based on the data captured in peak hours, as vessel operators only take consideration of safety distance to other vessels when in a crowded waterway.

Suppose there are many vessels passed a relatively close area of a waterway in peak hours, and the distribution (densities, frequencies) of distances ($-50 \sim 50$ metres) between any neighbouring vessels is presented as $\vec{k} = \{k_{-50}, \cdots k_{50}\}$. In particular, in peak hours, the distance between neighbouring vessels in crowded waterway is generally smaller than 50 metres, and they are mostly sailing in a straight line in this research. The maximum and minimum value of $\vec{k}$ are presented as $k_{\max} = \max(k_1, \cdots, k_{j-1})$, $k_{\min} = \min(k_1, \cdots, d_{j-1})$. The normalised distribution (densities) of distances is presented as, $\overrightarrow{k^*} = \{(k_1 - k_{\min})/(k_{\max} - k_{\min}), \cdots, (k_{j-1} - k_{\min})/(k_{\max} - k_{\min})\}$.

The undetermined coefficients of encountered vessel APF is denoted as $\overrightarrow{para_{vessel}} = \{\sigma_{speed}, \sigma_v, w_v, l_v\}$, all the positions of encountered vessels are denoted as $\{(x_1^v, y_1^v), \cdots, (x_{j-1}^v, y_{j-1}^v)\}$. In particular, the speed of monitored targets (probably vessels) can be measured by radar, which is presented as $\vec{v} = \{v_1, \cdots v_{j-1}\}$. Hence, the collision potentials caused by encountered vessel can be calculated with $\vec{v}$ and $\overrightarrow{para_{vessel}} = \{\sigma_{speed}, \sigma_v, w_v, l_v\}$, and given by,

$$P(v, x, y, \overrightarrow{para_{vessel}}) = \sum_{i=1}^{m} A_v \frac{\exp(-a_v \cdot K_v(x, y, x_i^v, y_i^v, w_v, l_v, \sigma_v))}{K_v(x, y, x_i^v, y_i^v, w_v, l_v, \sigma_v)} \tag{6.13}$$

In the moving direction, the normalised distribution of collision potential is presented as $\overrightarrow{p^*{}_{vessel}}$.

The undetermined coefficients of encountered vessels can be obtained with a nonlinear optimisation model, which is presented as,

$$\begin{aligned}\overrightarrow{para_{vessel}} &= \{\sigma_{speed}, \sigma_v, w_v, l_v\} \\ &= \text{argmin}_{feasiable} \sum_{i=-50}^{50} |[1 - P_{vessel}(x_i, y_i)] - (k_i - k_{\min})/(k_{\max} - k_{\min})|\end{aligned} \tag{6.14}$$

Since Eq. (6.14) is also continuously differentiable, the appropriate $\overrightarrow{para}$ can be obtained with the 'fmincon' function of MATLAB (Liu et al. 2003).

3.3 The Collision Probability Inference

The probability of a target being a real moving vessel and the collision potential of its position can be obtained by the above procedures. The next issue is to estimate the collision probability based on these two factors, which can be considered as two pieces of evidence. Considering the contribution in the risk recognition of manual operation, they can be regarded as being approximately independent of each other for simplicity. Hence, in this research, Dempster's rule is applicable for the evidence combination (Li and Pang 2013).

Suppose $\Theta = \{\theta_0, \theta_1\}$ is a set of mutually exclusive and collectively exhaustive propositions for the collision probability estimation of a blip. θ_0 is the *Collision* state, denoting a situation that the corresponding blip will collide with an obstacle; θ_1 is the *Non-collision* state, denoting a situation that the corresponding blip will not collide with any obstacle. Let Ø represent the empty set. In practice, the Unknown state θ_2 can be represented by the frame of discernment Θ itself, and it means the state that is neither θ_0 nor θ_1. Thus, the power set of Θ consists of 4 subsets of Θ, and is denoted by 2^{Θ} or $P(\Theta)$, as follows:

$$P(\Theta) = \{\varnothing, \theta_0, \theta_1, \theta_2\}$$

A *Basic Probability Assignment* (*bpa*) is a function p: $2^{\Theta} \rightarrow [0, 1]$ that satisfies,

$$p(\varnothing) = 0, \sum_{\theta \subseteq \Theta} p(\theta) = 1 \tag{6.15}$$

where the basic probability $p(\theta)$ is assigned exactly to a proposition θ and not to any smaller subset of θ. Then, the two factors discussed previously can be transformed to two pieces of evidence.

There is a target located at the position (x_k, y_k), and its probability of being a real moving vessel is estimated as p based on Bayesian Network (Ma et al. 2016a). Based on the authenticity of the target only, the basic probabilities about the

$\theta_0, \theta_1, \theta_2$ states can be obtained as follows, or a piece of evidence can be constructed,

$$e_1 : \{p(\theta_0), p(\theta_1), p(\theta_2)\} = \{p, (1-p) \tag{6.16}$$

In this area, there are M individual points $\{(x_1, y_1), (x_2, y_2), \cdots, (x_M, y_M)\}$. Their collision potentials to obstacles are presented as $\{P(x_1, y_1, \overrightarrow{para}), \cdots, P(x_M, y_M, \overrightarrow{para})\}$ based on Eq. (6.5), where $\overrightarrow{para}$ is obtained with the method presented in Sect. 3.2. The maximum collision potential is presented as $P_{\max}{}' = \max[P(x_1, y_1, \overrightarrow{para}), \cdots, P(x_M, y_M, \overrightarrow{para})]$; the minimum collision potential is presented as $P_{\min}{}' = \min[P(x_1, y_1, \overrightarrow{para}), \cdots, P(x_M, y_M, \overrightarrow{para})]$. Therefore, the normalised collision potential of position (x_k, y_k) is presented as,

$$P'_{normal}(x_k, y_k) = [P(x_k, y_k, \overrightarrow{para}) - P'_{\min}] / (P'_{\max} - P'_{\min}) \tag{6.17}$$

Based on the collision potential of the blip's position only, the basic probabilities about the $\theta_0, \theta_1, \theta_2$ states can be obtained as follows, or the piece of evidence is constructed as,

$$e_2 : \{p(\theta_0), p(\theta_1), p(\theta_2)\} = \{P'_{normal}(x_k, y_k), 1 - P'_{normal}(x_k, y_k), 0\} \tag{6.18}$$

Dempster's rule can then be used to combine the two pieces of evidence, which is presented as follows,

$$m(\theta) = [m_1 \bigoplus m_2] = \begin{cases} 0 & \theta = \varnothing \\ \dfrac{\sum_{B \cap C = \theta} m_1(B) m_2(C)}{1 - \sum_{B \cap C = \phi} m_1(B) m_2(C)} & \theta \neq \varnothing \end{cases} \tag{6.19}$$

where θ is a proposition that can be any subset of a set of hypotheses; $m(\theta)$ is the basic probability for θ; $m_1(B)$ is the basic probability for proposition B from the first piece of evidence; $m_2(C)$ is the basic probability for proposition C from the second piece of evidence; lastly, $\varnothing$ is the empty set. Therefore, the basic probability about the *Collision* state θ_0, or the collision probability of the blip based on the two pieces of evidence is presented as:

$$\begin{aligned} p(\theta_0) = {} & P'_{normal}(x_k, y_k) \times p / \{1 - P'_{normal}(x_k, y_k) \times (1-p) \\ & - [1 - P'_{normal}(x_k, y_k)] \times p \end{aligned} \tag{6.20}$$

4 A Case Study

4.1 Experimental Platforms

To validate the ER rule and APF-based method proposed in Sect. 3, an experiment was conducted at Wuzhou, Guangxin, China. In particular, the test radar is

FAR2127s, which is widely used in the coastal surveillance. The experiment lasted from 09:00 to 10:55 on the 21th May 2016. In total, 271 targets were captured, including 159 vessels and 112 stationary targets or noises. In the experiment, all the targets were verified manually. It is noted that many observations or blips were indeed from the same target since the radar scanned the area once per 2.4 s. In total, 30,367 individual observations (blips) have been captured. In these observations (blips), 22,955 observations are from moving vessels and 7412 observations are from noises or stationary targets. In the following research, all the stationary and noise targets are treated as noise samples for simplicity. Eventually, the verified samples have been divided into three parts by time. The first half of the samples are used to model the correlations between quantified attributes and the probabilities of states as discussed in Sect. 3.1, the other half of the samples are used for global validation.

Meanwhile, an AIS receiver was placed in the same area, which received 1,200,000 AIS messages from 15th March to 25th April 2016. Particularly, all the AIS messages are obtained from the same area as that of the blip recognition. These records will be used for obtaining the coefficients of collision potential fields as described in Sect. 3.2.

A software program is developed and shown in Fig. 6.4. As shown in this figure, radar images have already been overlapped on the e-chart of this waterway. Three typical verified objects were notified in Fig. 6.6. They are buoys No.17, No.35, and a moving vessel No.19. Intuitively, the moving vessel objects are different from noises in terms of the attributes of the velocity, course, and graphic shape.

Using the ER-based methods proposed in Sect. 3.1, these characteristics, velocity, vector and slenderness are quantified in the software program. All the blips in sequential images have been transformed to verified records that are presented in a text form with discrete values. A typical record is presented in Fig. 6.5. The record contains several fields, which are separated by commas and represent different types of discrete attribute values. In this way, the course (direction), velocity, and size are all stored in one record. Moreover, the verified vessel and noise records are saved separately.

4.2 *Probabilistic Inference and the Result Validation*

All the blip samples have been transformed to text forms shown in Fig. 6.5. Then, the probabilistic relationships between attribute values and probabilities or belief degrees of being moving vessels or noises can be obtained as follows.

- *Velocity*

  Using the first half of the verified samples captured in the experiment described in Sect. 4.1, the velocity distributions of the verified moving vessels and noise (or stationary) objects in 10 frames are presented in Figs. 6.6 and 6.7. The X axis denotes the velocity values, and the Y axis denotes the frequency in

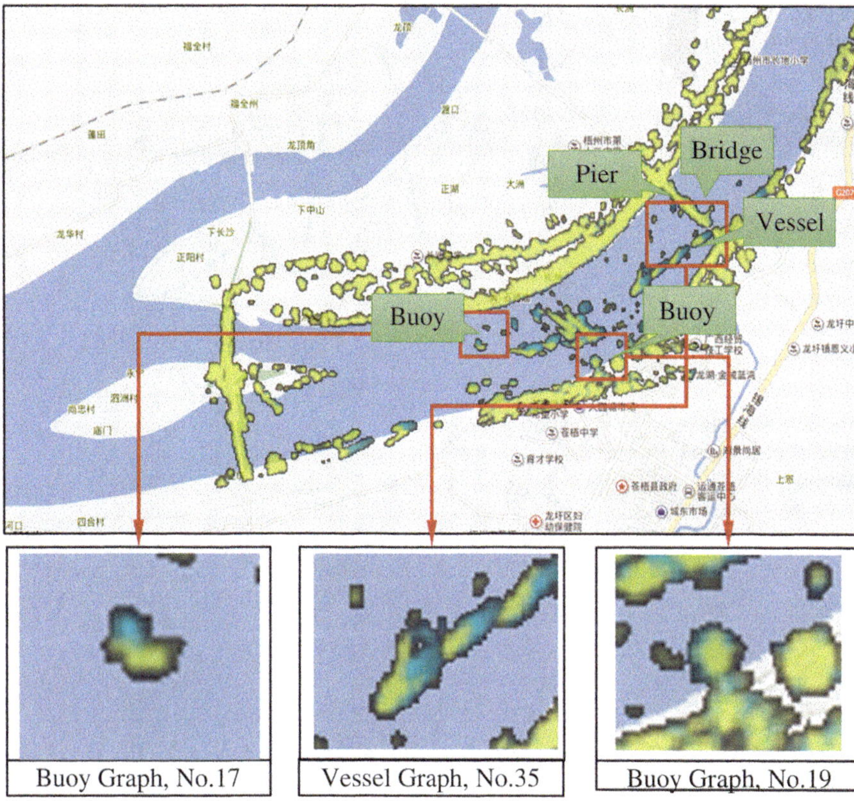

Fig. 6.4 Experimental software application program based on VC++

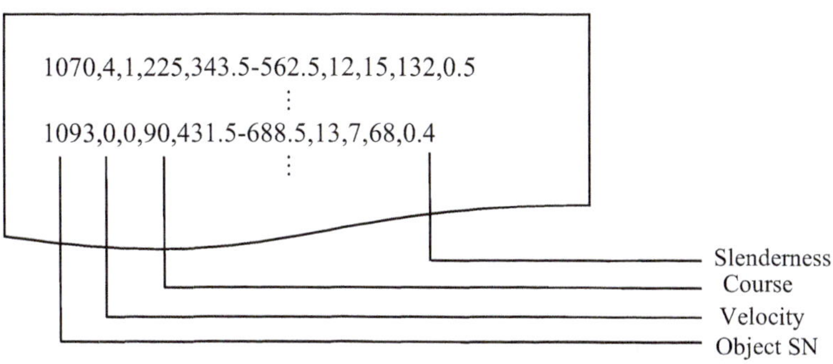

Fig. 6.5 Text record definitions

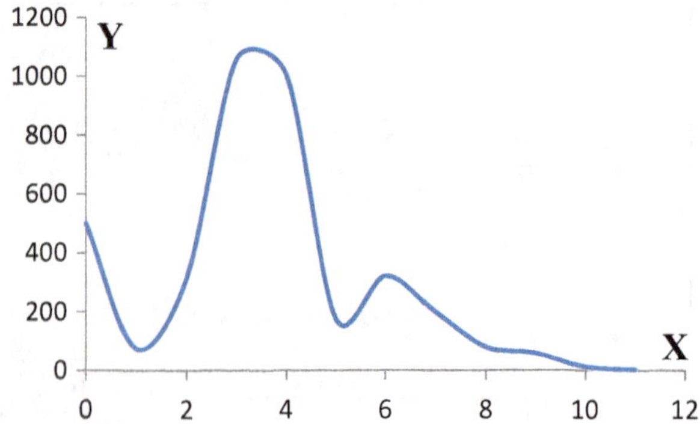

Fig. 6.6 Vessel blips' velocity distribution (True state)

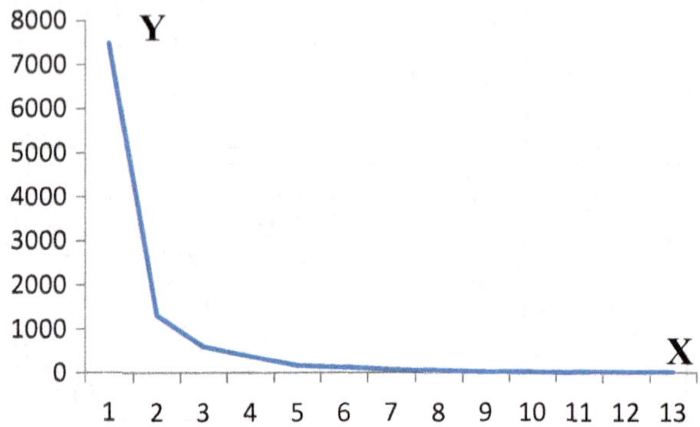

Fig. 6.7 Noise blips' velocity distribution (False state)

the first half of the samples. It can be inferred that the distribution differences are distinctive between of two states by contrasting Figs. 6.6 and 6.7.

In this case study, there are only two kinds of blips captured, True state (moving vessels) and False state (noises or stationary objects). No blips with the Unknown state have been captured.

Taking velocity 4 pixels per 10 frames as an example, the velocity evidence can be obtained as follows. The velocity of 4 pixels per 10 frames is the 5th value in the velocity attribute. According to Fig. 6.8, Fig. 6.9, and Eq. (3), the likelihoods of this blip being at the True state and False state are presented as,

$$c_5^0 = y_5^0/Q^0 = 0.0173 \tag{6.21}$$

$$c_5^1 = y_5^1/Q^1 = 0.2415 \tag{6.22}$$

As no blips with the Unknown state have been captured in this experiment, according to Eq. (4), the normalised likelihoods or probabilities of this blip being at the True state and False state can be presented as,

$$p_5^0 = c_5^0/\left(c_5^0 + c_5^1\right) = 0.0668 \tag{6.23}$$

$$p_5^1 = c_5^1/\left(c_5^0 + c_5^1\right) = 0.9331 \tag{6.24}$$

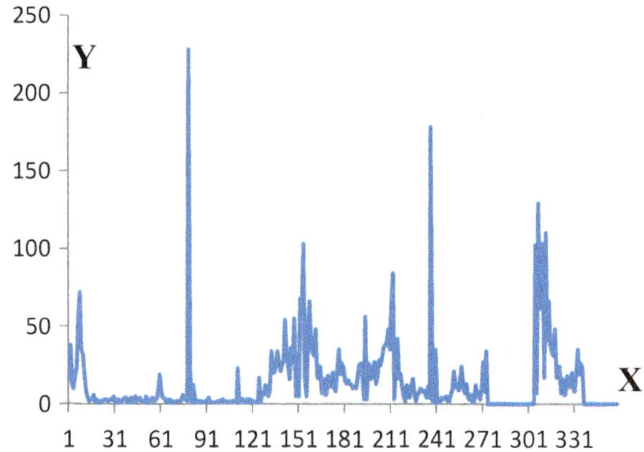

Fig. 6.8 Distribution of vessel blips' motion course in 10 frames (True state)

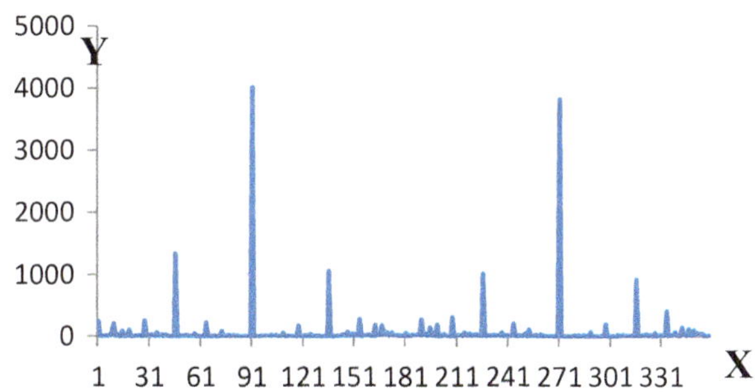

Fig. 6.9 Distribution of noise blips' motion course in 10 frames (False state)

In this occasion, $p_5^2 = 0$. Using this procedure, for any velocity value in Figs. 6.6 and 6.7, the probability or belief degree of each state (True, False, or Unknown) can be obtained. It is worth noting that the reliability and weight of this piece of evidence are still undetermined.

- *Motion course*

  Using the same verified samples as the velocity evidence, the motion course (direction) distributions in 10 frames of the verified moving vessels and noise (or stationary) objects are presented in Figs. 6.8 and 6.9. The X axis denotes the direction angle values, and the Y axis denotes the frequency in the first half of the samples. The regularity is also clear in the distributions. The noise objects are more likely to drift around a small area, and the centres might move in a short distance in 10 frames. Since the resolution of radar image is usually limited, the centres of such blips will drift 1 or 2 pixels vertically or horizontally in 10 frames after the binarization and segmentation. Therefore, the course values of noises (False state) crowd on 0, 45, 90, 135, 180, 225, 270 and 315 degrees. On contrary, the motion directions of moving vessels (True state) are different, and they crowd on the major directions of the waterway.

  The normalised likelihoods or probabilities to each state (True, False, or Unknown) with the motion direction can be computed using Eqs. (3) and (4).

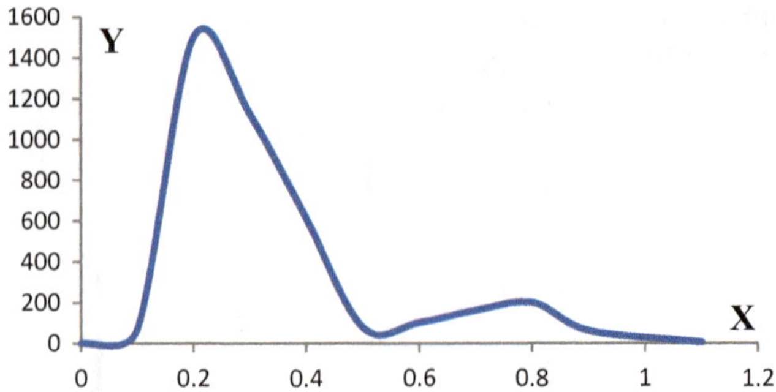

Fig. 6.10 Distribution of vessel blips' shape S_2/S_1 (True state)

Fig. 6.11 Distribution of noise blips' shape S_2/S_1 (False state)

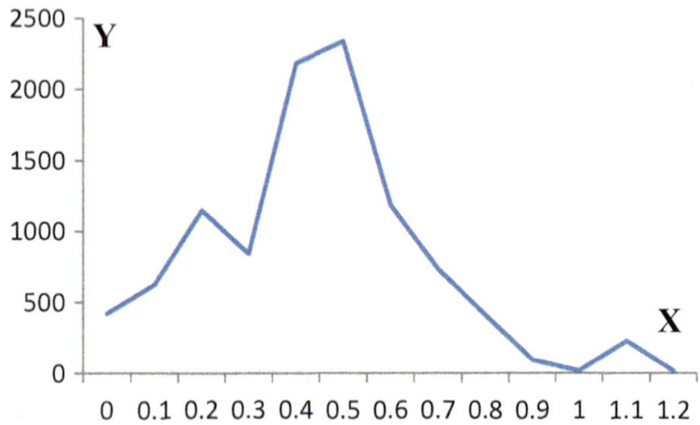

- *Object Shape S_2/S_1*

  The object graphic shape S_2/S_1, or slenderness distributions of the verified moving vessels and noise (or stationary) objects are presented in Fig. 6.10 and 6.11. The X axis denotes the slenderness values, and Y axis denotes the frequency in the first half of the samples. As the quotient of pixel size to blip circumcircle area, the slenderness value is continuous, meaning that there are infinite intervals. However, in radar images, the pixel size of a blip is limited; therefore the interval of 0.1 is considered to be sufficient to describe it accurately.

  In particular, the blip size is based on how many pixels it is occupying, and the value is denoted as an integer. The value of the circumcircle area is denoted as a float, which is based on the coordinates of outermost edge points. Therefore, if a blip is too small, the size is possible larger than the circumcircle area. In other words, the slenderness might be greater than 1. Similar to other evidence, the normalised likelihoods or probabilities to each state (True, False, or Unknown) with the slenderness can be computed using Eqs. (3) and (4).
- *Evidence combination*

  Through the procedures above, the correlations between the velocity, courses, slenderness and the belief degrees (probabilities) of blips being at the False,

Table 6.3 Results of analysis of the verified samples using the developed model, when $\{w_1, w_2, w_3\} = \{0.9500, 0.9500, 0.9500\}$

	Total	Correct identification	In-correct identification	Accuracy
Moving vessel	3706	3063	643	82.65%
Noises or stationary objects	11,477	10,475	1002	91.27%
Overall	15,183	13,538	1645	89.16%

True, and Unknown states can be established. With the methods described in Sects. 3.1 and 4.1, the verified samples can be transformed to text records. Then, such text records can be mapped to authenticity probabilities (belief degrees). Eventually, the ER rule is used to combine such evidence with corresponding weights and reliabilities, as discussed in Sect. 3.2.

As discussed, the reliability and weight coefficients of a piece of evidence should be equal when there is no verified sample or a specific optimisation objective. The observation area showed in Fig. 6.2 is open and clear, which is a common scenario. All the pieces of quantified evidence are from the same images. For all the evidence, reliability and weight can be considered as equal to 0.95.

As mentioned in Sect. 4.1, the last quarter of the verified samples are used for a global validation. Using these verified samples, the proposed approach is validated as follows. 50% is an intuitive and reasonable threshold. In other words, if the reasoning probability of a blip being a moving vessel is larger than 50%, the blip (observation) is considered as a true moving vessel. Otherwise, it can be considered as a noise or stationary object. The identification results are shown in Table 6.3.

This table shows the results obtained from the developed model. There are 2082 verified observations of being moving vessels and 15,183 verified observations of being noises or stationary objects in the analysis. As for the verified observations (blips) of being moving vessels, the developed model produced 3063 correct identifications out of 3706 observations, leading to an identification accuracy of 82.65%. As for the 11,477 verified observations of being noises or stationary objects, the model produces an identification accuracy of 91.27%. In total, the identification accuracy is 89.16%, and the ER rule turns out to be efficient. It is worth noting that mistakes are easily made by experienced operators.

4.3 The Modelling of Collision Potential Fields

The next step is to estimate the collision potentials of adjacent positions, which might be quantified in accordance with the distribution of passing vessels as described in Sect. 3.2. A survey region in Fig. 6.12 is chosen which contains

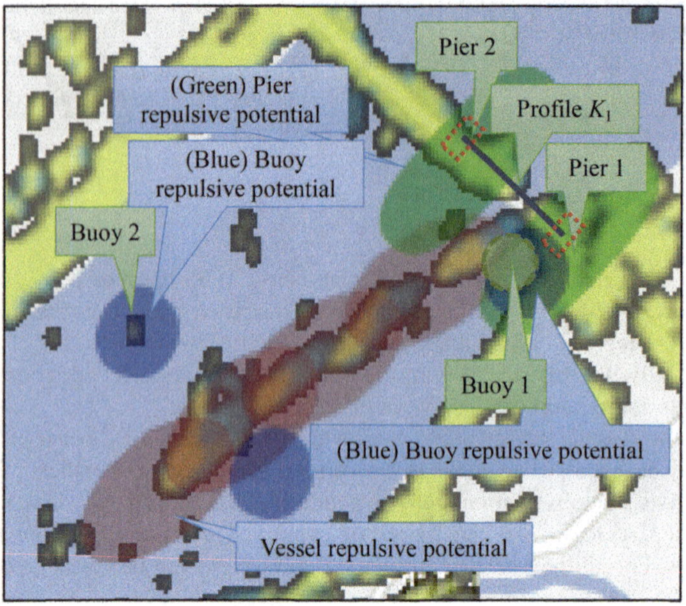

Fig. 6.12 The survey region

three piers, 3 buoys, and a major channel. In Fig. 6.12, the survey region is also represented with the e-chart format. Buoy 1 is represented as a green circle at the bottom; the blue solid line between the centres of Pier 1 and Pier 2 is selected to be the examined cross profile that has been described in Sect. 3.3, namely profile K_1, which can be used to train the coefficients of APF cuased by static obstacles (Ma et al. 2016b).

As described in Sect. 3.2, all the piers and buoys can be modelled as the sources of collision potential fields with the APF model, and the corresponding collision potential distribution can be obtained with the Yokawa potential function (Volpe and Khosla 1990). The red dotted rectangles in Fig. 6.10 indicate the *potential edge rectangles* of Pier 1 and Pier 2, defined in the reference (Ma et al. 2016b). The corresponding collision potential fields are represented as two highlighted green regions. In addition, the collision potential field of Buoy 1 represented as a highlighted blue eclipse (Ma et al. 2016b).

The distribution of passing vessels on profile K_1 can be inferred based on the collision potential fields of Pier 1, Pier 2, and Buoy 1. Apparently, the vessel distribution should be symmetrical on profile K_1 if Pier 1 and Pier 2 are the only static obstacles. Buoy 1 produces an extra collision potential field on the right side; in other words, Buoy 1 "repulses" passing vessels from the right side. Therefore, a conjecture is easy to be made that the peak value of the vessel distribution on profile K_1 should be slightly shifted to the left hand side due to the corresponding "repulsions" or collision potential, especially in a long time observation.

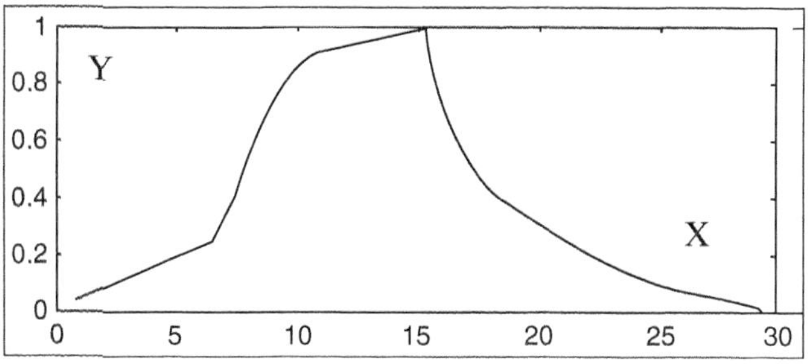

Fig. 6.13 The normalised distribution of passing vessels on profile K_1

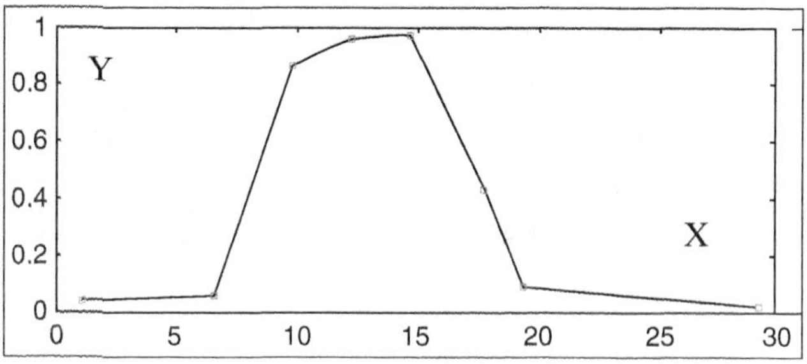

Fig. 6.14 The normalised distribution of safety degree on profile K_1

With the help of the software program described in Fig. 6.4, profile K_1 is analysed with 30 statistical individual points or sections in Fig. 6.13. Based on the AIS records, the distribution of passing vessels on profile K_1 can be normalised with Eq. (6.8) and presented in Fig. 6.11, where the X-axis represents the distance to Pier 2, and the Y-axis represents the normalised densities. Apparently, the densities follow a normal distribution, and the peak value is situated in the left side of profile K_1 between pier 2 and pier 3 as expected.

- *The coefficient training of static obstacles*

  The next task is to obtain the coefficients of these potential fields. Taking profile K_1 as an example, the coefficients should make the distribution of collision potentials consistent with the distribution of passing vessels. Therefore, the coefficients can be obtained in a nonlinear optimisation model, as discussed in the reference (Ma et al. 2016b). In this occasion, $i = 30$, the coefficients are obtained as $\overrightarrow{para} = \{\alpha_b, \xi, \sigma_b, \alpha_b, w, l, \sigma_p\} = \{0, 10.8322, 0.5228, 0, 12.444, 21.355, 0.0060\}$ using the '*fmincon*' function of MATLAB 2013b.

  Then, the collision potential distribution of profile K_1 can be estimated, and the normalised "safety distribution" is presented in Fig. 6.14, which is defined in Eq. (6.7) of the reference (Ma et al. 2016b). The X-axis represents the profile positions, and Y-axis represents the normalised "safety degree". By comparing Figs. 6.13 and 6.14, a good agreement can be found. In other words, the

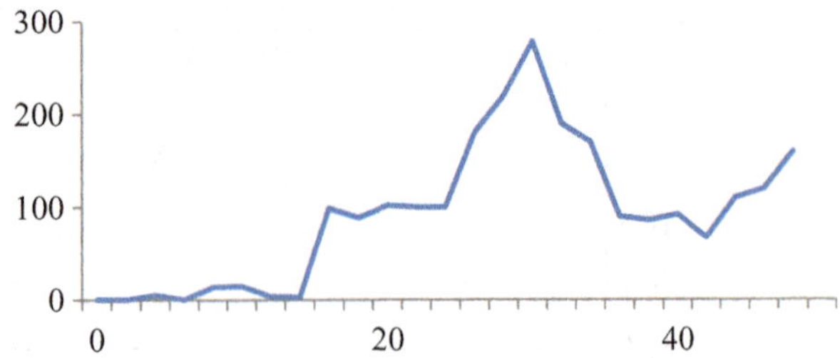

Fig. 6.15 The distribution (densities) of distances between neighbouring vessels

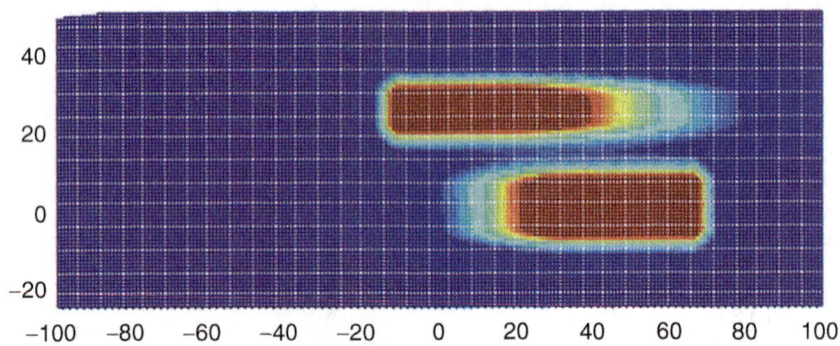

Fig. 6.16 The heat map of collision potentials caused by encountered vessels

distribution of collision potentials is consistent with the distribution of passing vessels on profile K_1.

The Bhattacharyya distance (Ma et al. 2016b) between Figs. 6.12 and 6.13 is 0.021, proving that the collision potentials are highly consistent with the real vessel distribution, and the coefficients obtained from the optimisation model are appropriate.

- *The coefficient training of encountered vessels*

  The distribution (densities) of distances between neighbouring vessels is presented in Fig. 6.15. Similar to the procedures above, the encountered vessel APF coefficients $\overrightarrow{para_{vessel}} = \{\sigma_{speed}, \sigma_v, w_v, l_v\} = \{0.8711, 0.1236, 10.2, 45.5\}$ based on Eqs. (6.10) and (6.11) using the 'fmincon' function of MATLAB 2013b. If there are two encountered vessels, the distribution of collision potential caused by them is presented in Fig. 6.16.

  For any individual target, its overall collision potential can be estimated by overlapping all the collision potential caused by static obstacles and encountered vessel.

4.4 Collision Probability Estimation

As discussed above, the probability of a blip being a real moving vessel, and the collision potential of its position are the two factors in determining whether it needs much attention in manual identification. In this research, the two factors are combined with Dempster's rule as described in Sect. 3.3. For instance, an object whose probability of being a moving vessel is 0.99 (99%) given by Bayesian Network (Ma et al. 2016b), and the normalised collision potential of this position is 0.37 (37%) given by Sect. 4.3. The two pieces of evidence are presented as $e_1 : \{p(\theta_0), p(\theta_1), p(\theta_2)\} = \{0.99{,}0.01{,}0\}$, $e_2 : \{p(\theta_0), p(\theta_1), p(\theta_2)\} = \{0.37{,}0.63{,}0\}$ based on Eqs. (6.15)–(6.19). Then, the basic probabilities about the θ_0, θ_1 and θ_2 states can be obtained as $\{p(\theta_0), p(\theta_1), p(\theta_2)\} = \{0.98{,}0.02{,}0\}$ by combining e_1 and e_2 based on Eq. (6.19). The collision probability of the target can be considered as $p(\theta_0) = 0.98$. In fact, $p(\theta_0)$ here represents a large belief degree about the *Collision* state for reminding the radar operators that the blip needs attention.

The efficiencies of the ER rule-based method and the APF model have been proved individually. Eventually, the proposed approach was tested with the verified samples, in order to prove its validity and reliability preliminarily. In the testing, the approach identified 37 objects that had the highest collision probability, and 31 of these objects were also inferred to be most dangerous by manual operation. In other words, a high agreement has been found.

5 Conclusions

To improve the efficiency and safety, this chapter proposed a probabilistic and APF-based methodology to estimate the collision probabilities of radar targets. The conclusions are given below. The APF model can be introduced to describe the collision potentials caused by obstacles. Moreover, the coefficients can be trained in a nonlinear optimisation model using AIS data records. According to manual operation, the collision probability of a radar target can be considered as the synthesis of the collision potential and the authenticity probability using probabilistic inference.

Stationary vessels were treated as noises in this research for simplicity. However, a new method may be needed to distinguish them from general noises. In manual judgments, for example, the continuous characteristics of a target are used as important evidence.

The concept of potential fields may need to be further investigated in order to fully realise the APF model's potential in ship collision assessment. This may be particularly useful for studying collision risks associated with berths and recommended channels.

The authenticity and collision potential of a blip were considered to be independent of each other and of equal weight in this research for simplification purposes. Further work may be useful to investigate how their dependency and their different weights would affect collision probability estimation.

References

Dellacherie C, Meyer PA (2011) Probabilities and potential, c: potential theory for discrete and continuous semigroups. Elsevier, North Holland

Dong Y, Frangopol DM (2015) Probabilistic ship collision risk and sustainability assessment considering risk attitudes. Struct Saf 53:75–84

Eleye-Datubo AG, Wall A, Wang J (2008) Marine and offshore safety assessment by incorporative risk modelling in a fuzzy-Bayesian network of an induced mass assignment paradigm. Risk Anal 28(1):95–112

Fujii Y, Oshima R, Yamanouchi H, Mizuki N (1974) Some factors affecting the frequency of accidents in marine traffic: I-the diameter of evasion for crossing encounters, II-the probability of stranding, III-the effect of darkness of the probability of collision and stranding. J Navig 27 (2):239–247

Guerriero M, Willett P, Coraluppi S, Carthel C (2008) Radar/AIS data fusion and SAR tasking for maritime surveillance. In: Proceedings of IEEE 11th international conference on information fusion. Cologne, pp 1–5

Hänninen M, Kujala P (2012) Influences of variables on ship collision probability in a Bayesian belief network model. Reliab Eng Syst Saf 102:27–40

IEC 62288 (2014) Maritime navigation and radio communication equipment and systems – Presentation of navigation-related information on shipborne navigational displays – General requirements, methods of testing and required test results

Khatib O (1986) Real-time obstacle avoidance for manipulators and mobile robots. Int J Robot Res 5(1):90–98

Kim JO, Khosla PK (1992) Real-time obstacle avoidance using harmonic potential functions. IEEE Trans Robot Autom 8(3):338–349

Kujala P, Hänninen M, Arola T, Ylitalo J (2009) Analysis of the marine traffic safety in the Gulf of Finland. Reliab Eng Syst Saf 94(8):1349–1357

Li HB, Shen YY, Liu YT (2007) Estimation of detection threshold in multiple ship target situations with HF ground wave radar. J Syst Eng Electron 18(4):739–744

Li B, Pang FW (2013) An approach of vessel collision risk assessment based on the D–S evidence theory. Ocean Eng 74:16–21

Liu GP, Yang JB, Whidborne JF (2003) Multiobjective optimisation and control. Res Stud Press, Baldock

Ma F, Wu Q, Yan X, Chu X, Zhang D (2015) Classification of automatic radar plotting aid targets based on improved Fuzzy C-Means. Transp Res Part C: Emerg Technol 51:180–195

Ma F, Chen YW, Yan XP, Chu XM, Wang J (2016a) A novel marine radar targets extraction approach based on sequential images and bayesian network. Ocean Eng 120:64–77

Ma F, Chen YW, Huang ZC, Yan XP, Wang J (2016b) A novel approach of collision assessment for coastal radar surveillance. Reliab Eng Syst Saf 155:179–195

Montiel O, Orozco-Rosas U, Sepúlveda R (2015) Path planning for mobile robots using bacterial potential field for avoiding static and dynamic obstacles. Expert Syst Appl 42(12):5177–5191

Montewka J, Goerlandt F, Kujala P (2012) Determination of collision criteria and causation factors appropriate to a model for estimating the probability of maritime accidents. Ocean Eng 40:50–61

Montewka J, Hinz T, Kujala P, Matusiak J (2010) Probability modelling of vessel collisions. Reliab Eng Syst Saf 95(5):573–589

Park MG, Jeon JH, Lee MC (2001) Obstacle avoidance for mobile robots using artificial potential field approach with simulated annealing. Proc IEEE Int Sym Ind Electron 3:1530–1535

Pêtrès C, Romero-Ramirez MA, Plumet F (2012) A potential field approach for reactive navigation of autonomous sailboats. Robot Auton Syst 60(12):1520–1527

Pedersen PT (2010) Review and application of ship collision and grounding analysis procedures. Mar Struct 23(3):241–262

Qu X, Meng Q, Suyi L (2011) Ship collision risk assessment for the Singapore Strait. Accid Anal Prev 43(6):2030–2036

Statheros T, Howells G, Maier KM (2008) Autonomous ship collision avoidance navigation concepts, technologies and techniques. J Navig 61(1):129–142

Volpe R, Khosla P (1990) Manipulator control with superquadric artificial potential functions: Theory and experiments. IEEE Trans Syst Man Cybern 20(6):1423–1436

Wang J (2001) The current status and future aspects in formal ship safety assessment. Saf Sci 38 (1):19–30

Yang JB, Xu DL (2014) A study on generalising Bayesian inference to evidential reasoning. In: International conference on belief functions. Springer, Cham, pp 180–189

Zhang D, Yan XP, Yang ZL, Wall A, Wang J (2013) Incorporation of formal safety assessment and Bayesian network in navigational risk estimation of the Yangtze River. Reliab Eng Syst Saf 118:93–105

Zhang Q, Chen D, Chen T (2012) An obstacle avoidance method of soccer robot based on evolutionary artificial potential field. Energy Procedia 16:1792–1798

Zhou D, Shen X, Yang W (2013) Radar target recognition based on fuzzy optimal transformation using high-resolution range profile. Pattern Recogn Lett 34(3):256–264

Chapter 7
Incorporating AHP and Evidential Reasoning for Quantitative Evaluation of Inland Port Performance

Chengpeng Wan, Di Zhang, and Hang Fang

Abstract Inland ports are core strategic resources promoting the development of regional economy. They are not only connections between production and consumption, domestic and foreign freight transport, but also critical infrastructure supporting the national and regional economic development, whose performance will inevitably influence the economy of the port city and even the relevant hinterlands. Thus, the measurement of inland port performance is of great significance in order to improve its performance and facilitate its development. Based on the literature review of port performance evaluation and brainstorms with domain experts, an assessment model for quantitatively measuring inland port performance is established in a three-level hierarchical structure composing of a goal level, a criteria level, and an index level. The relative importance of each criterion is obtained through the Analytic Hierarchy Process (AHP) method and evaluation results in the index level are then aggregated and calculated using the Evidential Reasoning approach. Based on a utility function, the performance of inland port is represented in the form of a crisp value for comparison in different years. The

C. Wan
Intelligent Transportation Systems Research Center (ITSC), Wuhan University of Technology, Wuhan, China

National Engineering Research Center for Water Transport Safety (WTSC), Wuhan University of Technology, Wuhan, China

Liverpool Logistics, Offshore and Marine (LOOM) Research Institute, Liverpool John Moores University, Liverpool, UK

D. Zhang (✉)
Intelligent Transportation Systems Research Center (ITSC), Wuhan University of Technology, Wuhan, China

National Engineering Research Center for Water Transport Safety (WTSC), Wuhan University of Technology, Wuhan, China
e-mail: zhangdi@whut.edu.cn

H. Fang
National Engineering Research Center for Water Transport Safety (WTSC), Wuhan University of Technology, Wuhan, China

P.T.-W. Lee, Z. Yang (eds.), *Multi-Criteria Decision Making in Maritime Studies and Logistics*, International Series in Operations Research & Management Science 260,
DOI 10.1007/978-3-319-62338-2_7

application of the assessment model is demonstrated using a case of the Port of Wuhan. The results of case study indicate that the assessment model developed in this chapter is able to provide insights for the evaluation of the performance of inland ports in different countries.

Keywords Inland port performance • Inland waterway transportation • Evidential reasoning • Hierarchical model • Analytic hierarchy process

1 Introduction

Inland waterway transportation plays a significant role in the cargo transport and resource distribution between the hinterland and coastal regions. It is irreplaceable among various transport modes in modern times due to its advantages such as low cost (e.g. investment cost and maintenance cost), low energy consumption, convenience (especially in the movement of heavy and bulky goods), and high transport capacity, when compared to rail and road transport. Thus, the acceleration of the development of inland water transport complies with the scientific development concept and the request of building an energy-efficient and environment-friendly society. The importance of inland waterway transportation has been seen in China and America, as well as a number of European countries (World Bank 2009). For example, the Yangtze River, as the longest inland waterway in China and the busiest one in the world since 2005 (Luo and Yang 2013), has contributed to more than 40% of the volume of freight traffic and about 35% of the total value of import and export trade in China (Xu 2014).

An inland port, as a crucial component of a multimodal transportation system, is essential to organise different types of transport patterns and support the logistic activities within the system. With the development of globalization and the evolution of logistics concepts, the role of inland ports has evolved, and its importance in promoting the economic development of the hinterland and integrating regional economy has been increasing. According to the Ministry of Transport (MoT) of China, the number of berths in China's inland ports in recent 5 years is shown in Table 7.1 (MoT 2011, 2012, 2013, 2014 and 2015).

Table 7.1 reveals a stable development of inland ports in China in which the number of berths with a capacity of more than 10,000 tons in 2015 has increased more than 20% compared to that 5 years ago. However, inland ports in China have suffered from an unbalanced development and for a long period their development has lagged behind that of coastal ports. Problems such as the insufficient utilisation of the available resources, distinct diversities existing in ports among different waters, and those in the ports between upstream and downstream in the same river result in the overall low efficiency of the inland waterway transportation networks (Wang and Bi 2010). Under the rapid development of world economic globalisation, the inter-port competition is becoming increasingly fierce, which further amplifies the negative effects of poor port performance on a region or even

Table 7.1 Number of berths with capacity of more than 10^4 tons in inland ports in China from 2011 to 2015

Year No. of Berth (pieces)	2011	2012	2013	2014	2015
With a capacity of $1*10^4$–$3*10^4$ tons	160	168	169	169	174
With a capacity of $3*10^4$–$5*10^4$ tons	95	103	102	104	103
With a capacity of $5*10^4$–$1*10^5$ tons	79	92	116	126	128
With a capacity more than 10^5 tons	6	6	7	7	9
Total number	340	369	394	406	414

Source: Compiled from the *Statistical Bulletin of Highway and Waterway Transportation Industry (2011–2015)*, published by the Ministry of Transport of China

country's trade (Esmer 2008). Therefore, it is of particular importance to measure the performance of ports in order to achieve the success of the port industry and the social welfare of citizens (Thomas and de Monie 2000).

The measurement of a port performance can provide port/terminal managers with following information: how efficiently the port or terminal is operating, what resources it needs to carry out its activities, whether it has achieved the expected or scheduled production targets or not, how its present performance is compared with its past performance, and how its performance is compared with that of its competitors. The above information helps port managers make rational and appropriate decisions to improve competitiveness of their ports. Although the importance of conducting port performance measurement has been widely accepted, the development of suitable Port Performance Indicators (PPIs) as well as the evaluation based on the selected PPIs are not always straightforward, due to the fact that a port is a complex dynamic system consisting of numerous interacting elements influencing various aspects of its operations (Esmer 2008).

A classic monograph on the port performance (UNCTAD 1976) classified PPIs into two main categories: financial indicators and operational indicators. The former category contains indicators like berth occupancy revenue per ton of cargo, cargo-handling revenue per ton of cargo, and capital equipment expenditure per ton of cargo, while the latter includes indicators such as waiting time, service time, and turn-around time. In 1999, UNCTAD suggested another two categories of PPIs, i.e., macro and micro performance indicators (UNCTAD 1999). The macro quantifies the influence of ports on economic activity and the micro measures input/output ratio of port operations (Bichou and Gray 2004). These have been used as a reference point (Marlow and Casaca 2003), and a lot of research has been conducted about measurement of productivity-related indicators, e.g. Talley (1988), Talley (1994), Tongzon (1995), and Tabernacle (1995), to name but just a few. Another traditional way of port performance measurement focuses on the efficiency aspect. In this field, two more complex, yet more appropriately holistic approaches that have been extensively studied and utilised are Data Envelopment Analysis (DEA) and Stochastic Frontier Analysis (SFA). They have promoted the progress of the efficiency measurement in terms of the evaluation of port performance (Tongzon 2001; Wang et al. 2003; Cullinane and Song 2006). Furthermore, a comprehensive comparison between two methods in the analysis of technical efficiency of container ports can be found in Cullinane et al. (2006).

Different ideas on the classification of PPIs have been proposed concerning the measurement of port performance. Hassan (1993) categorised complicated interconnected port operations into four types in order to obtain a better understanding of it, which assists in modelling and simulation: ship transport mode operations, cargo handling operations, warehousing operations, and inland transport mode operations. Kişi et al. (1999) analysed the port performance from four aspects, namely ship, cargo, berth, and labour. Trujillo and Nombela (1999) suggested that PPIs are classified into three broad categories: physical indicators, factor productivity indicators, and economic and financial indicators. While de Langen et al. (2007) concluded three types of PPIs as output indicators, upgrading indicators, and license to operate indicators. In the research of Thomas and Monie (2000), the measurement was mainly conducted from production, productivity, utilisation, and service measures. In which, production measures are reflected by throughput of ship, quay transfer, container yard, and receipt/delivery; productivity measures calculate the ratio of output to input; utilisation measures are used to determine how intensively the production resources are used; and service measures indicate the satisfaction of the customers in terms of reliability, regularity, and rapidity. Consequently, performance indicators considering environment protection, safety, and customer service are incorporated in the development of PPIs (de Langen et al. 2007). Most of the research on the port performance has been conducted, focusing mainly on the performance of seaports, or some certain parts of a port such as container terminals, and cargo terminals. Relatively less attention has been paid on the evaluation of the inland port performance (e.g. Arango et al. 2011; Cortés et al. 2007).

As a result, it leads to a research gap to be fulfilled, exploring appropriate PPIs for inland ports, as well as flexible methods for the evaluation of inland port performance based on the selected PPIs. This chapter aims to establish a novel model for the indication of port performance, and propose supporting methods for realising the quantitative measurement. To achieve the aim, the rest of the chapter is organised as follows. On the basis of the literature review, the evaluation model is developed and presented in the following section. In Sect. 3, the background information on the proposed methods is introduced. The Analytic Hierarchy Process (AHP) method is applied to determine the weight of individual indicators at each level of the model and the assessment results in the bottom level are aggregated to obtain the overall evaluation of port performance using the evidential reasoning (ER) approach. Furthermore, this chapter analyses assessment grades for each criterion and converts both quantitative and qualitative criteria to the same utility space by employing a series of fuzzy membership functions. A case study on one of the major inland ports in China is conducted in Sect. 4 to demonstrate the applicability of the proposed assessment model. Sect. 5 concludes this chapter with contributions and limitations being presented. It is worth noting that the concept of inland port in this chapter refers to the port on an inland waterway, such as a river or canal (no matter it has a connection to the ocean or not), rather than a cluster of distribution and logistics centres (Rahimi et al. 2008) which is also usually known as a dry port.

2 Inland Port Performance Assessment Model

The Inland port performance assessment model (IPPAM) is a dynamic complex system involving many indicators from various aspects. The indicators included in the model should be comprehensive, as well as exclusive without overlaps between each other. They should not only reflect the real-time changes accordingly to adapt to the needs of social development, but also maintain a relative stability for the evaluation of port performance in a period. Various criteria have been discussed in previous studies in terms of the selection of performance indicators, including the consistency with goals and objectives, conciseness, data availability, limitation of expense and time on data collection, and measurability (Talley 1994).

The IPPAM developed in this chapter consists of three levels. The top level reflects the goal of the model, which is to assess the performance of inland ports. The second level (criteria level) contains four main elements, namely infrastructure, operations and management, financial status, and development potential. These four elements are important, which enables the integration of different types of indicators reflecting the performance of a port from various aspects. They provide access to the analysis of both hard and soft power of a port considering the impacts from macro- as well as micro-environment when measuring port performance (e.g. Bichou and Gray 2004; de Langen et al. 2007; Metalla et al. 2015;). Thus, they are utilised in this research and the indicators in criteria level are set to be infrastructure (B1), operations and management (B2), financial status (B3), and development potential (B4).

The indicators in the bottom level (index level) are chosen in terms of their associated elements in the upper level along with the literature review and in-depth discussions with domain experts. Infrastructure is the basis of port operations and the indispensable foundation of port logistics, mainly including measurement on berths, yards, quays, machinery and equipment, and tugs. Infrastructure has a great influence on the operation efficiency and effectiveness in port logistics. Regarding the operations, the cargo throughput of port is the most straightforward one in the evaluation of the productivity of a port. Other indicators relate to the utilisation of storage yard, development of informatisation, safety management, employees, and logistics service quality (Lam and Su 2015). Financial status presents the port performance in a monetary term, especially the performance related to operational efficiency. High operational efficacy will lead to a healthy financial status, which in return will contribute to a better port performance (Gu 2012). Development potential indicates the potential of port development, which affects its performance in the future. Cargo throughput growth rate is an indicator directly reflecting its performance from its local level, while the growth rate of GDP and the support from government are influencing factors from a macro perspective.

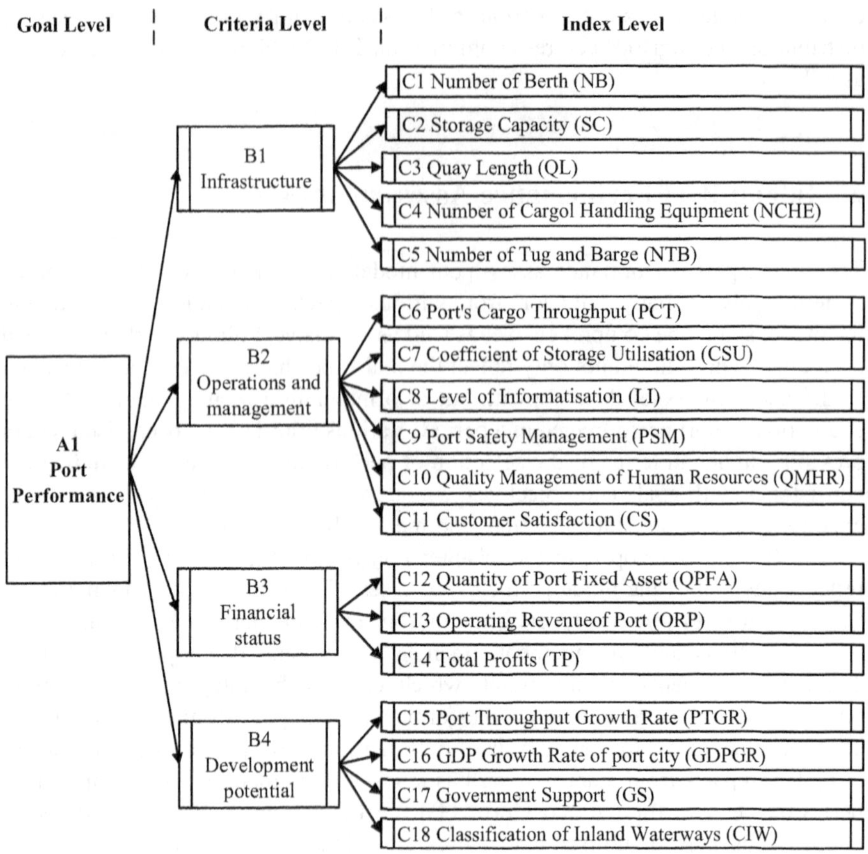

Fig. 7.1 Assessment model for inland port performance

Consequently, 18 indicators are identified and selected with respect to their significant roles reflecting the port performance, as shown in Fig. 7.1. They are further verified through extensive discussions with domain experts. Three domain experts are:

- Expert No.1: A senior officer who has worked in a port authority for more than 8 years,
- Expert No.2: An experienced port manager working in the Port of Wuhan, and
- Expert No.3: A professor engaged in inland waterway transportation and port research for more than 10 years.

Interpretations of each indicator in the index level are provided as follows:

(1) Infrastructure

C1, Number of Berth (NB) is one of the main indicators directly reflecting the scale of a port. The number of production berth of an inland port is counted here for analysis.

C2, Storage Capacity (SC) reflects the capacity of yard or warehouse to store goods or for turnover.

C3, Quay Length (QL) is the total length of quays in an inland port, reflecting the ability for loading or unloading of ships or boats.

C4, Number of Cargo Handling Equipment (NCHE): this indicator shows the ability of a port to move, storage, and control cargos.

C5, Number of Tug and Barge (NTB) indicates the how much tug assistance service can be offered for ships in a crowded harbour or a narrow channel.

(2) Operations and management

C6, Port's Cargo Throughput (PCT) is official statistics measuring the quantity of all kinds of cargo that can pass through a port. It is one of the important quantitative indicators to evaluate port productivity.

C7, Coefficient of Storage Utilisation (CSU) refers to how well the storage yard space is used during a certain period. It is calculated by Eq. (7.1).

$$C7 = \frac{Tons\ of\ cargo\ in\ storage\ per\ day}{Average\ warehouse\ capacity} \times 100\% \tag{7.1}$$

C8, Level of Informatisation (LI) refers to the extent to which a port is becoming information-based.

C9, Port Safety Management (PSM) is measured by the number of serious accident occurred in a port per year.

C10, Quality Management of Human Resources (QMHR) decides the employee competence of a port enterprise with respect to their age, education level, and skills.

C11, Customer Satisfaction (SC) mainly reflects the abilities to retain existing customers and attracting new customers. The level of customer satisfaction of a port will affect its share in the market. It can be judged from, for example, cargo on-delivery rate, value added service, and cargo complaint report ratio.

(3) Financial status

C12, Quantity of Port Fixed Asset (QPFA) refers to assets whose future economic benefit is probable to flow into the entity, whose cost can be measured reliably (IASC 1993). The more the quantity is, the better service can be acquired.

C13, Operating Revenue of Port (ORP) reflects all cash that flows into a port company from its primary operations including sales of goods and services. It can be obtained through the annual report of a company.

C14, Total Profits (TP) is the measure of a company's success equal to the net revenue that remains once all costs have been deducted. It can be calculated based on data in the annual report of a company.

(4) Development potential

C15, Port Throughput Growth Rate (PTGR):

This indicator reflects the development trend and speed of port's service capability which is calculated in Eq. (7.2),

$$C15 = \frac{Cargo throughput of the current period - Cargo throughput of the prior period}{Cargo throughput of the prior period} \times 100\% \tag{7.2}$$

C16, GDP Growth Rate of port city (GDPGR):

It is one of the main indicators reflecting the economic development trend of a port city which is calculated in Eq. (7.3),

$$C16 = \frac{GDP\ of\ the\ current\ period - GDP\ of\ the\ prior\ period}{GDP\ of\ the\ prior\ period} \times 100\% \tag{7.3}$$

C17, Government Support (GS):

This indicator reflects the support level (either financial or policy-related) from local government and country.

C18, Classification of Inland Waterways (CIW) shows the external navigation condition connecting to the port area, which affects its development potential in the future. It can be obtained through the official published standards.

3 Methodology

The following steps are developed in order to estimate inland port performance (Zhang et al. 2015).

Step 1. Carry out the pairwise comparisons in each level of the hierarchical structure in terms of the relative importance of the identified indicators to the port performance and calculate the weighting vectors of the indicators at the corresponding level.

Step 2. Develop a set of grading evaluation criteria and fuzzy membership functions to transform quantitative criteria into qualitative ones using an information transformation technique.

Step 3. The ER algorithm is used to carry out the synthesis of evaluation results of bottom indicators in the hierarchical structure.

Step 4. The results are prioritised and compared by using utility values for obtaining the quantitative results of inland port performance.

Step 5. The proposed model along with the research findings are discussed.

3.1 Analytic Hierarchy Process (AHP)

AHP was developed by Saaty and it was designed to solve complex multi-criteria decision problems (Saaty 1980). It requires decision makers to supply judgments about the relative importance of each criterion and then specify a preference for each decision alternative against each criterion. It is especially appropriate for complex decisions which involve the comparison of decision criteria that are difficult to quantify. AHP is based on the assumption that when dealing with a complex decision the natural human reaction is to cluster the decision criteria according to their common characteristics. Since the method was introduced three decades ago, many useful applications have been seen in the literature, including but not limited to, evaluation of green port (Maritz et al. 2014), port selections (Ugboma et al. 2006), industrial engineering application (Yang et al. 2003) and many more.

3.2 Evidential Reasoning (ER)

ER was developed in the 1990s to deal with Multiple Criteria Decision Making (MCDM) problems under uncertainty. The ER algorithm is based on the decision theory and the Dempster–Shafer (D-S) theory of evidence, which is well suited for handling incomplete assessment of uncertainty (Yang 2001). The algorithm can be used to aggregate criteria of a multilevel structure. ER is widely used in many applications such as engineering design, system safety, risk assessment, organizational self-assessment and supplier assessment (e.g. Chin et al. 2009; Liu et al. 2008; Wan et al. 2016).

The set $S(E) = \{(H_n, \beta_n), n = 1, \ldots, N\}$ represents a criterion E which is assessed to grade H_n with degree of belief β_n, $n = 1, \ldots, N$. Let $m_{n,i}$ be a basic probability mass representing the degree to which the ith basic criterion e_i supports the hypothesis that the criterion y is assessed to the nth grade H_n. To obtain the combined degrees of belief of all the basic criteria, $E_{I(i)}$ is firstly defined as the subset of the first i basic criteria as follows:

$$E_{I(i)} = \{e_1, e_2, \ldots, e_i\} \tag{7.4}$$

Let $m_{n,I(i)}$ be a probability mass defined as the degree to which all the i criteria in $E_{I(i)}$ support the hypothesis that E is assessed to the grade H_n and let $m_{h,I(i)}$ be the remaining probability mass unassigned to individual grades after all the basic criteria in $E_{I(i)}$ have been assessed. Equations (7.5) and (7.6) are obviously correct when $i = 1$.

$$M_{n,I(1)} = m_{n,1}, n = 1, 2, \ldots\ldots, N \tag{7.5}$$

$$M_{h,I(1)} = m_{h,1} \tag{7.6}$$

By using Eqs. (7.5) and (7.6), Eq. (7.7) can be constructed for $i = 1, 2, \ldots, L\text{-}1$ to obtain the coefficients $m_{n,I(L)}$, $\bar{m}_{H,I(L)}$ and $\tilde{m}_{H,I(L)}$ (Yang and Xu 2002):

$$K_{I(i+1)} = \left[1 - \sum_{t=1}^{N} \sum_{\substack{j=1 \\ j \neq t}}^{N} m_{t,I(i)} m_{j,i+1}\right]^{-1} \tag{7.7}$$

$K_{I(i+1)}$ is a normalizing factor.$\{H_n\}$:

$$m_{n,I(i+1)} = K_{I(i+1)} \left[m_{n,I(i)} m_{n,i+1} + m_{H,I(i)} m_{n,i+1} + m_{n,I(i)} m_{H,i+1}\right] \ n = 1,\ 2, \ldots, N \tag{7.8}$$

$$\tilde{m}_{H,I(i+1)} = K_{I(i+1)} \left[\tilde{m}_{H,I(i)} \tilde{m}_{H,i+1} + \bar{m}_{H,I(i)} \tilde{m}_{H,i+1} + \tilde{m}_{H,I(i)} \bar{m}_{H,i+1}\right] \tag{7.9}$$

$$\bar{m}_{H,I(i+1)} = K_{I(i+1)} \bar{m}_{H,I(i)} \bar{m}_{H,i+1} \tag{7.10}$$

$\{H\}$:

$$m_{H,I(i)} = \tilde{m}_{H,I(i)} + \bar{m}_{H,I(i)},\ i = 1,\ 2, \ldots, L\ - 1 \tag{7.11}$$

At last, the combined degrees of belief of all the basic criteria for the assessment to criterion E are calculated by:

$$\{H_n\} : \beta_n = \frac{m_{n,I(L)}}{1 - \bar{m}_{H,I(L)}}, n = 1,\ 2, \ldots, N \tag{7.12}$$

$$\{H\} : \beta_H = \frac{\tilde{m}_{H,I(L)}}{1 - \bar{m}_{H,I(L)}} \tag{7.13}$$

The ER approach is used in Step 3 of the proposed methodology for synthesising the evaluations of the basic indicators in the hierarchical structure.

3.3 Utility Value

It is difficult to rank the level of port performance by using belief degrees associated with linguistic terms because they are not sufficient to show the difference between the assessments. Numerical values (crisp values) are therefore generated from the obtained distributed assessments. The concept of expected utility is used to obtain a crisp value for each alternative in order to rank them in terms of development levels.

Suppose the utility of an evaluation grade H_n is denoted by $u(H_n)$ and $u(H_{n+1}) > u(H_n)$ if H_{n+1} is more preferred than H_n (Yang 2001). Therefore, the utility of the general criterion can be calculated using a linear distribution as Eqs. (7.14) and (7.15):

$$u(H_n) = \frac{n-1}{N-1}, \quad n = 1,2,\ldots, N \tag{7.14}$$

where, N denotes the number of the linguist terms.

$$u(E) = \sum_{n=1}^{N} \beta_n u(H_n) \tag{7.15}$$

Thus, a crisp value can be calculated based on the distribution generated via the ER technique and thus a comparison between alternatives can, therefore, be carried out.

3.4 *Degree of Evaluation Indicator Membership*

As both the quantitative and qualitative indicators are included in the evaluation system, they should be transformed and presented in certain grades respectively, as shown in Sects. 3.4.1 and 3.4.2.

3.4.1 Membership Degree of Quantitative Indicators

This study divides the development of inland port performance into four levels, namely, Excellent, Good, Fair and Poor. Membership degrees of quantitative indicators can be calculated through the fuzzy membership functions adapted from Chen (2014), as shown in Fig. 7.2. It is composed of fuzzy triangular and trapezoidal distributions.

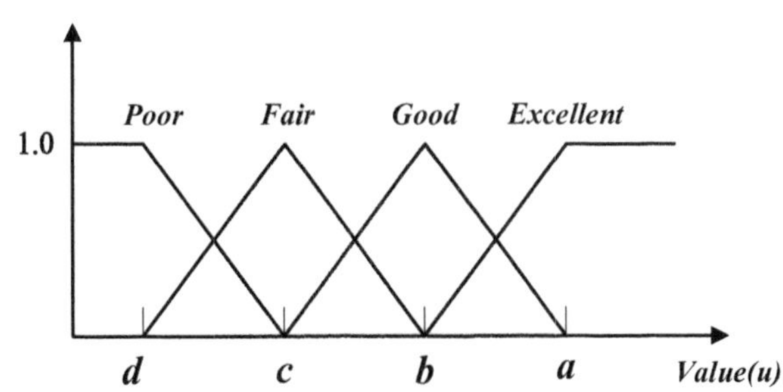

Fig. 7.2 Fuzzy membership of quantitative index

Here, a refers to the most possible value of Excellent, while d is the most possible value of Poor. b represents the most possible value of Good, and c indicates the most possible value of Fair. Supposing the value of an index is u, then.

(a) When $u \geq a$, the grading will be 100% Excellent;
(b) When $u \leq d$, the grading will be 100% Poor;
(c) When $b < u < a$, the grading will be $(u - b)/(a - b)$ Excellent and $(a - u)/(a - b)$ Good;
(d) When $c < u \leq b$, the grading will be $(u - c)/(b - c)$ Good and $(b - u)/(b - c)$ Fair.
(e) When $d < u \leq c$, the grading will be $(u - d)/(c - d)$ Fair and $(c - u)/(c - d)$ Poor.

In this study, the standards of grading for quantitative indexes are obtained from recent studies (Chen 2014; Xu 2014) and in-depth discussions with the experts described in Sect. 2. They are shown in Table 7.2.

3.4.2 Membership Degree of Qualitative Indicators

The grades of qualitative indexes are also described using Excellent, Good, Fair and Poor. Definitions of each grade for qualitative indicators (China Water Transportation Construction Association 2013; Xu 2014) are shown in Table 7.3.

Table 7.2 Grading for quantitative indicators

Quantitative indicator	a	b	c	d
C1 NB (berth number)	60	50	40	30
C2 SC (10^3 m^2)	500	400	300	200
C3 QL (km)	8	7	6	5
C4 NCHE (set)	350	300	250	200
C5 NTB (boat number)	200	150	100	50
C6 PCT (10^6 ton)	50	40	30	20
C7 CSU (%)	80	70	60	50
C9 PSM (accident number/year)	0	1	2	3
C12 QPFA (10^8 CNY[a])	16	12	8	4
C13 ORP (10^8 CNY)	16	12	8	4
C14 TP (10^6 CNY)	16	12	8	4
C15 PTGR (%)	40	30	20	10
C16 GDPGR (%)	70	60	50	40
C18 CIW (Level)	I	III	V	VII

[a]CNY = Chinese Yuan

Table 7.3 Standards of each grade for qualitative indicators

Qualitative indicators	Definition of each grade			
	Poor	Fair	Good	Excellent
C8 Level of Informatisation (LI)	Not be able to use information management technology such as electronic data interchange (EDI), cyber physical system (CPS), management information system (MIS), & etc.	Limited application of EDI technology with more than 10 types of relevant EDI documents. Initial realisation of office automation.	More than 50% of the files are managed using EDI technology, and partly usage of MIS.	More than 90% of the files are managed using EDI technology, and widely usage of CPS, MIS, and other information management technologies.
C10 Quality Management of Human Resources (QMHR)	Poor educated and inexperienced employees.	Average educated employees yet need more training on daily operations.	Average educated and skillful employees. But the employee structure of the company needs to be improved.	Higher educated and experienced employees. Rational employee structure of the company.
C11 Customer Satisfaction (CS)	Low cargo on-time delivery rate, no value-add service, and high complaint report rate	Normal cargo on-time delivery rate, poor value-add service, and several complaints reported	High cargo on-time delivery rate, fair value-add service, and low complaint report rate	High cargo on-time delivery rate, good value-add service, and almost no complaint reported
C17 Government Support (GS)	No official regulation on port management. No financial support from the government.	Unified regulation on port management, and financial support from the government.	Unified regulation on port management, and support from the government on both financial and political aspects.	Sound port management regulations, and strong support from government on both financial and political aspects.

4 Case Study

The National Plan for Inland Waterways and Ports (MOT 2007) presents a port network composing of 28 major inland ports that are distributed in four main waters (namely, the Yangtze River waters, the Pearl River waters, the Beijing-Hangzhou Grand Canal and Huaihe River waters, the Heilongjiang River and Song-Liao waters) in China. These inland ports are located in main regional areas radiating to the neighbouring cities, as shown in Fig. 7.3.

As an important waterway transport hub connecting Wuhan City with other regions, the Port of Wuhan has been developed into a considerable scale, three-dimensional traffic transportation system, with a flexible road network composed of three state roads, eight artery roads, and various other routes connecting Wuhan

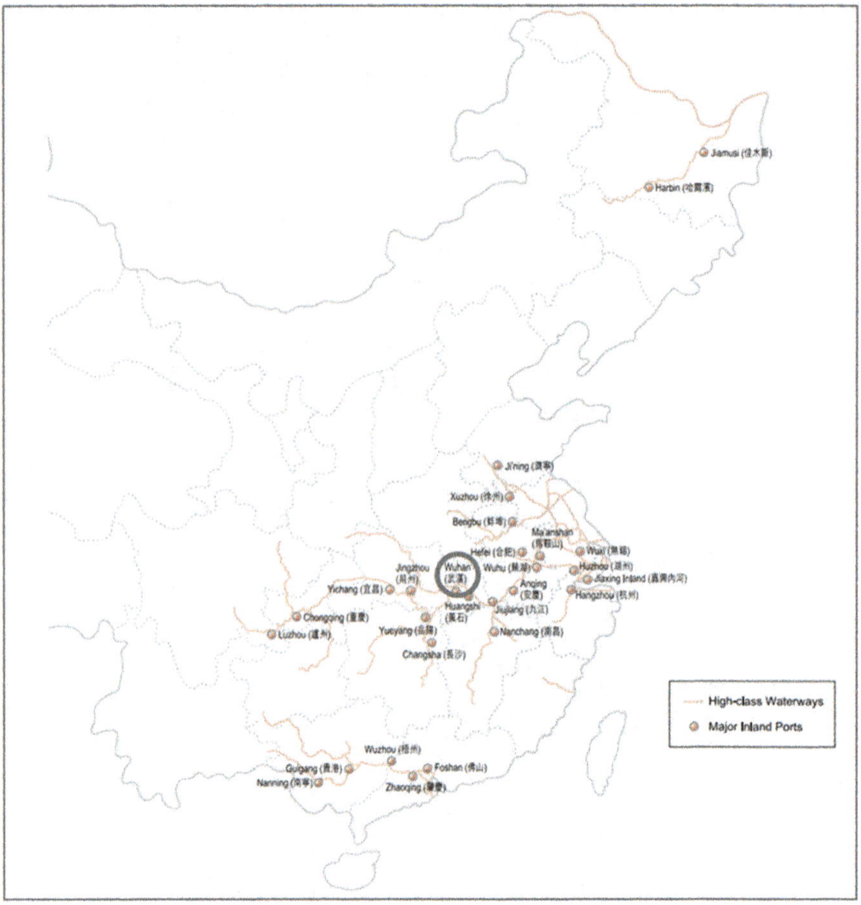

Fig. 7.3 The 28 major inland ports in China (Source: Li & Fung Research Centre 2009)

with other 195 cities located in eight neighbouring provinces (China Port Association 2011). Besides, it is the largest comprehensive inland port in the upper and middle Yangtze River of China with a total port area of 122.45 km^2. Thus, it has been selected as the case study in this research. As the Port of Wuhan is mainly operated by the Wuhan Port Group Co., Ltd., we take it as a representative when measuring the performance of the port. Therefore, the data from the company, such as its annual report, relevant statistical data, news, and company reports in 2007, 2010, and 2013 respectively, are collected and used for the case study.

4.1 Calculation of Weights of Evaluation Indicator

Pairwise comparisons are made by three domain experts mentioned in Sect. 2 through in-depth interviews and the weights of each index in the assessment model are obtained using the AHP method. Since the knowledge and experience of all three experts involved are considered as equivalent, the normalised relative weight of every expert is equally assigned while combining their judgments. A similar process can be implemented to each level and the weighting vectors of all pairwise comparison matrixes can be obtained to represent the local importance degree of each indicator. The weights of all indicators are shown in Table 7.4.

Table 7.4 Weights of each indicator of the assessment model

Goal Level	Criteria level	Index level	Relative weights	Global weights
Port performance	Infrastructures 0.243	C1	0.211	0.051
		C2	0.243	0.059
		C3	0.125	0.030
		C4	0.308	0.075
		C5	0.113	0.027
	Operations and management 0.317	C6	0.234	0.074
		C7	0.131	0.042
		C8	0.166	0.053
		C9	0.199	0.063
		C10	0.117	0.037
		C11	0.153	0.049
	Financial status 0.163	C12	0.374	0.061
		C13	0.183	0.030
		C14	0.443	0.072
	Development potential 0.277	C15	0.224	0.062
		C16	0.277	0.077
		C17	0.332	0.092
		C18	0.167	0.046

4.2 Evaluation Results of Index Level

Historical objective data used in the evaluation of quantitative indicators are obtained from official statistics, annual reports, statistical yearbooks (Shujuquan Online Forum 2014) and research literature, as shown in Table 7.5.

Evaluation results of each quantitative indicator can be calculated and expressed by Excellent, Good, Fair, and Poor, according to the membership functions and standards in Table 7.2, while those of qualitative indexes are determined, according to experts judgments by use of standards in Table 7.3. Judgments from the three experts described in Sect. 2 are merged together with equal weights. Taking the performance of Port of Wuhan in 2007 as an example, its GPDGR is 60.6%, locating between "a" (70) and "b" (60). As a result, 60.6% belongs to "60" (Good) with 94% degree of belief and "70" (Excellent) with 6% degree of belief. The evaluation results (grades) are shown in Table 7.6. It should be noted that the case study in this research is conducted with a time span of 3 years rather than one, which matters when calculating the throughput growth rate and GDP growth rate.

4.3 Evaluation of Inland Port Performance

In this section, the IDS software (Xu and Yang 2005) was used to compute the performance of the Port of Wuhan, employing the ER algorithm for synthesis of the criteria in the hierarchical structure. All the inputs with weightings of the relevant lowest level criteria are combined to determine the estimation of their corresponding higher level criteria. Consequently, the performance of the port in different years can be calculated and shown in Fig. 7.4.

Table 7.5 Values of quantitative indicators in 2007, 2010, and 2013

	Value		
Qualitative indexes	2007	2010	2013
C1 NB (berth number)	43	51	51
C2 SC (10^3 m^2)	272	434	457
C3 QL (km)	6.7	7.6	7.8
C4 NCHE (set)	217	318	318
C5 NTB (boat number)	83	105	140
C6 PCT (10^6 ton)	26.3	37.9	42.2
C7 CSU (%)	52	68	73
C9 PSM (accident number/year)	0	0	0
C12 QPFA (10^8 CNY)	2.8	6.1	11.3
C13 ORP (10^8 CNY)	4.1	6.9	13.4
C14 TP (10^6 CNY)	4.6	8.9	12.3
C15 PTGR (%)	35.6	44.1	11.9
C16 GDPGR (%)	60.6	75.6	64.1
C18 CIW (level)	I	I	I

Table 7.6 Evaluation of indicators on the performance of the Port of Wuhan in 2007

		Grade			
Criteria Level	Index level	Excellent	Good	Fair	Poor
Infrastructure	C1 NB		0.3	0.7	
	C2 SC			0.72	0.18
	C3 QL		0.7	0.3	
	C4 NCHE			0.17	0.83
	C5 NTB			0.66	0.34
Operations and management	C6 PCT			0.63	0.37
	C7 CSU		0.8	0.2	
	C8 LI		0.7	0.3	
	C9 PSM		1		
	C10 QMHR		0.4	0.6	
	C11 CS		0.8	0.2	
Financial status	C12 QPFA			0.1	0.9
	C13 ORP			0.03	0.97
	C14 TP			0.15	0.85
Development potential	C15 PTGR	0.56	0.44		
	C16 GDPGR	0.06	0.94		
	C17 GS		1		
	C18 CIW	1			

It can be seen from Fig. 7.4 that the performance of the Port of Wuhan in 2007 is evaluated as 25.86% poor, 18.22% fair, 49.80% good, and 6.12% Excellent. Thus, the utility value of the port in 2007 can be calculated as 0.4540 using Eqs. (7.11) and (7.12). Similarly, the utility values of the port in different years can be calculated and ranked as u (performance in 2007) $= 0.4540 < u$ (performance in 2010) $=0.7259 < u$ (performance in 2013) $= 0.7582$, as shown in Fig. 7.5. Meantime, the crispy values associated with each investigated port can be used as a benchmark to measure the performance of different inland ports in a cross-sectional study.

4.4 Discussion

The variation trends of indexes in Criteria Level can be virtually presented in Fig. 7.6.

Overall, the utility values of "Infrastructure", "Operations and management", and "Financial status" show an increasing trend during the research period, which is in accordance with the Goal level. The performance of Port of Wuhan in terms of infrastructure has been greatly improved from 2007 to 2010, while its improvement speed decreased in the following 3 years. Although the performance of both "operations and management" and "financial status" shows a continuous improving trend, the improvement of the latter one is more obvious. This is due to the great

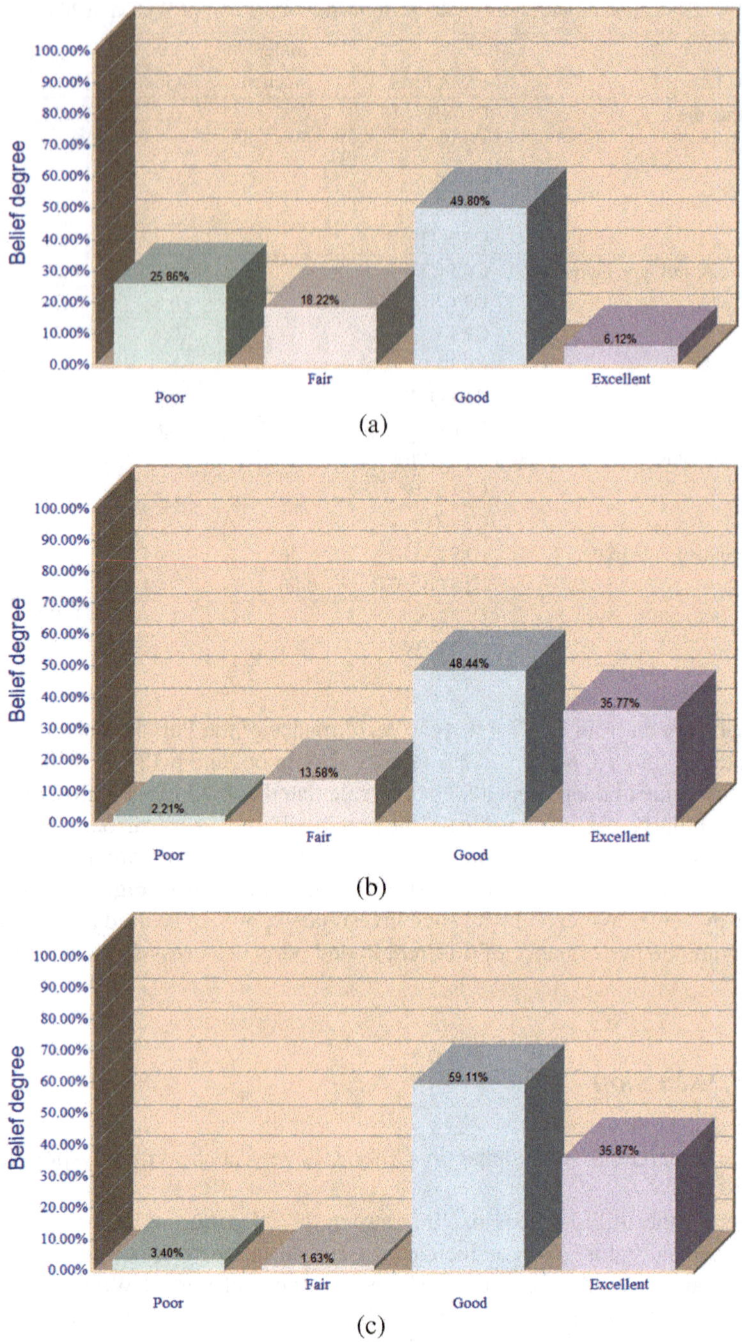

Fig. 7.4 The performance of the Port of Wuhan in 2007 (**a**), 2010 (**b**), and 2013 (**c**)

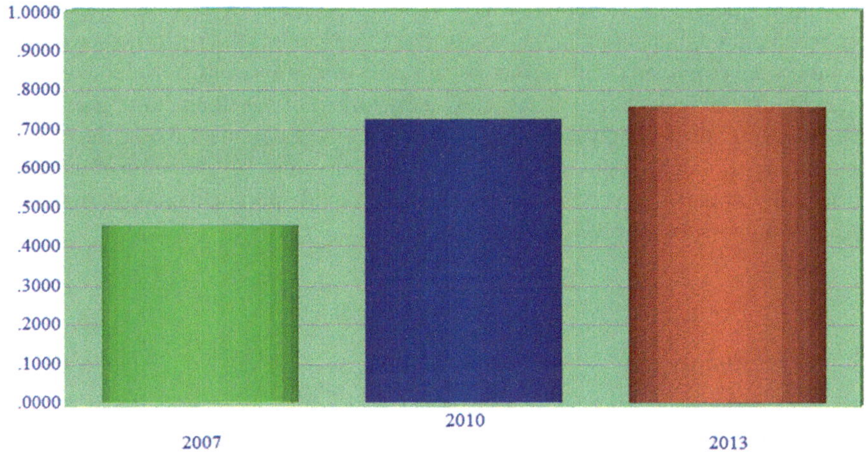

Fig. 7.5 The utility values of performance of the Port of Wuhan in 2007, 2010, and 2013

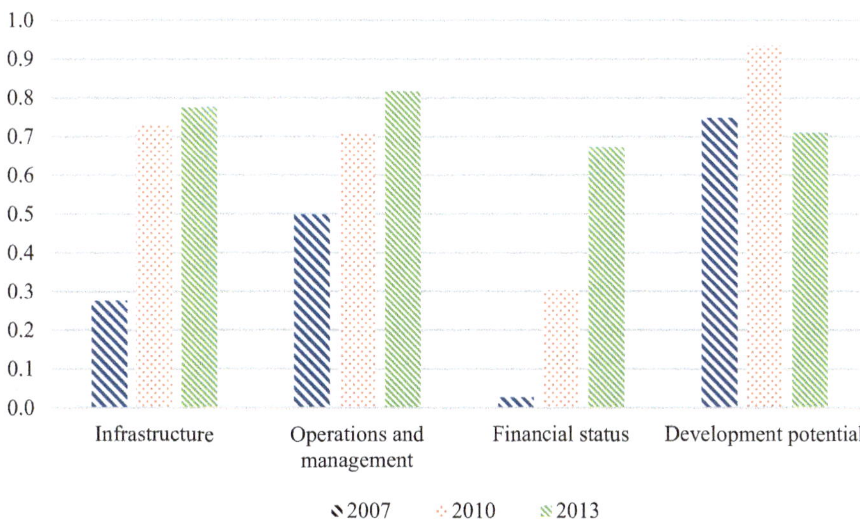

Fig. 7.6 The utility values of indicators in Criteria Level from by year

profits the Port of Wuhan during the period. Regarding the development potential of the port, its performance in the second period (from 2010 to 2013) is worse compared to the first one (from 2007 to 2010), which reveals a need for the growth of cargo throughput when further checking the indicators under this criterion.

Considering Fig. 7.5, it can be seen that the performance of Port of Wuhan holds a trend of improvement from 2007, which is in harmony with the real situation in the port, evidenced by the facts that a) great financial support from the local government and b) enormous investment from Wuhan Port Group Co., Ltd. In the port infrastructure construction, which to some extent verify the validity of the proposed model. Indicators in the assessment model reflect the performance of, for example, development of infrastructure and daily operations and management of an inland port, from which the deficiencies and the trend of development of green port can be identified and analysed as well. Therefore, the assessment model developed in this chapter can be applied not only in the evaluation of the present performance of inland port performance but also as a tool to provide port managers with certain insights on how to improve operations.

5 Conclusion

This chapter has developed a hierarchical model for evaluating the inland port performance from four main perspectives, namely, infrastructure, operations and management, financial status, and development potential. AHP and ER are used in the proposed model to calculate the relative importance of the relevant qualitative and quantitative criteria and to deal with synthesis in order to achieve the estimation of the top level goal. The proposed model is tested by a case study by evaluating the performance of the Port of Wuhan from 2007 to 2013. The novel model and flexible method presented in this chapter could be applied for evaluating performance of inland ports in other areas in order to formulate corresponding measures to improve its management and development.

The limitation of this study lies in the ignorance of possible relationships among the indexes in the model proposed in this chapter. Thus, advanced models capable of addressing the interdependencies among them should be generated in the future work. This chapter contributes to the existing literature of inland port performance evaluation. It also serves as an exploratory study to develop a set of suitable indexes system for the quantitative evaluation of inland port performance. The novel model and flexible method presented in this chapter will be not only applicable for evaluating the performance of inland ports, but also capable of providing useful insights and guides for port authorities and the stakeholders to formulate measurement of their management.

Acknowledgements The authors would like to thank the EU FP7 Marie Curie IRSES project "ENRICH" (612546) for its financial support to this research. The first author would also like to thank the China Scholarship Council (201506950023).

References

Arango C, Cortés P, Muñuzuri J, Onieva L (2011) Berth allocation planning in Seville inland port by simulation and optimisation. Adv Eng Inform 25(3):452–461

Bichou K, Gray R (2004) A logistics and supply chain management approach to port performance measurement. Marit Policy Manag 31(1):47–67

Chen G (2014) Research on the competence of Tianjin bulk and general cargo terminal. Master thesis. Dalian Maritime University, Dalian

China Water Transportation Construction Association (2013) Technical specification for port facility maintenance, JTS 310–2013, China

China Port Association (2011) China port yearbooks 2010, statistical databank on economic and social development. Accessible on: http://tongji.cnki.net/kns55/navi/YearBook.aspx?id=N2011030182&floor=1

Chin K, Yang JB, Guo M, Lam P (2009) An evidential-reasoning-interval-based method for new product design assessment. IEEE Trans Eng Manag 56(1):142–156

Cortés P, Muñuzuri J, Ibáñez JN, Guadix J (2007) Simulation of freight traffic in the Seville inland port. Simul Model Pract Theory 15(3):256–271

Cullinane K, Song DW (2006) Estimating the relative efficiency of European container ports: a stochastic frontier analysis. Res Transp Econ 16:85–115

Cullinane K, Wang TF, Song DW, Ji P (2006) The technical efficiency of container ports: comparing data envelopment analysis and stochastic frontier analysis. Transp Res A Policy Pract 40(4):354–374

de Langen P, Nidjam M, van der Horst M (2007) New indicators to measure port performance. J Mar Res 4(1):23–36

Esmer S (2008) Performance measurements of container terminal operations. Dokuz Eylul University (DEU) J GSSS 10(1):238–256

Gu HX (2012) Research on evaluation index of port logistics performance. Chongqing Jiaotong University, Chongqing

Hassan SA (1993) Port activity simulation: an overview. ACM SIGSIM Simulation Digest 23 (2):17–36

IASC (1993) International Accounting Standard No. 16: accounting for property, plant and equipment. IASC, London

Kişi H, Zorba Y Kalkan M (1999) Assessment of port performance: application on Port of Izmir. Strategic Approaches for Maritime Industries in Poland and Turkey, eds: Mustafa

Lam JSL, Su S (2015) Disruption risks and mitigation strategies: an analysis of Asian ports. Marit Policy Manag 42(5):415–435

Li & Fung Research Centre (2009) An update on the transport infrastructure development in China: water transportation. China Distrib Trading 58:1–8

Liu J, Yang JB, Ruan D, Martinez L, Wang J (2008) Self-tuning of fuzzy belief rule bases for engineering system safety analysis. Ann Oper Res 163(1):143–168

Luo XL Yang JQ (2013) Study on the development of Yangtze River shipping in china. World Conference on Transport Research (WCTR), Rio de Janeiro

Maritz A, Shieh CJ, Yeh SP (2014) Innovation and success factors in the construction of green ports. J Environ Prot Ecol 15(3A):1255–1263

Marlow PB, Casaca ACP (2003) Measuring lean ports performance. Int J Transp Manag 1 (4):189–202

Metalla O, Vyshka E, Lumi D (2015) Defining the most important port performance indicators: a case of Albanian ports. Int J Econ Comm Manag 3(10):808–819

MoT (2007) Statistical Bulletin of Highway and Waterway Transportation industry in 2007. Accessible on: http://www.moc.gov.cn/ftls/guihuaxunli/xiangguanziliao/201512/t20151225_1962294.html

MoT (2011) Statistical Bulletin of Highway and Waterway Transportation Industry in 2011. Accessible on: http://www.moc.gov.cn/fenxigongbao/hangyegongbao/201510/t20151013_1894758.html

MoT (2012) Statistical Bulletin of Highway and Waterway Transportation Industry in 2012. Accessible on: http://www.moc.gov.cn/fenxigongbao/hangyegongbao/201510/t20151013_1894759.html

MoT (2013) Statistical Bulletin of Highway and Waterway Transportation Industry in 2013. Accessible on: http://zizhan.mot.gov.cn/zfxxgk/bnssj/zhghs/201405/t20140513_1618277.html
MoT (2014) Statistical Bulletin of Highway and Waterway Transportation Industry in 2014. Accessible on: http://zizhan.mot.gov.cn/zfxxgk/bnssj/zhghs/201504/t20150430_1810598.html
MoT (2015) Statistical Bulletin of Highway and Waterway Transportation Industry in 2015. Accessible on: http://zizhan.mot.gov.cn/zfxxgk/bnssj/zhghs/201605/t20160506_2024006.html
Rahimi M, Asef-Vaziri A, Harrison R (2008) Integrating inland ports into the intermodal goods movement system for ports of Los Angeles and long Beach. METRANS Transportation Center, Project, Los Angeles, pp 07–01
Saaty TL (1980) The analytic hierarchy process. McGraw Hill International, New York
Shujuquan Online Forum (2014) Summary of China port yearbooks from 2000 to 2014, Accessible on: https://www.douban.com/note/449756663/
Tabernacle JB (1995) A study of the changes in performance of quayside container cranes. Marit Policy Manag 22(2):115–124
Talley WK (1988) Optimum throughput and performance evaluation of marine terminals. Marit Policy Manag 15(4):327–331
Talley WK (1994) Performance indicators and port performance evaluation. Log Transport Rev 30 (4):339
Thomas B, Monie G (2000) The measurement of port performance: with particular reference on ten container terminal operations. In International Labour Organization (ILO) Port Worker Development Programme (PDP). Cardiff, Antwerp
Tongzon JL (1995) Determinants of port performance and efficiency. Transp Res A Policy Pract 29(3):245–252
Tongzon JL (2001) Efficiency measurement of selected Australian and other international ports using data envelopment analysis. Transp Res A Policy Pract 35(2):107–122
Trujillo L, Nombela G (1999) Privatization and regulation of the seaport industry (Vol. 2181). World Bank Publications
Ugboma C, Ugboma O, Ogwude IC (2006) An analytic hierarchy process (AHP) approach to port selection decisions–empirical evidence from Nigerian ports. Marit Econ Log 8(3):251–266
UNCTAD (1976) Port performance indicators. United Nations Conference on Trade and Development, New York
UNCTAD (1999) Technical note: the fourth generation port. United Nations Conference on Trade and Development, Geneva
Wan C, Yan X, Zhang D, Pu B (2016) A novel model for development level analysis of LNG fueled ships. In: Transportation Research Board 95th Annual Meeting (No. 16–3325)
Wang L, Bi ZW (2010) Evaluating the operational efficiency of major inland ports in china using the 3-stage DEA model. Ind Econ Res 4:40–48
Wang TF, Cullinane K, Song DW (2003) Container port production efficiency: a comparative study of DEA and FDH approaches. J Eastern Asia Society for Transportation Studies 5:698–713
World Bank (2009) Sustainable development of inland waterway transport in China. Washington, DC: World Bank. Accessible on: http://documents.worldbank.org/curated/en/958141468218389761/Sustainable-development-of-inland-waterway-transport-in-China
Xu DL, Yang JB (2005) Intelligent decision system based on the evidential reasoning approach and its applications. Journal of Telecommunications and Information Technology 3:73–80
Xu LY (2014) Study of theory and method on comprehensive evaluation of inland hub development. Master thesis, Chongqing Jiaotong University, Chongqing
Yang JB (2001) Rule and utility based evidential reasoning approach for multiattribute decision analysis under uncertainties. Eur J Oper Res 131(1):31–61
Yang T, Lee RS, Hsieh C (2003) Solving a process engineer's manpower-planning problem using analytic hierarchy process. Prod Plan Control 14(3):266–272

Yang JB, Xu DL (2002) Nonlinear information aggregation via evidential reasoning in multiattribute decision analysis under uncertainty. IEEE Trans Syst Man Cybernetic Part A Syst Humans 32(3):376–393

Zhang D, Wan CP, Yang ZL (2015) Modeling and evaluation of green port development: a case study on Tianjin Port. In: Ng AKY et al. (ed) Climate change and adaptation planning for ports. Routledge, London. ISBN-13: 9781138797901

Chapter 8
Robust Evaluation of Risks in Ship-to-Ship Transfer Operations: Application of the STOCHASTIC UTA Multicriteria Decision Support Method

Dimitrios I. Stavrou, Eleftherios Y. Siskos, Nikolaos P. Ventikos, and John E. Psarras

Abstract Decision making in the maritime environment is a complex and difficult task. The stochastic nature of maritime operations, in combination with the complicated and often hostile sea setting, composes a scene where obvious solutions are not always the best choice. The assessment and evaluation of risks often lays on the exploitation of the experts' knowledge. However, this approach may suffer from uncertainty, stemming from potential biases inherent to the opinions of the experts. The aim of this chapter is to evaluate the risks of a ship-to-ship transfer of cargo operations, proposing a challenging approach to address the aforementioned uncertainty, by utilizing a sophisticated and targeted multi-criteria decision aid (MCDA) framework. The STOCHASTIC Utility Additives (UTA) method, adopting the philosophy of aggregation–disaggregation, is applied, coupled with a robustness control procedure. This methodological framework helped in the examination, management and reduction of uncertainty, and in the eventual attainment of robust and reliable evaluation results.

Keywords Multicriteria decision aid (MCDA) • STOCHASTIC Utility Additives (UTA) • Ship-to-ship cargo transfer • Risk assessment

---

D.I. Stavrou (✉) • N.P. Ventikos
School of Naval Architecture and Marine Engineering, National Technical University of Athens, Athens, Greece
e-mail: dstaurou@gmail.com

E.Y. Siskos • J.E. Psarras
School of Electrical and Computer Engineering, National Technical University of Athens, Athens, Greece

P.T.-W. Lee, Z. Yang (eds.), *Multi-Criteria Decision Making in Maritime Studies and Logistics*, International Series in Operations Research & Management Science 260,
DOI 10.1007/978-3-319-62338-2_8

1 Introduction

Maritime activities suffer inherently from risks that can significantly compromise the success of relevant operations. The origins of these risks can be located in several factors; the need to deal with the compound and often hostile sea environment, where an operation is conducted; the human element, due to the unavoidable possibility of human errors, and the potential failure of the means to complete the process, including the failure of vessels or relevant equipment during an operation. These risks jeopardize the success of a maritime operation and can lead to accidents with adverse effects to human life or the environment, in combination with the loss of property (Stavrou et al. 2015). In the context of such a complex and volatile environment, the decision-making process can be daunting. Undoubtedly, such decisions are usually based on subjective and irrational factors, such as instinct, intuition, inspiration, and prediction, instead of relying on the exploitation of all the available information and knowledge. A comprehensive understanding of the nature of the problem is a prerequisite to assess and rank all the alternative solutions in a sensible and realistic way. Experts with relevant experience in the field can provide such knowledge, in search of the optimal decision, which satisfies, to the highest degree, the different combinations of the attributes of the possible solutions and actions.

On the other hand, inevitably, a decision-making environment in the maritime sector consists of many types of uncertainty. This chapter focuses on the uncertainty resulting as part of the opinions of the experts, when they assess and evaluate safety issues. Potential biases of the experts can be initially found in the rational way that they interpret reality. Thus, when multiple experts evaluate and rank different risk scenarios, the most probable is to get as many different rankings, due to their subjective preferences. Disagreement also arises on the determination of the relative importance of the evaluation criteria. In both cases, the real challenge for the decision analyst is to account the experts' differentiated viewpoints and at the same time manage the instability of the results.

A promising way to address the challenges is the use of multicriteria decision aid (MCDA) methodologies under uncertainty. In this context, the additive utility models have a dominant position among the MCDA methods and can be especially implemented, when the analyst seeks to build a preference order of a set of actions, based on a consistent family of criteria (Hurson and Siskos 2014). Several different approaches for the development of an additive utility function have been presented over the last few years. Examples can be found in the multi-attribute utility theory (MAUT), which generalizes the single-attribute utility theory and employs a global preference utility function that expresses the decision maker's preferences. Models of this theory were developed by Keeney and Raiffa (1976), Farquhar (1984) and Fishburn (1967), as well as Figueira et al. (2005), who adopted the concept of the trade-offs between the criteria to build the additive value or utility function, as well as to determine the relative importance of the selected criteria.

In addition, Martel and D'Avignon (1982) proposed the use of confidence indices. D'Avignon and Vincke (1988) suggested a method to compare alternatives with preference indices and Zaras and Martel (1994) addressed stochastic dominance rules, consisting of pairwise comparisons of the actions and ranking of actions, according to the determined stochastic dominance relation. Fan et al. (2010) also proposed a method based on pairwise comparisons of alternatives with random evaluations using probability theory. However, the need to cope with uncertainty in a more efficient way led researchers to seek more sophisticated tools to overcome the impreciseness and ambiguity of a complex system. In this context, Zadeh (1965) introduced fuzzy set theory as a remedy for ill-defined systems, suffering from uncertainty. Fuzzy set theory has been successfully combined with the MCDA methodologies. An extensive overview of fuzzy MCDA applications can be found in Ashari Alias et al. (2008), as well as in Mardani et al. (2015).

An alternative way to build an additive utility is based on the philosophy of the aggregation-disaggregation (Siskos et al. 2005). The Utility Additives (UTA) family method of Jacquet-Lagrèze and Siskos (1982, 2001) are the most popular representatives of the disaggregation or analytical framework. The ability of these methods to exploit the decision maker's mechanism of evaluating and extract it implicitly renders them promising to deal with complex evaluation problems. Although the literature includes many different deterministic UTA methods for various evaluation problems, there is an evident absence of applications of UTA methods under a stochastic framework. As maritime activities inherently have a stochastic nature, the development of a stochastic UTA model is salient in order to assess the potential risks of maritime operations, such as ship-to-ship (STS) transfers.

This study focuses on the implementation of the STOCHASTIC UTA method for the risk assessment of the hazards of a STS transfer operation. STS transfer is the operation of transferring cargo at sea between ships. It has more than 50 years of continuous development and safety improvements, being always at the top of the agenda in the shipping business. STS transfers maintain a good safety record; however, the adverse effects of a potential accident cause great concern to stakeholders, who are always open to the development of new risk assessment techniques that will support them in making the right safety decisions in an operation.

In summary, the aim of this chapter is to develop and apply a model for the robust evaluation of risk scenarios under a multicriteria and stochastic framework, with a special focus on the STS transfer operation. The chapter is structured as follows: Section 2 provides an outlook of risks in STS transfer operations, and Sect. 3 presents relevant multicriteria modellings and related methodologies. Sect. 4 outlines the scenario evaluation problem, under consideration, while Sect. 5 presents the theoretical background of the STOCHASTIC UTA method and the bipolar robustness control procedure for robust evaluation. Sect. 6 refers to the implementation of the scenario evaluation and Sect. 7 includes a relevant discussion. Finally, Sect. 8 concludes the chapter.

2 An Outlook on the Risks in Ship-to-Ship (STS) Transfer Operations

The ship-to-ship (STS) transfer of cargo operation refers to the transfer of cargo between ships at sea, either stationary or underway. Currently, STS transfer operations are conducted in more than 50 dedicated locations around the world. STS transfer operations have more than 50 continuous years of evolution and development in the context of the optimization of the distribution plan between the source and the final consumers. STS transfer operations can be coordinated through different arrangements, according to specific operational conditions/characteristics. Thus, vessels can be positioned alongside to align their manifolds before continuing with the actual transfer or employ a tandem mooring configuration, in which vessels are positioned in line, one after the other (OCIMF 2013).

2.1 Phases of a STS Transfer Operation

A common STS transfer operation includes four discrete phases: the preparation phase, the mooring process, the transfer of cargo and, finally, the unmooring procedure. In the phase of preparation, certain conditions and requirements should be met to safeguard the entire operation. After the completion of this phase, the "run-in" procedure begins. During her approaching, the mother vessel moves in a steady direction and at a slow speed, whereas the manoeuvring vessel approaches the first vessel with the objective to bring their manifolds in line, to continue with the transfer process. The mooring phase consists of the necessary actions to securely berth the two vessels using mooring lines, fenders, and all other supporting equipment. Later, the actual transfer of cargo consists of the operational verification of various issues, such as control of the initial and maximum transfer rate, emergency shutdown transfers procedures, ballasting and de-ballasting processes, and etc. After the completion of the transfer phase, the unmooring procedure commences. During unmooring, the deviation method is determined and followed, to finally separate the vessels from each other. Stavrou and Ventikos (2014b) present a comprehensive description of the different phases of the STS transfer operation. Figure 8.1 shows a common STS transfer operation.

2.2 Types of Transferred Cargo

Nowadays, oil accounts for 32.6% and liquefied natural gas (LNG) 23.7% of global energy consumption (BP 2014). As such, the procedures of transhipment at sea form an essential link in the chain of the global transfer of energy.

Fig. 8.1 A typical STS transfer of oil cargo operation (Source: OCIMF 2013)

Oil STS transfers were initially used in the 1960s, due to the increased draft dimension that limited the access of large ships (VLCCs and ULCCs) to specific oil terminals at shallow rivers in the Gulf of Mexico (MAIB 2010). The first attempt at the European level was in the early of 1970s and due to the inability of tankers to cross the Suez Canal because of the Arab-Israeli wars. To overcome this issue, super tankers carried oil from the Middle East to European terminals via the Cape of Good Hope. The oil was delivered in parcels to small tankers due to the inability of the European ports to accommodate the super tankers. As an illustrative example, a typical VLCC can carry approximately two billion barrels and offload the cargo into four Aframax type vessels within 4 days. Amendments to the Annex of the Protocol of 1978 related to the International Convention for the Prevention of Pollution from Ships. The addition of Chap. 8 to MARPOL Annex I (MEPC 186 (59)/ 2009) was the first significant action by the International Maritime Organization (IMO) to establish common rules during STS transfer operations. The effort was performed in collaboration with other interested parties, such as the Oil Companies International Marine Forum (OCIMF) and the International Chamber of Shipping (ICS).

Although the initial use of such procedures concerned only the transfer of oil, contemporary STS operations have been extended to the transfer of cargoes, other than oil, such as liquefied gases (LPG, LNG) (Waterborne Energy 2010) or even dry cargoes, such as ore. Nevertheless, dry bulk STS operations are mostly conducted in the Middle and the Far East, where iron ore is transhipped to allow access to shallow ports, where melting mills are located (EC 2009).

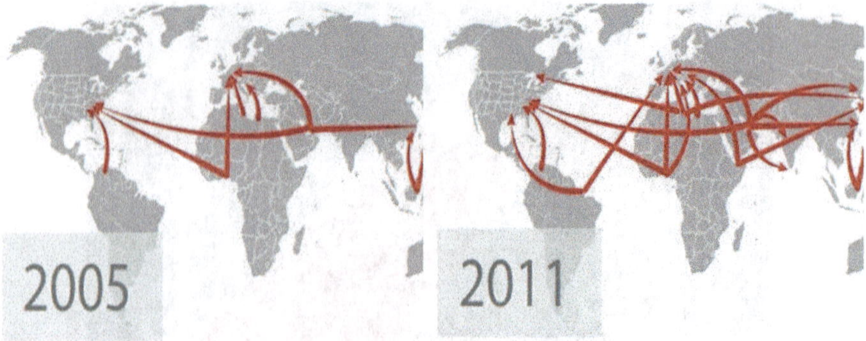

Fig. 8.2 LNG growth from 2005 to 2011

The concept of transferring liquefied gases from ship-to-ship arose due to the rapid growth of natural gas global energy consumption and the indisputable benefits, when LNG cargoes are transferred over long distances (over 1000 miles). Ever since the world's first commercial LNG plant, in the Algerian port city of Arzew, started delivering to the United Kingdom in 1964, under a 15-year contract, the pattern of LNG trade has been shifting and changing. This growth of the LNG network can be presented in numbers by the increase of LNG terminals from eight in 1996 to 15 in 2012, whereas today there are 26 with another 65 waiting to be completed (Gorstenko and Tikhomolova 2012). Figure 8.2 shows the development of the LNG network from 2005 to 2011.

Finally, bulk STS operations have found significant growth in the markets of the Far East because of China's global domination of goods, such as iron (70%), copper (42%), coal (47%), nickel (36%), lead (44%) and zinc (41%). The most representative type of bulk carriers, conducting dry cargo transfer operations, is the Capesize (150,000–400,000 DWT), whereas other types of vessels are the Handysize (20,000–35,000 DWT), the Handymax (35.000–50,000 DWT) and the Panamax (50,000–80,000 DWT). The Velamax (VLOC) type, with an enormous displacement of 380,000–400,000 DWT and a draft of over 22 m, is used for the transfer of dry cargoes from Brazil to Europe and Asia. Only a few ports can accommodate this type of vessel, and, thus, the transfer operation can be considered as a "one-way road." Lastly, the transfer of dry cargoes usually perquisites the existence of a third vessel dedicated to conducting the transfer operation (Arcelor Mittal 2013).

2.3 *Factors that Affect a STS Transfer Operation*

The execution of a STS transfer operation is based on specific and discrete concessive steps under the coordination of the Person of Overall Advisory Control (POAC) and his assistants. The POAC is responsible for the implementation of the different phases of the transfer operation; however, masters remain responsible for the safety of their vessels during the operation. The operation is conducted according to the guidelines and recommendations of the "STS transfer plan". The plan contains information about each phase of a STS transfer operation. It describes the mooring and unmooring phases; it provides cargo and ballast information; it specifies the duties of the involved personnel; it also gives a contingency plan in case of an emergency and other valuable information for operation issues (IMO 2010). STS transfer is a difficult and complex operation, of which success depends on various factors that interact during the operation. For example, the vessels and their technical specifications (dimensions, compatibility issues, propulsion installations, mooring and anchorage equipment, etc.) should be considered to ensure that the vessels are compatible, and their characteristics match each other. The use of special STS equipment that is necessary to conduct the operation (fenders, transfer hoses, etc.) is another factor that affects the operation due to the possibility of failure during the process (National Academy Press 1998). The skills and experience of the POAC and his assistants are also critical factors that determine the success of the procedure. Other external factors that interact during the operation are the selected transfer area (traffic density, geological factors, security threats, underwater pipelines and/or cables, water depth) and the weather and environmental conditions (sea state, the wind, tidal currents, visibility, etc.).

2.4 Risk Assessment of STS Transfers

STS transfer operations have a good safety record. However, the possibility of an accident with adverse effects on human beings, the environment, and property remains, forcing the stakeholders to pay special attention to safety matters during an operation. To mitigate or even eliminate the possibility of a potential accident, several classical and modern risk assessment methodologies have been developed to identify the hazards and evaluate the potential risks coming from them.

International organizations published several guidelines and recommendations (IMO 2010; OCIMF 2013). Also, papers or studies of independent researchers (Østvik and Grønstøl 2005; Skjong et al. 2007; Elsayed et al. 2013) have addressed the most probable high-level risks, during a STS operation (Stavrou et al. 2015):

- A *collision between the involved vessels*: Past accident records demonstrate the significance of the risk of collision/contact between the vessels involved in an STS transfer operation (Ventikos and Stavrou 2013). A rough analysis of Marico Marine (2004) on 1270 STS transfer operations addressed collision as the highest risk among all risks under consideration.
- A *fire on deck and/or an explosion*: An ignition or even explosion during a transfer procedure is a prevailing threat during the transfer process (Østvik and Grønstøl 2005; Skjong et al. 2007). A fire and/or an explosion may result in personnel fatalities or injuries and/or structural damage to the vessels involved in

the STS transfer operation. For this type of accident Stavrou and Ventikos (2015c) calculated a rate of 7.39×10^{-3} accidents per year.

- *Cargo leakage*: STS transfers are usually conducted with two hoses of 10″ in diameter and typically 24 m in length. These dimensions allow an oil transfer rate of approximately 2000 m^3/h (COWI 2011). In this context, a potential accident, due to the rapture of the transfer line, may lead to serious pollution through leakage at sea or a large amount of oil-on-board in the case of leakage on deck. Both risks can occur during the transfer of cargo, due to operational reasons. Cargo leakage can cause various adverse environmental effects and lead to injuries or fatalities of the personnel involved, an explosion, and hull damage.
- *Human Injury/Fatality*: A human injury or fatality can occur in various forms. Injuries or fatalities can also occur during the transfer of the personnel involved, such as crew members, mooring masters, surveyors, agents or customs officials, before and after the entire operation (Spenser 2010). Moreover, previous accident reports demonstrate the mooring operations relevant to the mooring and unmooring procedures are often related to injuries as well.
- *Damage to Cargo Tanks*: The guidelines of OCIMF (2013) refer to the risk of structural damage to cargo tanks, during the phase of the cargo transfer, due to environmental (e.g. abnormal weather conditions) or technical (e.g. overpressure) factors.

3 Multicriteria Modelling of STS Risks

Multicriteria decision aid (MCDA) is a scientific field that provides the tools, models and methodologies to effectively deal with decision problems. MCDA, with a view to supporting decision making, allows the analyst to compare and evaluate different actions /alternatives on certain criteria, and reach an efficient solution, based on the preferences of the decision maker. The first attempt to apply MCDA methodologies in the risk assessment of cargo transfers at sea can be found in the work of Elsayed et al. (2013). The authors proposed a fuzzy Technique of Order Preference by Similarity to the Ideal Solution (TOPSIS) modelling approach for the risk assessment of liquefied natural gas carriers during loading and offloading at terminals. The combination of fuzzy logic with the TOPSIS technique efficiently solved the problem, and interesting conclusions surfaced.

Stavrou and Ventikos (2014a) tested the effectiveness of the classical process failure mode and effect analysis (PFMEA) process, when different risk scenarios related to the hazards of an STS transfer operation were assessed and evaluated. The objective was to examine if the traditional approaches could adequately address the risks of STS transfers, and, if not, to identify the weaknesses that limited their performance. Risks were properly evaluated and managed; however, certain weaknesses elevated the uncertainty by the experts' bias when they evaluated the risk scenarios on three risk factors. These weaknesses were the inability to weight the risk factors, the inaction to deal effectively with the disagreements of the experts and, finally, the fact that different combinations of risk factors could result in the same risk priority number for scenarios, without however being able to perform a qualitative assessment of the combined factors.

The next attempt was in the work of Stavrou and Ventikos (2014b), who developed an assessment framework of the risks in STS transfer operations, using the Fuzzy Technique of Order Preference by Similarity to the Ideal Solution (TOPSIS) method. Specifically, a team of experts with experience in STS transfer operations evaluated and prioritized different risk scenarios, considering the occurrence of a risk and the severity of its consequences.

The above methodology was later expanded (Stavrou and Ventikos 2015b) through the inclusion of the risk factor of detectability for the risk scenarios evaluation. The addition of the new factor (in addition to the factors of occurrence and severity) provided meaningful conclusions regarding the potential interactions.

The authors also examined another multicriteria methodology (Stavrou et al. 2015) for STS transfer operations, which was based on the outranking relation concept using stochastic criteria evaluation and interval scaled importance weights. The proposed methodology, inspired by the work of Martel and D'Avignon (1982), used confidence indices to compare alternative risk scenarios for stochastic risk criteria evaluation. The same team of experts, as in in the other two studies, evaluated the scenarios based on three risk factors (likelihood, severity, and detectability). Finally, a fuzzy domination relation was applied to complete the ranking of the risks.

The classical PFMEA process had serious problems that needed to be solved. The employment of the multicriteria methodologies solved these problems and proved that MCDA methods could effectively deal with problems related to risk identification and evaluation in the maritime environment. The fuzzy TOPSIS approach made some assumptions to calculate the fuzzy weight numbers, whereas the stochastic outranking approach applied weight ranges to reduce uncertainty. However, each method had certain weaknesses that limited its effectiveness during implementation. The latter approach had questionable results in the case of weight evaluation disagreements among the experts, while the stochastic outranking approach demonstrated that it is extremely difficult to be implemented in the case of a large number (more than 10–12) of risk scenarios due to computational complexity.

The implementation of the multicriteria methods solved the problems of the classical risk assessment approaches. However, the rise of new problems leads us to seek a more effective way to deal with the uncertainty coming from the experts' opinion, when dealing with safety issues. An effective way to address the problem comes from the use of the method of a global criterion. This multi-criteria theory is based on the aggregation-disaggregation or analytical approach (Siskos 2008), with the objective of constructing one or more additive utility functions.

The analytical approach (Siskos 2008) is based on the use of the preference order of a preselected set of actions determined by the decision maker. Next, a model able to verify this preference order is extracted, and, finally, the model is used to extrapolate the results to the rest of the set of actions (Spyridakos and Yannacopoulos 2015). The most well-known analytical methods come from the family of the UTA methods initially proposed by Jacquet-Lagrèze and Siskos (1982, 2001). UTA methods refer to the philosophy of assessing a set of utility functions, assuming the axiomatic basis of MAUT and adopting the preference disaggregation principle (Siskos et al. 2005). These methods aim inductively to construct one or more additive utility functions using preference pre-ordered actions. The UTA method introduces an innovative way to analyse and rank different alternatives, due to its ability to extract and visualize the thinking mechanisms of the decision makers. No additional weight ranking is needed, and the challenge is to create the initial preference order, which is the base to extract the model that yields it. UTASTAR was the evolution of the initial UTA and introduced some significant improvements. UTASTAR is a deterministic MCDA approach, and it was successfully applied by the authors in a study for the selection of the most suitable STS transfer location (see Stavrou and Ventikos 2015e). However, the stochastic nature of the problem of risk evaluation by a team of experts makes UTASTAR weak and inappropriate. An effective way to deal with the stochastic nature coming from the experts' judgment is the use of the STOCHASTIC UTA (see Siskos et al. 2005). This method is the result of the expansion and modification of the UTASTAR method to be able to take probability distributions as input data instead of crisp criteria values. The analytical steps of the method are presented in detail in Sect. 5.

4 A Scenario Evaluation Problem

The STS transfer operation is a complex and very demanding process, which needs the special attention of the crews, involved from the initial phase of preparation until the final separation of the vessels. Section 2 has already covered the variety of risks that can compromise the success of a transfer operation. Risk assessment is necessary to identify the hazards and evaluate the risks of such operations. Historical data coming from previous accidents can, evidently, provide sufficient knowledge for the identification of the risks. However, in most cases, these data are not enough to address the entire range of the potential risks. This is due to a variety of reasons, such as the significantly low rate of accidents during STS transfers, the absence of an accepted and well-recognized mechanism, able to maintain accident records or near misses, changes in laws and regulations, etc. Thus, an effective way to study risks, during STS transfer operations, is the use of hypothetical risk scenarios, composed and evaluated by a team of experts with relative experience in such operations. Experts can provide adequate information regarding the risks of an operation, covering the majority of potential hazards and bridging the gaps coming from the use of historical data. Risk scenarios refer to narrated descriptions of causes and effects from different circumstances or combinations of events that may occur and lead to an accident. Risk scenarios have been successfully applied to

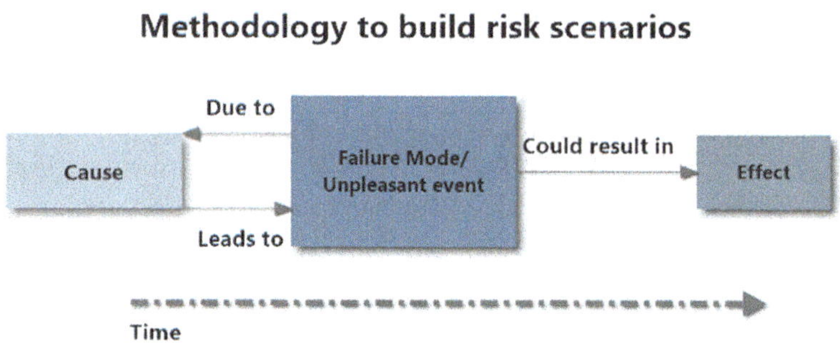

Fig. 8.3 Flow diagram of the inference technique proposed by Ford (2004)

solve MCDA problems in previous studies (Schoemaker 1995; Van der Heijden 1996; Islei et al. 1999; Durbach 2014). They can provide decision makers with a good understanding of the problem at hand and with valuable insight to the choices (Durbach and Stewart 2012). The inference technique to build risk scenarios comes from Ford's Handbook (2004). Figure 8.3 shows the flowchart diagram of the corresponding inference technique.

For the composition of risk scenarios, several guidelines are considered, as well as, recommendations, and studies relevant to the activity (Skjong et al. 2007; IMO 2010; OCIMF 2013). To this end, 43 risk scenarios were constructed including the most representative risks of all STS transfer phases, from preparation to the final separation of the vessels. Three scenarios surfaced from the phase of preparation. These scenarios were related to the transfer procedure of the personnel involved in the STS transfer operation (e.g., the POAC, the STS superintendents, etc.) that could result in injury or even fatality. The potential causes of the risks could be an inadequate compatibility study of the personnel landing area, defective personnel transfer equipment and unapproved transfer equipment that is not fit for the purpose. Moreover, 22 risk scenarios were related to the mooring/unmooring operations. These scenarios comprise the adequate navigational control of ships involved in STS transfers or third party passing vessels, which could lead to an uncontrolled event, resulting in property damage or injury/fatality of the personnel involved. Several causes were examined, e.g., a steering/propulsion failure, a fender defect or inadequate fendering, mismatched manoeuvring characteristics, inadequate incident management due to fatigue, pilotage errors.. Finally, 18 risk scenarios were identified, referring to the transfer process. These risks were cargo sloshing, due to ship motion under prevailing environmental conditions, hardware failure and overflow or overpressure, leading to an uncontrolled event, such as the loss of containment and the release of vapour, which could lead to ignition back to the source (resulting in fatalities, injuries, steelwork damage) or ignition with an explosion causing multiple fatalities/injuries and hull damage. Potential causes of such accidents could be inadequate training, procedures or experience, personnel fatigue, abnormal environmental conditions, the passing ship effect, etc.

Table 8.1 Scale of the occurrence number rate

Likelihood scale		
FI	Frequency	Definition
1	Extremely remote	Likely to occur once in the lifetime (20 years) of a world fleet of 5000 ships
2	Very remote	Likely to occur once per year in a fleet of 10,000 ships
3	Remote	Likely to occur once per year in a fleet of 1000 ships i.e. likely to occur in the total life of several similar ships
4	Unlikely	Likely to occur once per year in a fleet of 100 ships
5	Reasonably probable	Likely to occur once per year in a fleet of 10 ships i.e. likely to occur a few times during the ship's life
6	Probable	Likely to occur once per year in one ship
7	Frequent	Likely to occur once per month in one ship
8	Very frequent	Likely to occur once a week in one ship
9	Extremely frequent	Likely to occur twice a week in one ship
10	Very extremely frequent	Likely to occur every day in one ship

Table 8.2 Scale of the detection number rate

Detection scale	
1 or 2	Detection method is highly effective, and it is almost certain that the risk will be detected with adequate time
3 or 4	Detection method has moderately high effectiveness
5 or 6	Detection method has medium effectiveness
7 or 8	Detection method is unproven or unreliable, or effectiveness of detection method is unknown to detect in time
9 or 10	There is no detection method available or known that will provide an alert with enough time to plan for a contingency

After the determination of the risk scenarios, a team of six experts of relevant experience in STS transfer operations (three masters, one STS superintendent, one insurer and a STS provider) was formed. The experts had to survey and grade, for each scenario, its likelihood (occurrence number), impact (severity number) and detection (detection number) through predefined scales. Thus, each one of the 43 risk scenarios were characterized by a distribution of six evaluations, each one corresponding to the six members of the team of experts. Tables 8.1, 8.2 and 8.3 show the corresponding predefined scales. The scales come from IMO's guidelines, and they were previously applied successfully to other similar efforts (Østvik and Grønstøl 2005; Skjong et al. 2007). The occurrence number refers to the likelihood of a potential risk scenario occurring, and the severity number refers to the impact of the accident with regard to human safety, the environment, and the property. The detection number for the implementation of PFMEA refers to the ability of detection technique(s) or method(s) to detect the problem (Carbone and Tippett 2004).

Table 8.3 Scale of the severity number rate

Severity scale						
		Human safety		Property related		Environment related
SI	Severity	Description	Equiv. fatalities	Effect on ship	Economic losses	
1 or 2	Minor	Single or minor injuries	0,01	Local equipment damage (repair on board possible, downtime negligible)	US$ 30.000	Non-significant spill up to a few barrels of pollution to sea
		Small increase in operational duties of crew		Slight modifications of permissible operation conditions. Moderate degradation in handling characteristics		
3 or 4	Significant	Multiple or severe injuries	0,1	Non-severe ship damage (port stay required, downtime 1 day)	US$ 300.000	A few tonnes of pollution to sea. Situation is manageable
		Significant increase in operational duties of crew, but shall not be outside their capability		Significant modification of permissible operation conditions; not outside capability of competent crew. Significant degradation in handling characteristics.		
5 or 6	Severe	Single fatality or multiple severe injuries	1	Severe damage (yard repair required, downtime <1 week)	US$ 3 milion	Significant pollution demanding urgent measures for the control of situation and/or the cleaning of affected areas
		Dangerous increase in operational duties. Cannot reasonably be expected to cope with them without external assistance.		Marginal operation conditions. Essential need for outside assistance. Dangerous degradation in		

(continued)

Table 8.3 (continued)

Severity scale						
		Human safety		Property related		Environment related
SI	Severity	Description	Equiv. fatalities	Effect on ship	Economic losses	
				handling characteristics.		
7 or 8	Catastrophic	Multiple fatalities	10	Total loss (of, e.g. a medium size merchant ship)	US$ 30 milion	Major pollution with difficult control of situation and / or difficult cleaning to affected areas
9 or10	Disasterous	Large number of fatalities	100	Total loss (of, e.g. a large merchant ship)	US$ 300 milion	Uncontrolled pollution long-term effect on recipients' long-term disruption of the ecosystem

Table 8.4 Evaluation of the scenario P1 by the experts

Risk scenario	Scale number										Risk factor
	1	2	3	4	5	6	7	8	9	10	
P1	1	2	1	0	2	0	0	0	0	0	**Occurrence**
	1	0	1	1	0	1	1	1	0	0	**Detectability**
	0	1	0	0	0	3	0	1	0	1	**Impact to humans**
	0	4	0	1	0	0	0	1	0	0	**Impact to the environment**
	0	4	0	1	0	0	0	1	0	0	**Impact to the property**

Appendix A shows the experts' evaluations on all the risk factors. Table 8.4 shows indicatively the ranking of scenario P1 by the six experts.

5 The Preference Disaggregation Methodological Framework

5.1 Introduction to the UTA Methods

The UTA method is the result of the effort of Jacquet-Lagrèze and Siskos (1982), to re-determine the classical "cause and effect" approach to the MCDA problems. They supported the idea that the analyst can thoroughly examine the mechanism under which a decision is made and subsequently apply it to evaluate other actions under the same evaluation context.

Figure 8.4 shows the general methodological framework of the analytical/disaggregation approach. The first step is the collection of data and the determination of a set of reference actions, evaluated and ranked, based on a consistent family of criteria by the decision maker (DM). Next, a model, which is fully consistent with the DM's preference order, is constructed. If the consistency of the DM's ranking is verified, the model is extrapolated to the entire set of actions; otherwise, the reference actions are re-examined and returned to be re-evaluated. The UTA family methods are the most representative of the analytical methods (Jacquet-Lagrèze and Siskos 2001).

According to the UTA methods, the analyst takes for granted the preference order of the DM for a predefined set of reference actions A_R. These actions can be

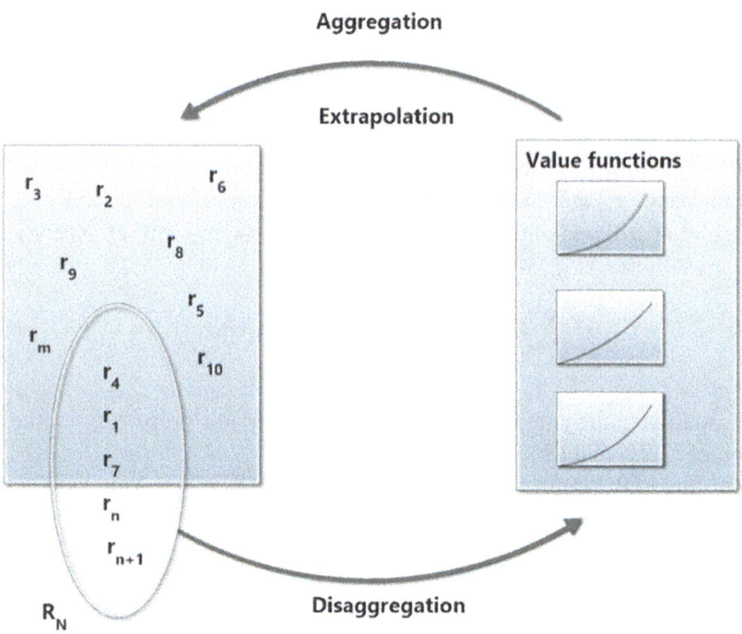

Fig. 8.4 The concept of the analytical approach through the UTA methodologies

either a sub-set of real actions or fictitious actions, constructed solely for the facilitation of the DM. Given the preference order of this set of actions, UTA employs linear programing techniques to determine the utility functions, which result to the same order (see Appendices B.1 and B.2 for the mathematical framework of the additive value model and the UTA method).

5.2 The UTA Method Under Uncertainty: STOCHASTIC UTA

The stochastic nature of maritime activities and their environment usually sets limitations on the implementation of deterministic decision-making models, which are based on criteria determined with certainty. The inherent stochastic nature of marine activities and the environment practically means that the information of the evaluation criteria cannot be known with certainty. Within the framework of multicriteria decision-aid under uncertainty, Jacquet-Lagreze and Siskos (1983) developed a specific version of UTA (STOCHASTIC UTA), in which the aggregation model to be inferred from a reference ranking is an additive utility function of the following form:

$$u(\boldsymbol{\delta}^a) = \sum_{i=1}^{n} \sum_{j=1}^{\alpha_i} \delta_i^a\left(g_i^j\right) u_i\left(g_i^j\right) \tag{8.1}$$

subject to normalization constraints (8.24) and (8.25), where δ_i^a is the distributional evaluation of action a on the $i-th$ criterion, $\delta_i^a\left(g_i^j\right)$ is the probability that the performance of action a on the $i-th$ criterion is g_i^j, $u_i\left(g_i^j\right)$ is the marginal value of the performance g_i^j, $u(\boldsymbol{\delta}^a)$ is the global utility of action a (see Fig. 8.5).

This global utility is of the von Neumann-Morgenstern form (cf. Keeney 1980), in the case of discrete g_i, where:

$$\sum_j \delta_i^a\left(g_i^j\right) = 1 \ (\textit{for discrete scales}) \tag{8.2}$$

The additive utility function of (8.1) maintains the properties as the value function of (8.21) and (8.22) (see Appendix B.1).

According to the implementation of the STOCHASTIC UTA method, there are three consecutive algorithmic steps:

Step 1 Express the global expected utilities of reference actions $u(\delta^{a_k}), k = 1, 2, \ldots, m$, in terms of variables:

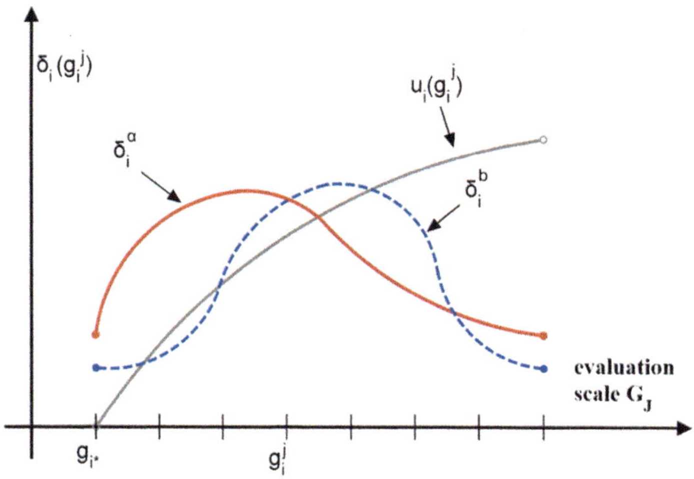

Fig. 8.5 Distributional evaluation and marginal value function (Siskos et al. 2005)

$$w_{ij} = u_i\left(g_i^{j+1}\right) - u_i\left(g_i^j\right) \geq 0, \forall i \ and \ j \tag{8.3}$$

The above transformation also considers the monotonicity of the marginal utility functions. It should be noted that, for $j = 1, w_{i1} = u_i\left(g_i^2\right) = 0$, because $u_i\left(g_i^1\right) = u_i(g_{i*}) = 0, \forall i = 1, 2, \ldots, n.$ Vice-versa:

$$u_i\left(g_i^j\right) = \sum_{t=1}^{j-1} w_{it}, \forall i \ and \ j > 1 \tag{8.4}$$

The number of the different degrees of g_i^j of the scale of criterion g_i is determined in relation to the distribution of the available information within the scale $\left[g_{i*}, g_i^*\right]$.

Step 2 The preference order of the reference actions, as expressed by the decision maker, is run from the top to the bottom, and the equations coming from the pairs (first, second), (second, third), etc. are written accordingly:

$$\begin{aligned} \Delta\left(\delta^a, \delta^b\right) &= u(\delta^a) - u\left(\delta^b\right) + \sigma^-(a) - \sigma^+(a) - \sigma^-(b) + \sigma^+(b) \\ &\geq \delta, if \ (a \succ b) \end{aligned} \tag{8.5}$$

$$\begin{aligned} \Delta\left(\delta^a, \delta^b\right) &= u(\delta^a) - u\left(\delta^b\right) + \sigma^-(a) - \sigma^+(a) - \sigma^-(b) + \sigma^+(b) \\ &= 0, if \ (a \sim b) \end{aligned} \tag{8.6}$$

where $\sigma^+(a) \geq 0$ *and* $\sigma^-(a) \geq 0$ are the error functions of the underestimation and overestimation of the reference actions accordingly, whereas δ corresponds to a small positive value dedicated to discriminating the different order positions.

Step 3 Next, the linear program is solved:

$$[\min]z = \sum_{a \in A_R} [\sigma^+(a) + \sigma^-(a)] \tag{8.7}$$

subject to the following constraints:

$$\begin{cases} \textit{Constraint } (B.15) \textit{ or } (B.16) \textit{ for each pair of succesive actions}: \\ \sum_i \sum_j w_{ij} = 1 \\ w_{ij} \geq 0 \forall i \textit{ and } j, \sigma^+(a) \geq 0, \sigma^-(a) \geq 0 \forall a \in A_R \end{cases}$$

(8.8and8.9)

After the solution of the linear program and the verification that the objective function (8.7) is zeroed, the additive value model is applied on the whole entity of the actions under evaluation, using the obtained values of the variables w_{ij}.

However, since the stability of the decision model and the obtained results is not guaranteed, it is advisable to proceed to a post-optimality analysis, which focuses on verifying the robustness of the results and increasing it to acceptable levels, if possible.

Recently Siskos and Psarras (2016) proposed an interactive bipolar robustness control procedure in order to manage robustness in both phases/poles of interactive decision support, namely the disaggregation and the aggregation one. Through this integrated procedure, the analyst has the possibility to examine, measure and analyze in a systematic way the robustness of both the decision model's parameters and its results (see next paragraph) .

5.3 Bipolar Robustness Control of the Evaluation Model

The disaggregation inference engine showed that the DM's preference model might not be a unique additive utility function but a set of functions, all compatible with the holistic preference statements. This infinite set of functions/models comprises a polyhedral set, confined under certain linear constraints.

Robustness analysis and control is a methodological framework to manage multiple decision models and multiple results that surface as outputs of the implementation of these models (Siskos and Psarras 2016). Consequently, robustness should be considered as a pivotal and essential decision support tool. Although bipolar robustness control is coupled perfectly to the UTA type methods, it can be flawlessly implemented in a synergy with several other analytical MCDA methods.

When it comes to the UTA family methods, the robustness control algorithm is initiated after the disaggregation of the holistic preferences on the set of reference actions (complete or partial ranking, pairwise comparisons, etc.). It then proceeds to the analysis of the robustness of the additive value model, with the option of discontinuing the modeling process, if the results are not satisfactory. In this case, the analyst asks the DM to enrich the reference set with additional reference actions or add other new preference statements.

In other way, the process moves to the aggregation pole, where the MCDA model is implemented (application of the additive value model) and the ranking of the real actions is obtained. Robustness is again measured in this pole, in terms of the stability of the ranking positions of each action. If the robustness of the results is adequate enough to support a safe decision, the algorithm is terminated, otherwise the analyst returns to the disaggregation pole and asks the DM for the acquisition of additional preferential information. Figure 8.6 depicts the algorithm followed during the implementation of bipolar robustness control.

The robustness control framework, when coupled with any UTA family method, uses two separate sets of robustness indices to judge: (i) the efficacy of the additive model in the disaggregation pole and (ii) the robustness of the final ranking results, achieved after the extrapolation of the model on the whole set *A* in the aggregation pole. The calculation of certain of these indices requires the implementation of certain techniques and standalone methods, in parallel with the decision support procedure.

5.3.1 Robustness Indices in the Disaggregation Pole

The indices, related to the disaggregation pole of the robustness control framework, focus on the stability of the model and its potentiality to produce results that are stable and not misleading or ambiguous. The objective of these indices is to build a robust and meaningful decision model that accurately reflects the preferences of the DM and assures that the results to be obtained are safeguarded against imprecision, which stems from the inefficient elicitation of information from the DM. On top of that, these indices have a practical meaning, since they prevent the analyst from performing heavy pointless computations, which are certain to lead to results of low quality. The whole computation burden is therefore decreased and the goals of the DM are reached by spending fewer resources.

The robustness indices proposed by Siskos and Psarras (2016) are categorized, based on which pole they applied.

The major indices, applying on the modeling of the evaluation problem while in the disaggregation pole - 1st pole - are distinguished in two separate categories: (i) the preferential parameters variance indices, (ii) and the statistical indices. All these indices can be subsequently visualized, in order to provide a more insightful view on the model's robustness.

Two indices can be recognized under this category. The use of these indices presupposes the production of multiple sets of preferential parameters. A usual way to achieve this, when implementing the UTA-type methods is the Max-Min LPs technique. During this technique all different parameters are successively minimized and maximized, under the set of feasibility constraints.

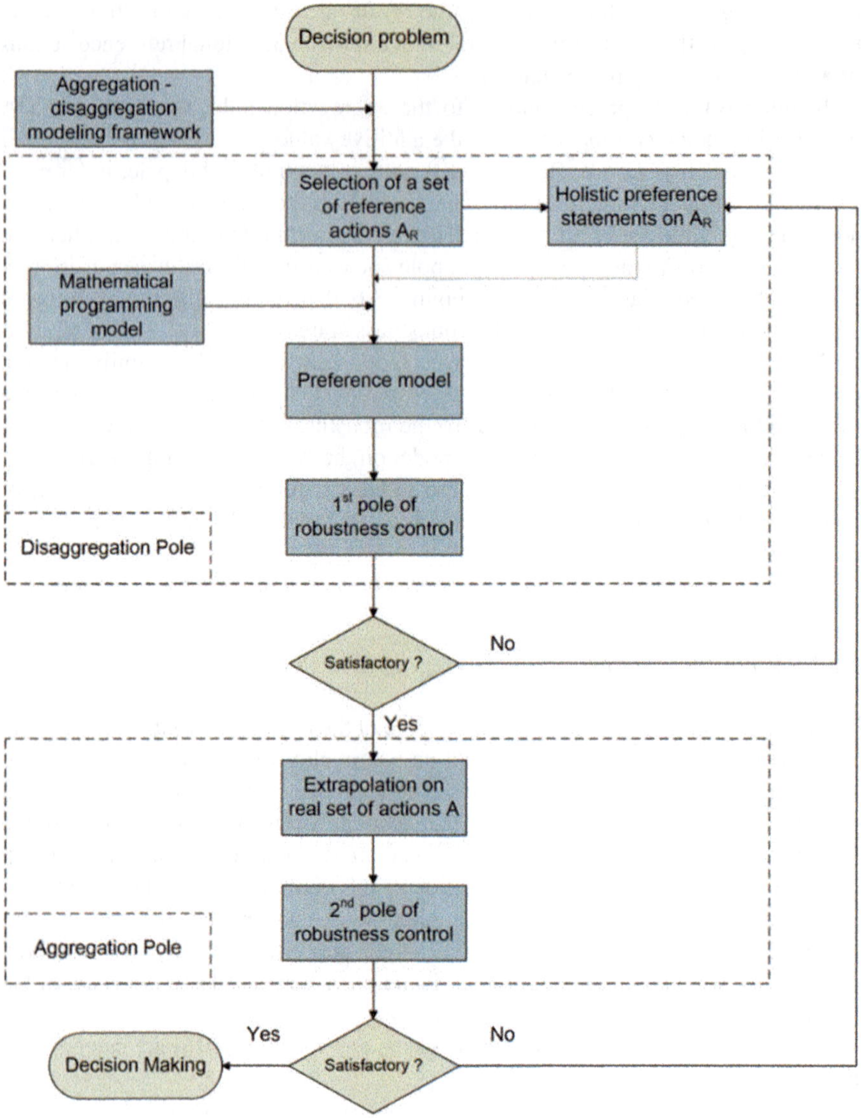

Fig. 8.6 Algorithm of the bipolar robustness control coupled with the UTA-type methods

Let p_{ij} are the set of the model's parameters produced by a robustness disaggregation technique, where i denotes a specific parameter of the model and j is an instance in which the parameter is estimated. In STOCHASTIC UTA method i varies from 1 to $\sum_{i=1}^{n} (\alpha_i - 1)$.

Average Range of the Preferential Parameters (ARP)

This index reveals the range of an average preferential parameter, after considering the preference information extracted by the DM. The calculation of the ARP requires the a priori implementation of the Max-Min LPs technique.

$$ARP = \frac{1}{p_m} \sum_{i=1}^{p} \sum_{j=1}^{m} \frac{\left(\max p_{ij} - \min p_{ij}\right)}{\max p_{ij}} \tag{8.10}$$

where m denotes the number of all instances of different preferential parameters p_{ij}.

This index ranges from [0–1] and receives lower values when the robustness of a model increases. *ARP* receives the value of 0 when a unique preference model is reflected on the mathematical formulation of the decision model.

Average Stability Index (ASI)

The average stability index is a robustness index proposed by Grigoroudis and Siskos (2002) and indicates the average value of the normalized standard deviation of the preferential parameters.

$$ASI = 1 - \frac{\frac{1}{n}\sum_{i=1}^{n} \sqrt{m \sum_{j=1}^{m} p_{ij}^2 - \left(\sum_{j=1}^{m} p_{ij}\right)^2}}{\frac{m}{n}\sqrt{n-1}} \tag{8.11}$$

where m denotes the number of different weighting instances produced.

ASI ranges in [0–1] and returns the value of 1 when perfect robustness has been achieved. Experimentations and applications to real life problems have shown that ASI should at least exceed 0.95.

The statistical robustness indices act as measures that aid the analyst to explore and gain insight on the feasible area of the preferential parameters – compatible with the DM's viewpoints. These indices again are coupled with certain analytic techniques.

Usually, the decision analyst considers that the control is satisfactory when at least one of the acceptable levels has been achieved. Of course, the analyst may also simultaneously proceed to the visualization of the variability of the model parameters.

5.3.2 Robustness Indices in the Aggregation Pole

The exploitation of the indices of the disaggregation pole offers to the analyst a comprehensive view of the robustness of the decision model. However, this fact does not guarantee the acquisition of robust results after the implementation of the decision model. The examination of the robustness, and the proposition of corresponding indices, in the aggregation pole – 2nd pole – is therefore necessary. Again, these indices work under the condition that certain auxiliary techniques and methods are implemented.

Average Range of the Ranking and Ratio of the Average Range of the Ranking

The Average Range of the Ranking Index (ARRI) and Ratio of the Average Range of the Ranking (RARR) are coupled with the Extreme Ranking Analysis technique, proposed by Kadziński, Greco and Słowiński (2012). ARRI depicts the possible number of positions that an average action can occupy in the whole ranking, while RARR offers a percentage of the aforementioned deviation on the whole number of the alternatives under evaluation. The optimal values of ARRI and RARR are 1 and 0% respectively. In general, a value of 5% for the latter index is usually acceptable by the decision analyst and the DM.

$$ARRI = \frac{1}{n} \sum_{i=1}^{n} \{|R_*(i) - R^*(i)| + 1\} \tag{8.12}$$

$$RARR = \frac{ARRI - 1}{N_n - 1} \cdot 100\% \tag{8.13}$$

where $R_*(i)$ and $R^*(i)$ are the worst and best possible ranking positions for an alternative i. N_n is the number of all the alternatives under evaluation.

Statistical Preference Relations Index (SPRI)

The statistical preference relations index (SPRI) offers a comprehensive way to examine the stability of all the ranking positions achieved by the whole entity of alternatives. It performs in synergy with random sampling techniques, the Manas-Nedoma algorithm (Manas and Nedoma 1968), and generally methods that provide a statistically adequate number of sets of preferential parameters, within the feasible area. SPRI calculates a separate probability, that each alternative occupies a single ranking position in the final ranking, and forms a meaningful measure, which gives a clear insight of the robustness of the final results.

Specifically, the estimation of the statistical probability, that an alternative i gets ranked in the t-th position, is performed using the following relation:

$$P_t^i = \frac{c_t^i}{V_T} \cdot 100\% \tag{8.14}$$

Where c_t^i is the number of samples/instances that position the alternative i in the t-th position, V_T is the number of all the samples/instances.

The statistical preference relations index is then calculated using the equation below:

$$SPRI = \frac{1}{T} \sum_{i}^{N_n} \sum_{T} P_t^i \tag{8.15}$$

SPRI reaches the optimal value of 100% when all the alternatives occupy a single ranking position with the statistical probability of 100%. In other words, the exactly same ranking occurs for all the preferential parameters samples/instances under consideration.

6 Application to the Scenario Evaluation

Risk scenarios have been developed and assigned values on 10-point scales of the 5 criteria, by the team of six experts, in cooperation with the decision analyst. According to the STOCHASTIC UTA method, the experts were asked to rank certain reference risk scenarios from the one bearing the highest risk to the less risky one. The target of the analysis is to achieve a robust and acceptable ranking of the 43 risk scenarios, through successive applications of the STOCHASTIC UTA method. In each successive iteration, the number of reference actions, which are given to the experts to rank, increases until robust ranking results are achieved.

As part of the robustness control procedure, the analyst decided to focus on the aggregation pole (2nd pole of robustness control), involving at the same time the team of experts to judge the quality of the results. Specifically, the extreme ranking analysis (ERA) in the disaggregation pole was performed, to measure and visualize the robustness of the ranking. The indices of ARRI and RARR were therefore used and presented to the team of experts, along with a visualization of the extreme ranking analysis, until they endorsed the adequacy of the results. These two indices were gladly greeted by the experts, due to their comprehensiveness and the fact that they are easily visualized.

Initially, the experts expressed their viewpoints on a subset of 5 risk scenarios (P2, M20, M24, T30 and T38), which were selected by the decision analyst among the 43 different ones (see Table 8.9 in the Appendix for the description of these risk scenarios). The mathematical formulations of the Stochastic UTA, in the four consecutive algorithmic steps of the method, are comprehensively presented below for the case of the 5 risk scenarios.

Step 1 In the first step, the global expected utilities of the five reference actions $u(\delta^{a_k}), k = 1, 2, \ldots, 5$, are calculated and then expressed in terms of variables w_{ij}, according to step 1 of the stochastic UTA algorithm. Considering that $u_{1j} = 0$, the following expressions are calculated:

$$u(\delta^{P2}) = \frac{2}{6} \cdot u_1(2) + \frac{2}{6} \cdot u_1(5) + \frac{1}{6} \cdot u_1(6) + \frac{1}{6} \cdot u_2(3) + \frac{3}{6} \cdot u_2(4) + \frac{1}{6} \cdot u_2(7) + \frac{1}{6} \cdot u_3(2)$$
$$+ \frac{3}{6} \cdot u_3(6) + \frac{2}{6} \cdot u_3(10) + \frac{4}{6} \cdot u_4(2) + \frac{1}{6} \cdot u_4(4) + \frac{1}{6} \cdot u_4(8) + \frac{4}{6} \cdot u_5(2) + \frac{1}{6} \cdot u_5(4) + \frac{1}{6} \cdot u_5(8)$$
$$= \frac{2}{6} \cdot w_{11} + \frac{2}{6} \cdot (w_{11} + w_{12} + w_{13} + w_{14}) + \frac{1}{6} \cdot (w_{11} + w_{12} + w_{13} + w_{14} + w_{15})$$
$$+ \frac{1}{6} \cdot (w_{21} + w_{22}) + \frac{3}{6} \cdot (w_{21} + w_{22} + w_{23}) + \frac{1}{6} \cdot (w_{21} + w_{22} + w_{23} + w_{24} + w_{25} + w_{26})$$
$$+ \frac{1}{6} \cdot w_{31} + \frac{3}{6} \cdot (w_{31} + w_{32} + w_{33} + w_{34} + w_{35})$$
$$+ \frac{2}{6} \cdot (w_{31} + w_{32} + w_{33} + w_{34} + w_{35} + w_{36} + w_{37} + w_{38} + w_{39}) + + \frac{4}{6} \cdot w_{41}$$
$$+ \frac{1}{6} \cdot (w_{41} + w_{42} + w_{43}) + \frac{1}{6} \cdot (w_{41} + w_{42} + w_{43} + w_{44} + w_{45} + w_{46} + w_{47})$$
$$+ \frac{4}{6} \cdot w_{51} + \frac{1}{6} \cdot (w_{51} + w_{52} + w_{53}) + \frac{1}{6} \cdot (w_{51} + w_{52} + w_{53} + w_{54} + w_{55} + w_{56} + w_{57})$$
$$= \frac{5}{6} \cdot w_{11} + \frac{3}{6} \cdot (w_{12} + w_{13} + w_{14}) + \frac{1}{6} \cdot w_{15} + \frac{5}{6} \cdot w_{21} + \frac{5}{6} \cdot w_{22} + \frac{4}{6} \cdot w_{23}$$
$$+ \frac{1}{6} \cdot (w_{24} + w_{25} + w_{26}) + w_{31} + \frac{5}{6} \cdot (w_{32} + w_{33} + w_{34} + w_{35})$$
$$+ \frac{2}{6} \cdot (w_{36} + w_{37} + w_{38} + w_{39}) + w_{41} + \frac{2}{6} \cdot (w_{42} + w_{43})$$
$$+ \frac{1}{6} \cdot (w_{44} + w_{45} + w_{46} + w_{47}) + w_{51} + \frac{2}{6} \cdot (w_{52} + w_{53})$$
$$+ \frac{1}{6} \cdot (w_{54} + w_{55} + w_{56} + w_{57}).$$

In the same way:

$$u(\delta^{M20}) = w_{11} + \frac{5}{6} \cdot w_{12} + \frac{3}{6} \cdot (w_{13} + w_{14}) + \frac{2}{6} \cdot w_{15} + \frac{1}{6} \cdot (w_{16} + w_{17}) + w_{22}$$
$$+ \frac{5}{6} \cdot w_{23} + \frac{4}{6} \cdot w_{24} + \frac{2}{6} \cdot w_{25} + \frac{1}{6} \cdot (w_{26} + w_{27}) + w_{31} + w_{32} + \frac{5}{6} \cdot w_{33}$$
$$+ \frac{2}{6} \cdot (w_{34} + w_{35}) + w_{43} + \frac{2}{6} \cdot (w_{44} + w_{45}) + w_{51} + \frac{5}{6} \cdot (w_{52} + w_{53}) + \frac{2}{6} \cdot (w_{54} + w_{55})$$

$$u\left(\delta^{M24}\right)=w_{12}+\frac{4}{6}\cdot\left(w_{13}+w_{14}\right)+\frac{3}{6}\cdot w_{15}+\frac{2}{6}\cdot w_{16}+w_{21}+w_{22}+\frac{5}{6}\cdot\left(w_{23}+w_{24}\right)$$
$$+\frac{3}{6}\cdot\left(w_{25}+w_{26}\right)+\frac{1}{6}\cdot w_{27}+\frac{4}{6}\cdot w_{31}+\frac{3}{6}\cdot\left(w_{32}+w_{33}+w_{34}+w_{35}\right)$$
$$+\frac{1}{6}\cdot\left(w_{36}+w_{37}\right)+w_{41}+\frac{4}{6}\cdot\left(w_{42}+w_{43}\right)+\frac{2}{6}\cdot\left(w_{44}+w_{45}\right)$$
$$+\frac{1}{6}\cdot\left(w_{46}+w_{47}\right)+w_{51}+w_{52}+w_{53}+\frac{3}{6}\cdot w_{54}+\frac{1}{6}\cdot\left(w_{55}+w_{56}+w_{57}\right)$$

$$u\left(\delta^{T30}\right)=w_{11}+w_{12}+\frac{4}{6}\cdot\left(w_{13}+w_{14}\right)+\frac{2}{6}\cdot w_{15}+\frac{1}{6}\cdot\left(w_{16}+w_{17}\right)+w_{21}+w_{22}$$
$$+\frac{5}{6}\cdot w_{23}+\frac{4}{6}\cdot w_{24}+\frac{2}{6}\cdot\left(w_{25}+w_{26}\right)+\frac{1}{6}\cdot\left(w_{27}+w_{28}+w_{29}\right)+w_{31}+w_{32}+w_{33}$$
$$+\frac{5}{6}\cdot\left(w_{34}+w_{35}\right)+\frac{2}{6}\cdot\left(w_{36}+w_{37}\right)+\frac{1}{6}\cdot\left(w_{38}+w_{39}\right)+w_{41}+w_{42}+w_{43}$$
$$+\frac{3}{6}\cdot\left(w_{44}+w_{45}\right)+\frac{2}{6}\cdot\left(w_{46}+w_{47}\right)+\frac{1}{6}\cdot\left(w_{48}+w_{49}\right)+w_{51}+w_{52}+w_{53}$$
$$+\frac{4}{6}\cdot\left(w_{54}+w_{55}\right)+\frac{3}{6}\cdot\left(w_{56}+w_{57}\right)+\frac{1}{6}\cdot\left(w_{58}+w_{59}\right)$$

$$u\left(\delta^{T38}\right)=w_{11}+\frac{3}{6}\cdot w_{12}+\frac{2}{6}\cdot\left(w_{13}+w_{14}\right)+\frac{1}{6}\cdot\left(w_{16}+w_{17}\right)+w_{21}+w_{22}$$
$$+\frac{4}{6}\cdot w_{23}+\frac{3}{6}\cdot w_{24}+\frac{1}{6}\cdot\left(w_{25}+w_{26}\right)+w_{31}+w_{32}+w_{33}+\frac{5}{6}\cdot\left(w_{34}+w_{35}\right)$$
$$+\frac{2}{6}\cdot\left(w_{36}+w_{37}\right)+\frac{1}{6}\cdot\left(w_{38}+w_{39}\right)+w_{41}+w_{42}+w_{43}+\frac{5}{6}\cdot\left(w_{44}+w_{45}\right)$$
$$+\frac{2}{6}\cdot\left(w_{46}+w_{47}\right)+\frac{1}{6}\cdot\left(w_{48}+w_{49}\right)+w_{51}+w_{52}+w_{53}+\frac{5}{6}\cdot\left(w_{54}+w_{55}\right)$$
$$+\frac{2}{6}\cdot\left(w_{56}+w_{57}\right)+\frac{1}{6}\cdot\left(w_{58}+w_{59}\right)$$

Step 2 In this step the preference order of the actions, provided by the experts is run from the top to the bottom, and the inequalities coming from the four consecutive pairs of actions are written accordingly.

The ranking of the 5 alternative risk scenarios from the one bearing the highest risk to the least risky one, as decided and articulated by the experts, is presented in Table 8.5.

Based, on the rank order of the reference risk scenarios, the following four inequalities arise:

Table 8.5 Preference order of the 5 references risk scenarios

Descending risk order of the 5 reference scenarios
M24
T30
M20
T38
P2

$$\Delta\left(\delta^{M24},\delta^{T30}\right)=u\left(\delta^{M24}\right)-u\left(\delta^{T30}\right)+\sigma^{-}(M24)-\sigma^{+}(M24)-\sigma^{-}(T30)+\sigma^{+}(T30)\geq\delta$$
$$\Delta\left(\delta^{T30},\delta^{M20}\right)=u\left(\delta^{T30}\right)-u\left(\delta^{M20}\right)+\sigma^{-}(T30)-\sigma^{+}(T30)-\sigma^{-}(M20)+\sigma^{+}(M20)\geq\delta$$
$$\Delta\left(\delta^{M20},\delta^{T38}\right)=u\left(\delta^{M20}\right)-u\left(\delta^{T38}\right)+\sigma^{-}(M20)-\sigma^{+}(M20)-\sigma^{-}(T38)+\sigma^{+}(T38)\geq\delta$$
$$\Delta\left(\delta^{T38},\delta^{P2}\right)=u\left(\delta^{T38}\right)-u\left(\delta^{P2}\right)+\sigma^{-}(T38)-\sigma^{+}(T38)-\sigma^{-}(P2)+\sigma^{+}(P2)\geq\delta$$

Step 3 & Step 4 In Step 3, the Stochastic UTA linear program (8.7, 8.8 and 8.9) is modelled and solved. The objective function should reach the value of zero, in order to ensure the mathematical feasibility and logic of the experts' ranking of the five alternatives. The value of δ is set at the commonly used value of 0.01.

$$[\min]z=\sigma^{-}(M24)+\sigma^{+}(M24)+\sigma^{-}(M24)+\sigma^{+}(M24)+\sigma^{-}(M24)+\sigma^{+}(M24)$$
$$+\sigma^{-}(T38)+\sigma^{+}(T38)+\sigma^{-}(P2)+\sigma^{+}(P2)$$

under the constraints:

$$\begin{cases}\Delta\left(\delta^{M24},\delta^{T30}\right)=u\left(\delta^{M24}\right)-u\left(\delta^{T30}\right)+\sigma^{-}(M24)-\sigma^{+}(M24)-\sigma^{-}(T30)+\sigma^{+}(T30)\geq0.01\\ \Delta\left(\delta^{T30},\delta^{M20}\right)=u\left(\delta^{T30}\right)-u\left(\delta^{M20}\right)+\sigma^{-}(T30)-\sigma^{+}(T30)-\sigma^{-}(M20)+\sigma^{+}(M20)\geq0.01\\ \Delta\left(\delta^{M20},\delta^{T38}\right)=u\left(\delta^{M20}\right)-u\left(\delta^{T38}\right)+\sigma^{-}(M20)-\sigma^{+}(M20)-\sigma^{-}(T38)+\sigma^{+}(T38)\geq0.01\\ \Delta\left(\delta^{T38},\delta^{P2}\right)=u\left(\delta^{T38}\right)-u\left(\delta^{P2}\right)+\sigma^{-}(T38)-\sigma^{+}(T38)-\sigma^{-}(P2)+\sigma^{+}(P2)\geq0.01\\ \sum_i\sum_j w_{ij}=1\\ w_{ij}\geq0\,\forall i\,and\,j,\sigma^{+}(a)\geq0,\sigma^{-}(a)\geq0\,\forall a\in A_R\end{cases}$$

However, the solution of the linear programming model, albeit feasible and with zero errors, gave unsatisfactory results with regard to their robustness, due to the enormous variations of the utilities $u_i\left(g_i^j\right)$. Indicatively, the implementation of the ERA at the aggregation pole revealed that an average risk scenario could potentially occupy 34.5 positions in the ranking, which is translated to a possible occupancy of 79.8% of the total ranking positions.

The inadequacy of the results forced the analyst to add two more actions to the reference preference set and ask the experts to include them in the previous 5-scenarios ranking. The addition of two more preference actions increased the

Table 8.6 Iterative process for the preference order of the risk scenarios

	Preference order for 5 risk scenarios	Preference order for 7 risk scenarios	Preference order for 9 risk scenarios	Preference order for 11 risk scenarios	Preference order for 13 risk scenarios
Higher risk	M24	M24	M24	M24	M24
	T30	T30	T30	T30	T30
	M20	T36	T36	T36	T36
	T38	M20	M9	M9	M9
	P2	T38	M20	M20	M20
		P2	T38	T38	M4
		M5	T40	M12	T38
			P2	T27	M12
			M5	T40	T27
				P2	T43
				M5	T40
					P2
Lower risk					M5

robustness of the results significantly, however, they were still not deemed as acceptable by the team of experts. This cyclic procedure was performed three additional times, including the addition of 2 reference risk scenarios in each run (9, 11 and 13 risk scenarios respectively). The procedure ended after the implementation of the ERA for the case of the 13 reference risk scenarios, the results of which were endorsed by the team of experts. The reference preferences of the five iterative processes, as articulated by the team of experts are shown in Table 8.6.

Figure 8.7 depicts the marginal value functions of the five criteria, reflecting the preferences of the experts, as inferred by the STOCHASTIC UTA method at the final iteration (13 reference risk scenarios). The Figure presents the variation between the minimum and the maximum possible values at each interior point, as well as the mean values.

These results show an orientation of the experts towards the detection number rate criterion, clearly considering it as the most important one. The rest of the criteria are similarly weighted. The experts' preference to the detection criterion indicates the importance of the ability to detect a risk on time to prevent it from happening. Even the worst hazard can be affordable, in case of the existence of a mechanism able to detect it on time. The experts are persons with adequate experience to STS transfer operations and the fact that detectability is the most important criterion for them, means that they deal with STS transfer risks with a more sophisticated point of view.

The results obtained by the ERA for the cases of 7, 9, 11 and 13 risk scenarios are presented in Fig. 8.8. The figure comparatively shows a significant improvement on

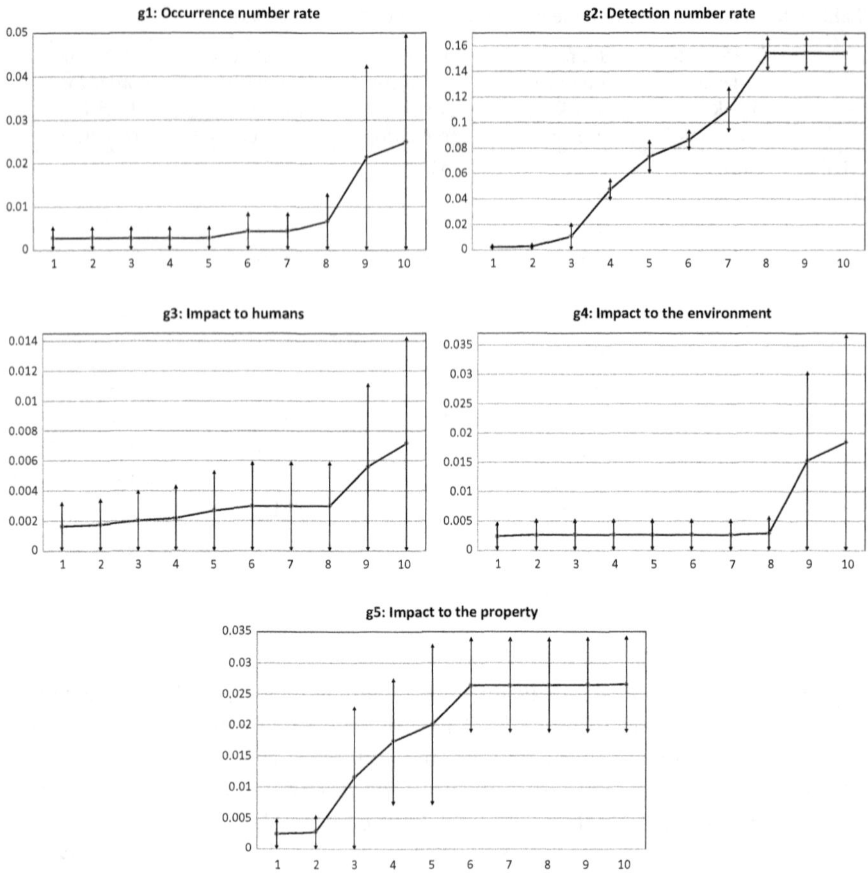

Fig. 8.7 Marginal value functions of the proposed representative additive value model and parameters variation

the robustness of the results. The robustness indices of ARRI and RARR, illustrated in Table 8.7 also highlight the dramatic increase in the stability of the results.

The ARRI, applied for the five different iterations of the application of the STOCHASTIC UTA and the additive value model, displayed an escalation of 489% throughout the whole robustness control procedure, while the Ratio of the Average Range of the Ranking increased by 588%.

Specifically, ARRI displays that at the last iteration, an average alternative can occupy 5.86 possible ranking positions, which, along to the fact that especially the head of the ranking, as shown in Fig. 8.8, is stable, led the team of experts to endorse the results.

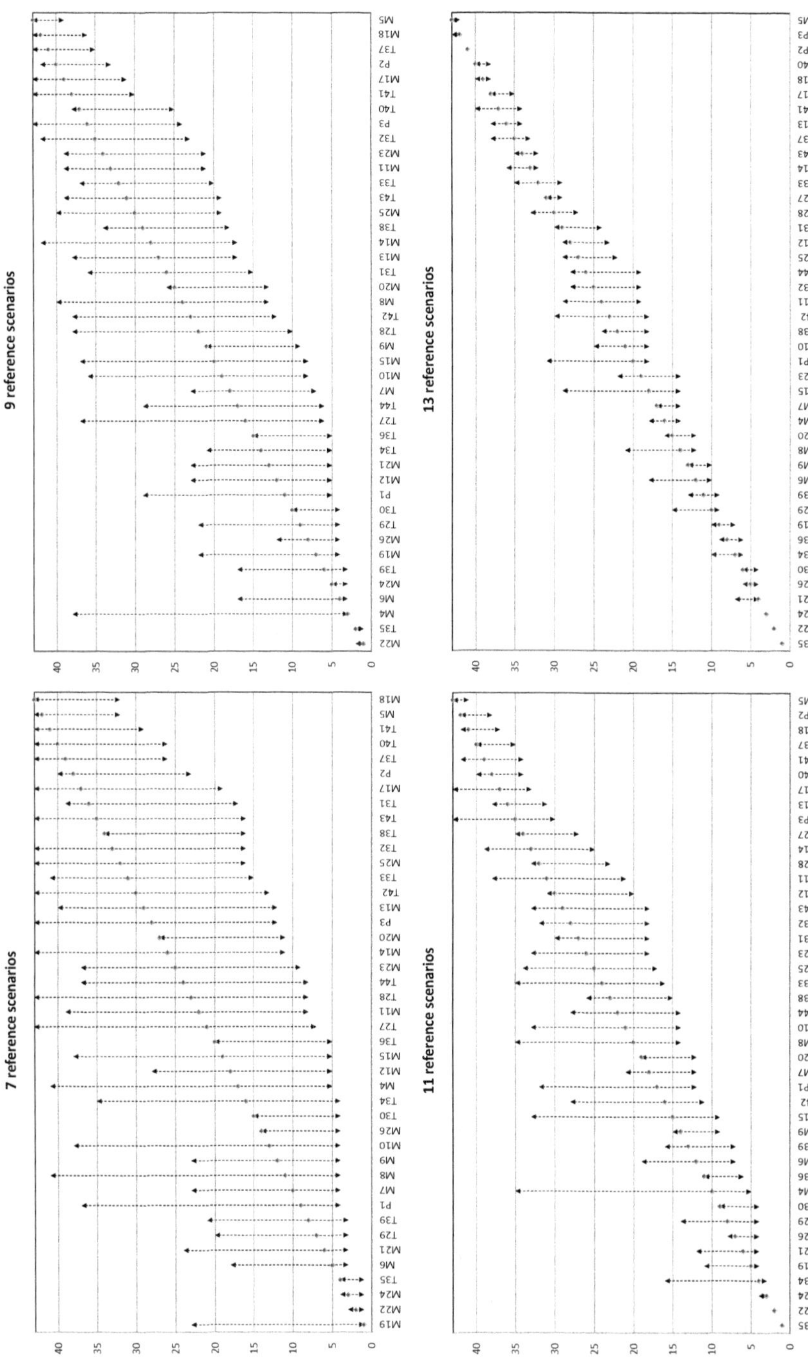

Fig. 8.8 Graphical comparison of the four successive applications of the extreme ranking analysis

Table 8.7 ARRI and RARR indices for all the successive applications of the additive value model

	5 risk scenarios	7 risk scenarios	9 risk scenarios	11 risk scenarios	13 risk scenarios
ARRI	34.51	23	17.53	11.74	5.86
RARR	79.8%	52.4%	39.0%	26.0%	11.6%

7 Discussion

This chapter presents a novel approach about risk evaluation with quantitative characteristics of accident scenarios, related to the phases of a STS transfer operation of petroleum products.

The results of the risk scenario ranking with respect to the preselected preference orders are shown in Table 8.6. Regarding the high-ranking risk positions the following can be observed by the obtained results:

- The scenarios T30 and M26, which refer to fatigue during the transfer phase, as well as the mooring/unmooring phases, indicate a prevailing threat. It is important that fatigue is a factor with low detectability, which means that it is a potential hazard that needs special consideration. The long duration of the STS transfer operations, on top of the limited crew sizes on-board the involved vessels, may be identified as the major contributing factors for the problem of human fatigue. The lightering operation of a VLCC to Aframax size "parcels" can last about 4 days for the entire operation (OSG 2009). On the other hand, the designated personnel according the OCIMF (2013) for the scheduled operation may be limited from the existing numbers of the crew.
- The scenario T35 is a prominent threat. It is relevant to inadequate control during transfer, due to different roll periods. This may cause loss of containment and/or vapour release leading to ignition back to source resulting in fatalities/injury/hardware/steelwork damage. The detection method for this scenario is unreliable, which means that the effectiveness of detection method is unknown to detect in time. To this end special consideration should be given to the compatibility characteristics during the preparation time and continuous checking of the weather conditions and changes should be conducted during the STS transfer operation.
- The scenarios M22 and M24 are highly risky, too. Scenario M22 is related with inadequate control during berthing/unberthing and side-by-side operations due to tug/support vessel failure, whereas scenario M24 is related to mooring equipment failure. Both M22 and M24 scenarios may lead to hull-to-hull contact/collision resulting in low energy collision and physical damage to one or both ships. The detection method for these scenarios has medium effectiveness.
- Other scenarios of special consideration, related in a direct or indirect way with human error are the inadequate planning for emergency breakaway (M21), the choice of inadequate tug/support vessel (M19), the inadequate experience (T29)

during the transfer process or a pilotage error during the mooring/unmooring operations. These scenarios may lead to hull-to-hull contact/collision resulting in energy collision and physical damage to one or both ships and their detection method is of medium effectiveness.

- Scenarios of non-human factor in high position ranking are the passing ship effect (T36), the Cargo sloshing during cargo transfer (T34), the damaged mooring due to chafing and cyclic loading (T39) and the fender defect (M6). The scenarios T36 and T34 may lead to the damage of the cargo tank structure, rendering the ship unable to make transfers, whereas the scenario T39 may lead to the loss of containment, leading to vapour collection in a confined space, resulting in ignition with explosion causing fatalities, injury and hull damage. Finally, scenario M6 may lead to hull-to-hull contact/collision resulting in low energy collision and physical damage to one or both ships.

All the above scenarios have a detection method of medium effectiveness. On the other hand, regarding the low-ranking risk positions the following observations can be obtained:

- Risks related to the transfer of personnel prior the transfer operation (P1, P2 and P3) does not create great concerns to the experts. Only scenario P1, which is related to inadequate compatibility study for the personnel landing area has a medium ranking position, whereas scenarios of defective personnel transfer equipment (P2) and equipment for the transfer of personnel unapproved/not fit for purpose (P3) are of less importance. This observation is in contradiction with previous studies (Spenser 2010), which focus exhaustively on the transfer operations of the personnel involved, such as crew members, mooring masters, surveyors, agents or customs officials, prior to and after the entire operation.
- Insignificant risks related with the human factor are the inadequate equipment inspection, testing and maintenance (M13), the inadequate site selection (M18), the inadequate training (M13), the use of level measurement and overfill protection systems inadequate for open water operation (T41), the inadequate procedures (M14) and the selection of transfer of cargo equipment not fit for purpose. The insignificant importance of these risks indicate that the experts are convinced that the personnel involved to STS transfer operations have the necessary safety culture for such operations. Nonetheless, this claim is also proven by the significantly low accident records during STS transfer operations.

Finally, other risks of less importance which are not related to the human factor are the defective overfill protection (T43), the poor visibility (M17) and the steering/propulsion failure (M5), which may lead to hull-to-hull contact/collision resulting in low energy collision and physical damage to one or both ships.

8 Conclusions

Evaluating the risks using marine accident scenarios remains a significant link within the chain of marine safety for all maritime activities, part of which are the STS transfer operations. The severely adverse effects of a potential accident, during a transfer operation, have led several researchers to focus their studies on this topic and develop models to evaluate hazardous scenarios of interest. Nevertheless, a serious weakness of these studies is the inability to combine effectively the subjectivity of experts' opinion.

The aim of this chapter was to evaluate risks related to STS transfer of cargo operations, by reducing the uncertainty coming from experts' viewpoints. To do so, a novel approach, based on the STOCHASTIC UTA method and a robustness control procedure, was presented and discussed. To the best of the authors knowledge, this is the first time that the STOCHASTIC UTA method is applied to a maritime operation to evaluate risk scenarios. In effect, the method can be flawlessly employed to activities where experts' judgment is exploited, such as the STS transfer operations. More specifically, this work ranked and clustered the relevant hazards according to their strength/potential to cause adverse consequences to humans, the environment and the property.

The applied methodology offered to the STS transfer operators an alternative way to conduct risk assessment using experts' knowledge. Moreover, the addition of the bipolar robustness control procedure to manage robustness of the model eliminated the weaknesses of the classical STOCHASTIC UTA approach improving the effectiveness of the extracted results. This approach is further ahead from the existing ones, providing operators a systematic tool to evaluate risks and safety of maritime activities. The employment of the proposed method can also be examined in the context of other new or current procedures with inadequate or lack of sufficient information (failure or accident data), which can be modelled and assessed in terms of risk and safety. Thus, operators can be equipped with a decision tool able to provide them the necessary knowledge to deal with risks coming from the stochastic nature of maritime environment. Another factor that supports the preference to the method lies in the analysis and study of the STS operations in an analytical and structured manner. Hence, each phase is thoroughly analyzed, each hazard and/or cause is carefully identified and all possible consequences are presented and evaluated.

The future perspectives of the STOCHASTIC UTA methodology may include synergies with other MCDA or mathematical programming techniques with the objective to determine more effective the utility functions mitigating or even eliminating the uncertainty coming inherently due to the stochastic nature of the real-case problems.

Appendix A Experts Evaluations on the Risk Alternatives on the Criteria Under Evaluation

Appendix B

B.1 Theoretical Background of the UTA Family Methodologies

A utility function u of a set of actions $A_R = \{a_1, a_2, \ldots, a_m\}$, based on a family of criteria $g_1, g_2, \ldots, g_n$, is expressed by the relationships:

$$u(\mathbf{g}) = u(g_1, g_2, \ldots, g_n) \tag{8.16}$$

$$\mathbf{g}(a) \rightarrow u[\mathbf{g}(a)] \tag{8.17}$$

where $u[\mathbf{g}(a)]$ corresponds to a real number, which expresses the global utility of action a.The criteria aggregation model in UTA is assumed to be an additive value function of the following form (Jacquet-Lagrèze and Siskos 1982, Fig. 8.9):

$$u(\mathbf{g}) = \sum_{i=1}^{n} p_i u_i(g_i) \tag{8.18}$$

under the constraints of normalization:

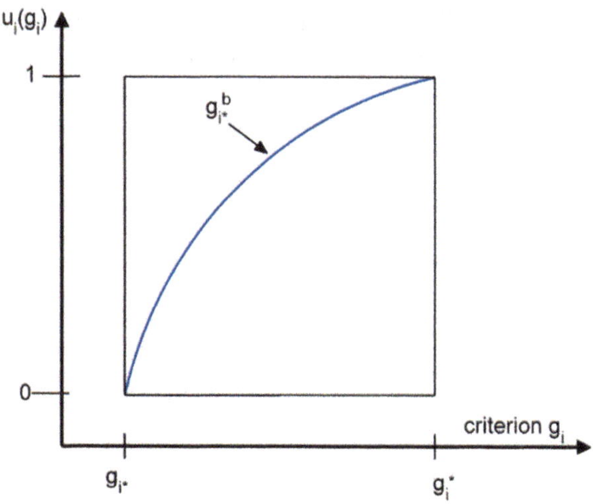

Fig. 8.9 Normalized marginal utility function

$$\sum_{i=1}^{n} p_i = 1 \tag{8.19}$$

$$u_i(g_{i*}) = 0, u_i(g_i^*) = 1 \ \forall i = 1, 2, \ldots n \tag{8.20}$$

where $u_i, i = 1, 2, \ldots n$ is the non-dicreased real valued functions of g_i, generally known as the marginal value or utility function, normalized between 0 and 1, g_{i*} and g_i^* correspond to the worst and the best values of criterion g_i, and p_i, is the weight of u_i.Both the marginal and the global value functions have the monotonicity property of the true criterion. For instance, in the case of the global value function the following properties hold:

$$u[\mathbf{g}(a)] > u[\mathbf{g}(b)] \Leftrightarrow a \succ \text{b, action } a \text{ is preferred to action b.} \tag{8.21}$$

$$u[\mathbf{g}(a)] = u[\mathbf{g}(b)] \Leftrightarrow a \sim \text{b, action } a \text{ is indifferent to action b.} \tag{8.22}$$

In UTA methods an equivalent unweighted formula is inferred (Jacquet-Lagrèze and Siskos 1982):

$$u(\mathbf{g}) = \sum_{i=1}^{n} u_i(g_i) \tag{8.23}$$

under the constrains of normalization:

$$\sum_{i=1}^{n} u_i(g_i^*) = 1 \tag{8.24}$$

$$u_i(g_{i*}) = 0, \forall i = 1, 2, \ldots n \tag{8.25}$$

The existence of such a preference model assumes the preferential independence of the criteria for the DM (Keeney and Raiffa 1976), while other conditions for additivity have been proposed by Fishburn (1966, 1967). This assumption does not pose significant problems in a posteriori analyses such as disaggregation analyses.

B.2 UTA Method Under Certainty

Given the additive model (8.23), (8.24), (8.25) and under the preference conditions (8.21), (8.22) the value of each alternative $a \in A_R$ may be written as:

$$u'[\mathbf{g}(a)] = \sum_{i=1}^{n} u_i[g_i(a)] + \sigma(a) \quad \forall a \in \mathrm{A}_R, \tag{8.26}$$

where $\sigma(a)$, is the potential error relative to $u'[\mathbf{g}(a)]$.The marginal value functions can be estimated in a piecewise linear form according to Jacquet-Lagrèze and Siskos (1982) by the use of linear interpolation. Thus, for each criterion, the interval $\left[g_{i*}, g_i^*\right]$ is cut into $(a_i - 1)$ equal intervals, and the end points g_i^j are given by the formula:

$$g_i^j = g_{i*} + \frac{j-1}{a_i - 1}\left(g_i^* - g_{i*}\right) \quad \forall j = 1, 2, \ldots, a_i \tag{8.27}$$

The marginal value of an action a is approximated by a linear interpolation, which means that for $g_i(a) \epsilon \left[g_i^j, g_i^{j+1}\right]$

$$u_i[g_i(a)] = u_i\left(g_i^j\right) + \frac{g_i(a) - g_i^j}{g_i^{j+1} - g_i^j}\left[u_i\left(g_i^{j+1}\right) - u_i\left(g_i^j\right)\right] \tag{8.28}$$

The set of reference actions $A_R = \{a_1, a_2, \ldots, a_m\}$, is supposed to be "rearranged" in such a way that a_1, is the head of the ranking and a_m its tail. Since the ranking has the form of a weak order R, for each pair of consecutive actions (a_k, a_{k+1}) it holds either $a_k \succ a_{k+1}$ (preference) or $a_k \sim a_{k+1}$ (indifference).Thus, if

$$\Delta(a_k, a_{k+1}) = u'[\mathbf{g}(a_k)] - u'[\mathbf{g}(a_{k+1})] \tag{8.29}$$

Then one of the following holds:

$$\begin{cases} \Delta(a_k, a_{k+1}) \geq \delta \textit{ if } a_k \succ a_{k+1} \\ \Delta(a_k, a_{k+1}) = 0 \textit{ if } a_k \sim a_{k+1} \end{cases}, \tag{8.30}$$

where δ is a small positive number so as to discriminate significantly two successive equivalence classes of R. Taking into account the hypothesis on monotonicity of preferences, the marginal values $u_i(g_i)$ must satisfy the set of the following constraints:

$$u_i\left(g_i^{j+1}\right) - u_i\left(g_i^j\right) \geq s_i \quad \forall j = 1, 2, \ldots, a_i - 1, i = 1, 2, \ldots, n \tag{8.31}$$

with $s_i \geq 0$ being indifference thresholds defined on each criterion g_i. Jacquet-Lagrèze and Siskos (1982) urge that it is not necessary to use these thresholds in the UTA model ($s_i = 0$), but they can be useful in order to avoid phenomena such as $u_i\left(g_i^{j+1}\right) = u_i\left(g_i^j\right)$, when $g_i^{j+1} \succ g_i^j$.The marginal value functions are finally estimated by means of the following linear program (LP) with (8.23), (8.24), (8.25), (8.30), and (8.31) as constraints and with an objective function depending on the $\sigma(\alpha)$ and indicating the amount of total deviation:

$$\begin{cases} [\min]F = \sum_{\alpha \in A_R} \sigma(\alpha) \\ \textit{subject to} \\ \Delta(a_k, a_{k+1}) \geq \delta \textit{ if } a_k \succ a_{k+1} \\ \Delta(a_k, a_{k+1}) = 0 \textit{ if } a_k \sim a_{k+1} \\ u_i\left(g_i^{j+1}\right) - u_i\left(g_i^j\right) \geq 0 \\ \sum_{i=1}^{n} u_i\left(g_i^*\right) = 1 \\ u_i(g_{i*}) = 0, \quad u_i\left(g_i^j\right) \geq 0, \quad \sigma(\alpha) \geq 0 \ \forall \alpha \in A_R, \forall i \textit{ and } j \end{cases} \tag{8.32}$$

The stability analysis of the results provided by LP (8.32) is considered as a post-optimality analysis problem. As Jacquet-Lagrèze and Siskos (1982) noted, if the optimum $F^* = 0$, the polyhedron of admissible solutions for $u_i(g_i)$ is not empty and many value functions lead to a perfect representation of the weak order R. Even when the optimal value F^* is strictly positive, other solutions, less good for F, can improve other satisfactory criteria, like Kendall's correlation coefficient τ.As shown in Fig. 8.10, the post-optimal solutions space is defined by the polyhedron:

$$\begin{cases} F \leq F^* + k(F^*) \\ \textit{all the constraints of LP } (5.17) \end{cases}, \tag{8.33}$$

where $k(F^*)$ is a positive threshold which is a small proportion of F^*.

The algorithms which could be used to explore the polyhedron (5.18) are branch and bound methods, like reverse simplex method (Van de Panne 1975), or techniques dealing with the notion of the labyrinth in graph theory, such as Tarry's method (Charnes and Cooper 1961) or the method of Manas and Nedoma (1968).

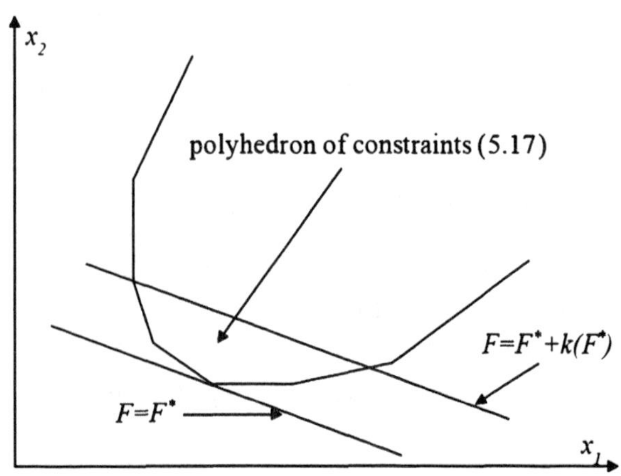

Fig. 8.10 Post-optimality analysis (Jacquet-Lagrèze and Siskos 1982)

Table 8.8 Experts evaluations on the risk alternatives on the criteria of occurrence, detectability, impact to humans, impact to the environment and impact to the property

	Occurrence										Detectability										Impact to human										Impact to the environment										Impact to the property									
	1	2	3	4	5	6	7	8	9	10	1	2	3	4	5	6	7	8	9	10	1	2	3	4	5	6	7	8	9	10	1	2	3	4	5	6	7	8	9	10	1	2	3	4	5	6	7	8	9	10
P1	1	2	1	0	2	0	0	0	0	0	1	0	1	1	0	1	1	1	0	0	0	1	0	0	0	3	0	1	0	1	0	4	0	1	0	0	0	1	0	0	0	4	0	1	0	0	0	1	0	0
P2	1	2	0	0	2	1	0	0	0	0	1	0	1	3	0	0	1	0	0	0	0	1	0	0	0	3	0	0	0	2	0	4	0	1	0	0	0	1	0	0	0	4	0	1	0	0	0	1	0	0
P3	1	1	0	0	1	3	0	0	0	0	1	0	2	2	0	0	1	0	0	0	0	0	0	0	0	3	0	1	0	2	0	5	0	0	0	0	0	1	0	0	0	4	0	1	0	1	0	0	0	0
M4	2	2	1	0	1	0	0	0	0	0	0	1	2	1	1	0	0	0	1	0	0	0	0	1	0	4	0	0	0	1	0	0	0	0	0	0	0	3	0	3	0	0	0	0	0	0	0	4	0	2
M5	0	2	1	0	1	1	1	0	0	0	0	3	1	0	1	0	1	0	0	0	0	0	0	5	0	1	0	0	0	0	0	2	0	3	0	0	0	1	0	0	0	2	0	2	0	0	0	2	0	0
M6	1	0	0	0	3	1	1	0	0	0	0	1	0	1	0	1	3	0	0	0	0	3	0	2	0	1	0	0	0	0	0	3	0	2	0	1	0	0	0	0	0	2	0	2	0	0	0	2	0	0
M7	0	0	1	0	1	3	1	0	0	0	0	1	0	2	0	1	1	0	0	1	0	2	0	3	0	1	0	0	0	0	0	3	0	2	0	1	0	0	0	0	0	2	0	2	0	0	0	2	0	0
M8	0	2	1	0	3	0	0	0	0	0	0	0	0	4	0	0	2	0	0	0	0	1	0	4	0	1	0	0	0	0	0	3	0	1	0	2	0	0	0	0	0	0	0	4	0	1	0	1	0	0
M9	0	0	3	0	1	1	1	0	0	0	0	1	1	1	1	0	1	0	0	1	0	0	0	2	0	2	0	1	0	1	0	0	0	1	0	4	0	0	0	1	0	0	0	0	0	4	0	1	0	1
M10	0	3	0	0	1	0	2	0	0	0	0	0	1	2	1	1	1	0	0	0	0	1	0	4	0	0	0	1	0	0	0	1	0	2	0	3	0	0	0	0	0	1	0	1	0	2	0	2	0	0
M11	0	2	2	0	2	0	0	0	0	0	0	2	0	2	0	0	1	1	0	0	0	2	0	2	0	1	0	1	0	0	0	1	0	3	0	1	0	1	0	0	0	1	0	2	0	1	0	2	0	0
M12	0	0	1	0	4	0	1	0	0	0	0	1	1	1	0	1	2	0	0	0	0	4	0	0	0	1	0	1	0	0	0	2	0	2	0	1	0	1	0	0	0	1	0	3	0	0	0	1	0	1
M13	0	0	1	0	4	0	1	0	0	0	0	2	1	0	1	1	1	0	0	0	0	3	0	1	0	1	0	1	0	0	0	1	0	3	0	1	0	1	0	0	0	1	0	2	0	1	0	2	0	0
M14	0	2	1	0	1	1	1	0	0	0	0	2	1	1	1	1	0	0	0	0	0	0	0	1	0	2	0	2	0	1	0	0	0	1	0	3	0	1	0	1	0	0	0	0	0	2	0	2	0	2
M15	0	1	2	0	1	2	0	0	0	0	0	0	3	0	1	0	1	0	0	1	0	1	0	2	0	2	0	1	0	0	0	1	0	2	0	2	0	1	0	0	0	1	0	1	0	1	0	3	0	0
M17	1	1	2	0	2	0	0	0	0	0	0	2	1	2	0	0	0	1	0	0	0	3	0	1	0	2	0	0	0	0	0	2	0	2	0	1	0	1	0	0	0	1	0	2	0	1	0	1	0	1
M18	1	1	1	0	3	0	0	0	0	0	0	1	2	1	1	1	0	0	0	0	0	3	0	1	0	1	0	1	0	0	0	2	0	2	0	1	0	1	0	0	0	0	0	3	0	1	0	2	0	0
M19	0	1	1	0	1	2	1	0	0	0	0	0	1	0	3	1	0	1	0	0	0	2	0	2	0	0	0	2	0	0	0	1	0	2	0	2	0	1	0	0	0	0	0	3	0	1	0	2	0	0
M20	0	1	2	0	1	1	0	1	0	0	0	0	1	1	2	1	0	1	0	0	0	1	0	3	0	2	0	0	0	0	0	0	0	4	0	2	0	0	0	0	0	1	0	3	0	2	0	0	0	0
M21	0	0	1	0	3	1	1	0	0	0	0	0	0	1	3	1	0	1	0	0	0	2	0	2	0	2	0	0	0	0	0	1	0	3	0	1	0	1	0	0	0	0	0	4	0	0	0	2	0	0
M22	0	1	2	0	1	1	1	0	0	0	0	0	0	1	0	1	2	2	0	0	0	2	0	2	0	1	0	1	0	0	0	1	0	2	0	1	0	2	0	0	0	0	0	3	0	1	0	1	0	1
M23	0	1	0	0	4	0	1	0	0	0	0	1	0	2	2	0	0	1	0	0	0	2	0	2	0	2	0	0	0	0	0	1	0	4	0	1	0	0	0	0	0	0	0	4	0	1	0	1	0	0
M24	0	0	2	0	1	1	2	0	0	0	0	0	1	0	2	0	2	1	0	0	0	1	0	2	0	2	0	1	0	0	0	2	0	2	0	1	0	1	0	0	0	0	0	3	0	2	0	1	0	0
M25	0	2	1	0	3	0	0	0	0	0	0	1	0	1	2	2	0	0	0	0	0	2	0	2	0	2	0	0	0	0	0	1	0	3	0	1	0	1	0	0	0	1	0	3	0	1	0	1	0	0
M26	0	1	1	0	2	1	0	1	0	0	0	0	1	1	1	1	1	0	0	1	0	1	0	3	0	1	0	1	0	0	0	0	0	3	0	2	0	1	0	0	0	0	0	2	0	2	0	2	0	0
T27	1	0	2	0	2	1	0	0	0	0	0	2	1	0	1	2	0	0	0	0	0	1	0	0	0	3	0	1	0	1	0	1	0	0	0	3	0	0	0	2	0	0	0	2	0	0	0	2	0	2
T28	0	1	2	0	2	1	0	0	0	0	0	2	1	0	1	2	0	0	0	0	0	1	0	0	0	2	0	2	0	1	0	1	0	0	0	3	0	1	0	1	0	0	0	1	0	1	0	3	0	1
T29	0	1	2	0	1	1	1	0	0	0	0	0	2	1	1	0	1	0	0	1	0	0	0	1	0	2	0	2	0	1	0	0	0	1	0	3	0	2	0	0	0	0	0	0	0	2	0	4	0	0
T30	0	0	2	0	2	1	0	1	0	0	0	0	1	1	2	0	1	0	0	1	0	0	0	1	0	3	0	1	0	1	0	0	0	3	0	1	0	1	0	1	0	0	0	2	0	1	0	2	0	1
T31	2	0	3	0	1	0	0	0	0	0	1	0	1	1	1	1	1	0	0	0	0	2	0	2	0	0	0	1	0	0	0	2	0	1	0	0	0	3	0	0	0	0	0	2	0	2	0	2	0	0

(continued)

Table 8.8 (continued)

	Occurrence										Detectability										Impact to human										Impact to the environment										Impact to the property									
T32	1	2	1	0	2	0	0	0	0	0	1	0	1	1	2	0	0	0	1	0	0	2	0	3	0	1	0	0	0	0	0	2	0	1	0	2	0	1	0	0	0	1	0	1	0	2	0	2	0	0
T33	0	3	0	0	3	0	0	0	0	0	0	1	0	2	1	1	1	0	0	0	0	3	0	2	0	1	0	0	0	0	0	0	0	1	0	4	0	1	0	0	0	1	0	1	0	2	0	0	0	0
T34	0	2	2	0	1	1	0	0	0	0	0	0	0	1	2	2	1	0	0	0	0	4	0	1	0	1	0	0	0	0	0	2	0	2	0	1	0	1	0	0	0	0	0	3	0	2	0	1	0	0
T35	1	0	1	0	2	1	1	0	0	0	0	0	0	1	0	1	1	2	1	0	0	1	0	3	0	2	0	0	0	0	0	0	0	1	0	4	0	1	0	0	0	0	0	2	0	3	0	1	0	0
T36	2	1	0	0	1	2	0	0	0	0	1	0	0	1	0	2	1	1	0	0	0	4	0	1	0	0	0	0	0	0	0	2	0	2	0	1	0	1	0	0	0	1	0	1	0	2	0	2	0	0
T37	0	1	2	0	3	0	0	0	0	0	0	0	2	2	2	0	0	0	0	0	0	0	0	3	0	2	0	1	0	0	0	0	0	1	0	4	0	1	0	0	0	0	0	3	0	2	0	1	0	0
T38	0	3	1	0	2	0	0	0	0	0	0	0	2	1	2	0	1	0	0	0	0	0	0	1	0	3	0	1	0	1	0	0	0	1	0	3	0	1	0	1	0	0	0	1	0	3	0	1	0	1
T39	1	3	0	0	0	1	1	0	0	0	0	0	1	2	0	1	2	0	0	0	0	0	0	2	0	2	0	1	0	1	0	1	0	1	0	2	0	1	0	1	0	0	0	2	0	2	0	1	0	1
T40	0	3	2	0	1	0	0	0	0	0	0	1	2	1	0	2	0	0	0	0	0	2	0	3	0	1	0	0	0	0	0	1	0	0	0	2	0	1	0	1	0	2	0	2	0	1	0	0	0	1
T41	1	3	1	0	1	0	0	0	0	0	0	1	1	2	1	1	0	0	0	0	0	4	0	1	0	1	0	0	0	0	0	0	0	2	0	2	0	1	0	1	0	1	0	4	0	0	0	0	0	1
T42	1	1	1	0	2	1	0	0	0	0	0	1	0	1	2	2	0	0	0	0	0	4	0	1	0	1	0	0	0	0	0	0	0	1	0	3	0	1	0	1	0	2	0	1	0	2	0	1	0	0
T43	1	3	1	0	0	1	0	0	0	0	1	0	1	0	2	2	0	0	0	0	0	3	0	2	0	1	0	0	0	0	0	0	0	2	0	2	0	2	0	0	0	2	0	2	0	1	0	1	0	0
T44	1	2	0	0	3	0	0	0	0	0	0	1	1	1	0	2	1	0	0	0	0	0	0	3	0	1	0	2	0	0	0	0	0	2	0	1	0	3	0	0	0	0	0	2	0	1	0	3	0	0

Table 8.9 Description of risk scenarios under evaluation for STS transfer of cargo operation

Risk ID number	STS phase	Description
P1	Preparation	Failure or improper use during personnel transfer due to inadequate compatibility study - personnel landing area resulting in loss of control during transfer resulting in fatality or injury.
P2	Preparation	Failure or improper use during personnel transfer due to defective personnel transfer equipment resulting in loss of control during transfer resulting in fatality or injury.
P3	Preparation	Failure or improper use during personnel transfer due to equipment for the transfer of personnel unapproved/not fit for purpose resulting in loss of control during transfer resulting in fatality or injury.
M4	Mooring/ unmooring	Inadequate navigational control by ships involved in STS or third party passing vessel due to steering/propulsion failure resulting in a high-energy collision resulting in significant equipment damage, fatalities, and loss of containment.
M5	Mooring/ unmooring	Inadequate control during berthing/underthing and side-by-side operations due to steering/propulsion failure resulting in the hull to hull contact/collision leading to low energy collision and physical damage to one or both ships.
M6	Mooring/ unmooring	Inadequate control during berthing/underthing and side-by-side operations due to fender defect resulting in the hull to hull contact/ collision leading to low energy collision and physical damage to one or both ships.
M7	Mooring/ unmooring	Inadequate control during berthing/underthing and side-by-side operations due to inadequate fendering resulting in the hull to hull contact/collision leading to low energy collision and physical damage to one or both ships.
M8	Mooring/ unmooring	Inadequate control during berthing/underthing and side-by-side operations due to mismatched manoeuvring characteristics resulting in the hull to hull contact/collision leading to low energy collision and physical damage to one or both ships.
M9	Mooring/ unmooring	Inadequate control during berthing/underthing and side-by-side operations due to pilotage error resulting in high energy collision and personal injury or fatality.
M10	Mooring/ unmooring	Inadequate control during berthing/underthing and side-by-side operations due to insufficient manoeuvring room resulting in hull to hull contact/collision leading to low energy collision and physical damage to one or both ships
M11	Mooring/ unmooring	Inadequate control during berthing/underthing and side-by-side operations due to inadequate watchkeeping resulting in hull to hull contact/collision leading to low energy collision and physical damage to one or both ships
M12	Mooring/ unmooring	Inadequate control during berthing/underthing and side-by-side operations due to inadequate communication resulting in hull to hull contact/collision leading to low energy collision and physical damage to one or both ships

(continued)

Table 8.9 (continued)

Risk ID number	STS phase	Description
M13	Mooring/ unmooring	Inadequate control during berthing/underthing and side-by-side operations due to inadequate training resulting in hull to hull contact/collision leading to low energy collision and physical damage to one or both ships
M14	Mooring/ unmooring	Inadequate control during berthing/underthing and side-by-side operations due to inadequate procedures resulting in high energy collision and personal injury or fatality
M15	Mooring/ unmooring	Inadequate control during berthing/underthing and side-by-side operations due to inadequate experience resulting in hull to hull contact/collision leading to low energy collision and physical damage to one or both ships
M17	Mooring/ unmooring	Inadequate control during berthing/underthing and side-by-side operations due to poor visibility resulting in hull to hull contact/ collision leading to low energy collision and physical damage to one or both ships
M18	Mooring/ unmooring	Inadequate control during berthing/underthing and side-by-side operations due to inadequate site selection resulting in hull to hull contact/collision leading to low energy collision and physical damage to one or both ships
M19	Mooring/ unmooring	Inadequate control during berthing/underthing and side-by-side operations due to inadequate tug/support vessel resulting in hull to hull contact/collision leading to low energy collision and physical damage to one or both ships
M20	Mooring/ unmooring	Inadequate control during berthing/underthing and side-by-side operations due to inadequate operational planning and control resulting in hull to hull contact/collision leading to low energy collision and physical damage to one or both ships
M21	Mooring/ unmooring	Inadequate control during berthing/underthing and side-by-side operations due to inadequate planning for emergency breakaway resulting in hull to hull contact/collision leading to low energy collision and physical damage to one or both ships
M22	Mooring/ unmooring	Inadequate control during berthing/underthing and side-by-side operations due to tug/support vessel failure resulting in low energy collision leading to significant equipment damage, fatalities and loss of containment
M23	Mooring/ unmooring	Inadequate control during berthing/unberthing and side-by-side operations due to inadequate compatibility study- bridge wing, separation and parallel body length resulting in the hull to hull contact/collision leading to low energy collision and physical damage to one or both ships.
M24	Mooring/ unmooring	Inadequate control during berthing/unberthing and side-by-side operations due to mooring equipment failure resulting in the hull to hull contact/collision leading to low energy collision and physical damage to one or both ships.
M25	Mooring/ unmooring	Inadequate control during berthing/unberthing and side-by-side operations due to inadequate incident management resulting in the hull to hull contact/collision leading to low energy collision and physical damage to one or both ships.

(continued)

Table 8.9 (continued)

Risk ID number	STS phase	Description
M26	Mooring/ unmooring	Inadequate control (e.g. master, helmsman, POAC) during berthing/unberthing and side-by-side operations leads to an uncontrolled event due to fatigue resulting in the hull to hull contact/collision leading to low energy collision and physical damage to one or both ships.
T27	Cargo transfer	Inadequate control during transfer leads to an uncontrolled event due to inadequate training resulting in loss of containment and vapour release leading to ignition back to source resulting in fatalities/injury/hardware/steelwork damage.
T28	Cargo transfer	Inadequate control during transfer leads to an uncontrolled event due to inadequate procedures resulting in loss of containment and vapour release leading to ignition back to source resulting in fatalities/injury/hardware/steelwork damage.
T29	Cargo transfer	Inadequate control during transfer leads to an uncontrolled event due to inadequate experience resulting in loss of containment and/or vapour release leading to ignition back to source resulting in fatalities/injury/hardware/steelwork damage.
T30	Cargo transfer	Inadequate control during transfer leads to an uncontrolled event due to fatigue resulting in loss of containment and/or vapour release leading to ignition back to source resulting in fatalities/ injury/hardware/steelwork damage.
T31	Cargo transfer	Cargo sloshing due to ship motions in the prevailing swell conditions due to abnormal metecean conditions resulting in damage to cargo tank structure resulting in ship being unfit to trade
T32	Cargo transfer	Cargo sloshing due to ship motions in the prevailing swell conditions due to inadequate weather forecasting resulting in damage to cargo tank structure resulting in ship being unfit to trade
T33	Cargo transfer	Inadequate control, hardware failure, overflow or overpressure during transfer leads to an uncontrolled event due to inadequate information on vessel motion limits for all filling levels resulting in loss of containment and cargo release
T34	Cargo transfer	Cargo sloshing due to ship motions in the prevailing swell conditions due to large roll angles resulting in damage to cargo tank structure resulting in ship being unfit to trade
T35	Cargo transfer	Inadequate control during transfer leads to an uncontrolled event due to different roll periods resulting in loss of containment and/ or vapour release leading to ignition back to source resulting in fatalities/ injury/ hardware/ steelwork damage
T36	Cargo transfer	Cargo sloshing due to ship motions in the prevailing swell conditions due to passing ship effect resulting in damage to cargo tank structure resulting in ship being unfit to trade
T37	Cargo transfer	Inadequate control during transfer leads to an uncontrolled event due to inadequate equipment inspection, testing and maintenance resulting in loss of containment and/or vapour release leading to ignition back to source resulting in hardware/steelwork damage

(continued)

Table 8.9 (continued)

Risk ID number	STS phase	Description
T38	Cargo transfer	Inadequate control during transfer leads to an uncontrolled event due to inadequate equipment inspection, testing and maintenance resulting in loss of containment leading to vapour collection in a confined space, leading to ignition with explosion causing fatalities, injury and hull damage
T39	Cargo transfer	Inadequate control during transfer leads to an uncontrolled event due to damaged mooring due to chafing and cyclic loading resulting in loss of containment leading to vapour collection in a confined space, leading to ignition with explosion causing fatalities, injury and hull damage
T40	Cargo transfer	Inadequate control, hardware failure, overflow or overpressure during transfer leads to an uncontrolled event due to inadequate transfer of cargo equipment (type approval/ fit for purpose) resulting in loss of containment and cargo release
T41	Cargo transfer	Inadequate control, hardware failure, overflow or overpressure during transfer leads to an uncontrolled event due to level measurement and overfill protection systems inadequate for open water operation resulting in loss of containment and cargo release
T42	Cargo transfer	Inadequate control, hardware failure, overflow or overpressure during transfer leads to an uncontrolled event due to transfer rate too high resulting in loss of containment and cargo release
T43	Cargo transfer	Inadequate control, hardware failure, overflow or overpressure during transfer leads to an uncontrolled event due to defective overfill protection resulting in loss of containment and cargo release
T44	Cargo transfer	Inadequate control during transfer leads to an uncontrolled event due to Indequate contingency planning resulting in loss of containment leading to vapour collection in a confined space, leading to ignition with explosion causing fatalities, injury and hull damage

References

ArcelorMittal (2013) Enviromental appraisal of the transhipment of iron ore offshore from the port of Buchanan, Liberia

Ashari Alias M, Mohd Hashim SZ, Samsudin S (2008) Multi-criteria decision making and its applications: a literature review. J Teknologi Maklumat, Jilid 20, Bil.2, pp. 129–152

BP PLC (2014) Statistical Review of World Energy, 2014. Online available: http://www.bp.com/statisticalreview. Accessed 20 Dec 2016

Carbone TA, Tippett DD (2004) Project risk management using the project risk FMEA. Eng Manag J 16(4):28–35

Charnes A, Cooper W (1961) Management models and industrial applications of linear programming, vol 1. Wiley, New York

COWI (2011) Spill risk in the Baltic Sea from ship collisions with fixed objects, offshore platforms and offshore. Kongens Lyngby

D'Avignon GR, Vincke P (1988) An outranking method under uncertainty. Eur J Oper Res 36:311–321
Durbach IN (2014) Outranking under uncertainty using scenarios. Eur J Oper Res 232(1):98–108
Durbach I, Stewart T (2012) Modelling uncertainty in multi-criteria decision analysis. EJOR 223 (1):1–14
Elsayed T, Marghanya K, Abdulkadera S (2013) Risk assessment of liquefied natural gas carriers using fuzzy TOPSIS. Ships Offshore Struct 9(4):355–364
European Commission (EC) (2009) Technical support for European action to reducing GHG emissions, Annex I, Ship-to-Ship transfer. DG ENV.C3/ATA/2008/0016, Delf
Fan Z, Liu Y, Feng B (2010) A method for stochastic multiple criteria decision making based on pairwise comparisons of alternatives with random evaluations. Eur J Oper Res 207(2):906–915
Farquhar PH (1984) Utility assessment methods. Manag Sci 30(11):1283–1300
Figueira J, Greco S, Ehrgott M (2005) Multiple criteria decision analysis. State of the art surveys. Springer, New York
Fishburn P (1966) A note on recent developments in additive utility theories for multiple factors situations. Oper Res 14:1143–1148
Fishburn PC (1967) Methods for estimating additive utilities. Manag Sci 13(7):435–453
Ford (2004) Failure mode and effects analysis, FMEA handbook (with robustness linkages), USA
Gorstenko O, Tikhomolova V (2012) Shale gas: more complications for price forecast developers. Part II: LNG
Grigoroudis E, Siskos Y (2002) Preference disaggregation for measuring and analysing customer satisfaction: the MUSA method. Eur J Oper Res 143(1):148–170
Hurson C, Siskos Y (2014) A synergy of multicriteria techniques to assess additive value models. Eur J Oper Res 238:540–551
International Maritime Organization (IMO), MEPC, 186(59) (2009) Amendments to the annex of the protocol of 1978 relating to the international convention for the prevention of pollution from ships 1973
International Maritime Organization (IMO), MEPC (2010) Manual on oil pollution, section I, prevention of pollution. IMO Publication, London
Islei G, Lockett G, Naude P (1999) Judgemental modelling as an aid to scenario planning and analysis. Omega 27(1):61–73
Jacquet-Lagrèze E, Siskos J (1982) Assessing a set of additive utility functions for multicriteria decision-making: the UTA method. Eur J Oper Res 10:151–164
Jacquet-Lagreze E, Siskos Y (eds) (1983) Methode de decision multicritere. Hommes et Techniques, Paris
Jacquet-Lagrèze E, Siskos Y (2001) Preference disaggregation: 20 years of MCDA experience. Eur J Oper Res 130(2):233–245
Kadziński M, Greco S, Słowiński R (2012) Extreme ranking analysis in robust ordinal regression. Omega 40(4):488–450
Keeney R, Raiffa H (1976) Decisions with multiple objectives: preferences and value Tradeoffs. Wiley, New York
Keeney RL (1980) Sitting energy facilities. Academic Press, New York
Manas M, Nedoma J (1968) Finding all vertices of a convex polyhedron. Numer Math 12:226–229
Mardani A, Jusoh A, Zavadskas EK (2015) Fuzzy multiple criteria decision-making techniques and applications – two decades' review from 1994 to 2014. Expert Syst Appl 42:4126–4148. http://dx.doi.org/10.1016/j.eswa.2015.01.003
Marine Accident Investigation Branch. MAIB (2010) Report on the investigation of the collision between MTSaetta and MTConger on completion of a ship to ship transfer 9.5 miles south east of Southwold. UK
Marico Marine (2004) "Falmouth Harbour Commissioners. Ship-to-ship Oil Transfer Preliminary Risk Assessment." IJBAS-IJENS 11(5). Report number: 04RB383. Southampton
Martel J, D'Avignon G (1982) Project ordering with multicriteria analysis. Multicriteria analysis. Part II: applications. Eur J Oper Res 10(1982):56–69

National Academy Press (1998) Oil spill risks from tank vessel lightering. OSG, Washington, DC. 2009. Lightering-The back-story, Newsletter Article on http://www.osg.com/

Oil Companies International Marine Forum (OCIMF) (2013) Ship to ship transfer guide for petroleum, chemical and liquefied cases, First edition 2013, Scotland

OSG (2009) Lightering-The back-story, Newsletter Article on http://www.osg.com/

Østvik I, Grønstøl E (2005) HAZID for LNG tankers, design, operation, and regulation for safety. Project co-funded by the European Commission within the Sixths Framework Programme (2002–2006). Document ID. SAFEDOR-D-4.3.1–2005–11-29- LMG-HAZID LNG Tankers–rev-03

Schoemaker P (1995) Scenario planning: a tool for strategic thinking. Sloan Manag Rev 36(25):25

Siskos I (2008) Decision models, 1st edn Athens, ISBN: 9789606759109

Siskos E, Psarras J (2016) Bipolar robustness control methodology in disaggregation MCDA approaches: application to European e-government evaluation, 28th European Conference on Operational Research, Poznan, 3–6 July 2016

Siskos Y, Grigoroudis E, Matsatsinis FE (2005) UTA methods, multiple criteria decision analysis: state of the art surveys, Vol. 78 of the series international series in operations research & management science, pp 297–334

Skjong R, Loer K, Hamann R (2007) Hazid of tanker operations. SAFEDOR Project. No IP-516278. Document Id. SAFEDOR-D-04.07.01–2007–12-06-GL-Tanker HAZID– rev-1.2

Spenser C (2010) Personnel transfer using ship's cranes, J of Standard Safety, Charles Taylor & Co. Limited

Spyridakos A, Yannacopoulos D (2015) Incorporating collective functions to multicriteria disaggregation-aggregation approaches for small group decision making. Ann Oper ResResearch 227(1):119–136. doi:10.1007/s10479-014-1609-3

Stavrou D, Ventikos N (2014a) Ship-to-Ship transfer of crude oil and oil products: a risk assessment applying fuzzy TOPSIS. In: Proceedings of the 2nd International Symposium on Naval Architecture and Maritime, Istanbul, pp 711–722

Stavrou DI, Ventikos N (2014b) Ship to ship (STS) transfer of cargo operations: risk assessment applying a fuzzy inference system (FIS). Int J Risk Anal Crisis Response 4(4):214–227

Stavrou D, Ventikos N (2015b) Risk assessment for oil transshipment applying PFMEA and TOPSIS. In: Proceedings of the 5th International Symposium on Ship Operations, Management and Economics, Athens, pp 1–9

Stavrou D, Ventikos N (2015c) Multi-criteria decision theory in the marine environment: application to the selection of the STS transfer location. In: Proceedings of the 4th student Conference of the EEEE, NTUA, Athens, pp 30–37, ISBN: 978–618–80361-5-4

Stavrou D, Ventikos N, Siskos Y (2015) Risk assessment of ship-to-ship transfer operations based on stochastic criteria evaluation and intervally scaled importance weights, Special Issue "OR and New Technologies in Management and Engineering" of the International Journal of Decision Support Systems (IJDSS), (in press)

Stavrou D, Ventikos N (2015) Multi-criteria decision theory in the marine environment: application to the selection of the STS transfer location. In: Proceedings of the 4th student Conference of the EEEE, NTUA, Athens, pp 30–37, ISBN: 978–618–80361-5-4

Van de Panne C (1975) Methods for linear and quadratic programming. North-Holland Publishing Company, Amsterdam

Van der Heijden K (1996) Scenarios: the art of strategic conversation. Wiley, New York

Ventikos NP, Stavrou ID (2013) "Ship to ship (STS) transfer of cargo: latest developments and operational risk assessment" SPOUDAI. JEB 63(3):172–180

Waterborne Energy (2010) LNG industry now OK with ship-to-ship transfer [Online]. Available at: http://blog.chron.com/newswatchenergy/2010/02/lng-industry-now-ok-with-ship-to-shiptransfer/. Accessed 10 Nov 2016

Zadeh LA (1965) Fuzzy Sets. Inf Control 8:338–353

Zaras K, Martel JM (1994) Multi-attribute analysis based on stochastic dominance. In: Munier B, Machina MJ (eds) Models and experiments in risk and rationality. Kluwer Academic Publishers, Dordrecht, pp 225–248

Chapter 9
Financial Performance Evaluation of Shipping Companies Using Entropy and Grey Relation Analysis

Paul Tae-Woo Lee, Cheng-Wei Lin, and Sung-Ho Shin

Abstract The international trade of Korea and Taiwan has been heavily dependent upon international sea transportation owing to geo-political aspects. Therefore, the two countries have promoted ocean-going shipping industry in order to support their export-oriented economies. Recent financial crisis in together with the economic slowdown has reduced seaborne trade cargoes, which resulted in remarkably deteriorated revenues of the container shipping sector. Major container shipping companies of both countries such as Evergreen, Yang Ming, Hyundai, and Hanjin under our study are no exception. This chapter aims to achieve two-fold aims. The first applies Entropy to find the relative weights of financial ratios of the four companies each year. In so doing, we can find the weights variance for the period of 1999–2009 based on the financial performance of the above companies. The second aim is to evaluate financial performance of the companies in the period by grey relation analysis. The findings in this chapter help shipping managers to mitigate impacts of the financial crisis on their companies.

Keywords Entropy • Grey relation analysis • Financial ratio analysis • Container shipping

---

This chapter is a reprint of Lee TW, Lin CW, Shin SH (2012) A Comparative Study on Financial Positions of Shipping Companies in Taiwan and Korea Using Entropy and Grey Relation Analysis. *Expert Systems with Application*, 39(5): 5649–5657. The authors have slightly revised it. They would like to express their sincere thanks to the Elsevier Publisher for giving them its copyright permission for this book chapter.

P.T.-W. Lee (✉)
Ocean College, Zhejiang University, Zhoushan 316021, Zhejiang Province, China
e-mail: paultaewoo.lee@zju.edu.cn

C.-W. Lin
Department of Logistics and Shipping Management, Kainan University, Taoyuan City, Taiwan

S.-H. Shin
Department of Logistics and Maritime Studies, The Hong Kong Polytechnic University, Kowloon, Hong Kong, SAR, China

P.T.-W. Lee, Z. Yang (eds.), *Multi-Criteria Decision Making in Maritime Studies and Logistics*, International Series in Operations Research & Management Science 260,
DOI 10.1007/978-3-319-62338-2_9

1 Introduction

International sea transportation plays an important role for supporting transaction of global trade. Container shipping in particular has been central to the maritime logistics transportation and supply chain management in the globalized economy (Heaver 2002; Levinson 2006; Panayides 2006; Notteboom and Rodrigue 2008; Song and Lee 2009). But since the financial crisis in association with economic slow-down in the United States in 2008, container shipping companies are pulling more and more ships out of service, due to a lack of demand, and placing them at anchor indefinitely.[1] This trend was again confirmed by Drewry Publishing (2009) comparing the second quarter of 2008 and the second quarter of 2009, carrier revenue declined by the following estimated percentages: −31% in Transpacific route, −63% in Far East-North Europe-Far East route, and −38% in Transatlantic route, recording average −43% of the three main trades.

The international trade of South Korea and Taiwan has been dependent upon international sea transportation owing to geo-political aspects. Major container shipping companies of both countries such as Evergreen, Yang Ming, Hyundai, and Hanjin are no exception under this worldwide financial crisis impact. The water transportation sector of the two countries has high rates of backward and forward linkages so that there is a multi-dimensional influence on the national economy in a way, having tight inter-relationship between the production structure and the water transport sector (Chang et al. 2006). Having said that, we find a significant topic to conduct a comparative study on the financial positions of container shipping companies listed on stock markets in Korea and Taiwan. Therefore, this chapter focuses on exploring practical procedure of financial ratio analysis to identify the various features that need to reflect financial crisis by looking into financial statements. On the basis of such analysis, this chapter aims to achieve two-fold aims. The first applies Entropy to find the relative weights of financial ratios of the four companies each year. In so doing, we can find the weights variance for the period of 1999–2009 based on the financial performance of the four companies. The second is to evaluate financial performances of the four companies by grey relation analysis (GRA), which is introduced by Deng (1982). The findings in this chapter help shipping managers to mitigate impacts of the financial crisis on their companies.

[1]For example, as of November, 2009, prices were falling throughout the industry. In the spring, large commercial shipping companies like Maersk and Hapag-Lloyd were still charging about $2,000 (€1,600) to ship a container from Asia to Europe. Some companies collected only $500 (€400) for the same service. In the spring in 2008, it cost $30,000 (€24,000) a day to charter a ship containing 2,500 standard containers (TEU). As of 5th December 2008, that price dropped to less than $12,000 (€9,600). Alexander Jung, Thomas Schulz and Wieland Wagner (2009), "Shipping industry drowning in financial woes", Der Spiegel, August 14 2009.

2 Literature Review

A container shipping company is sensitive to capital structure, profitability, and cost reduction in turbulent times (Guzhva and Pagiavlas 2003; Chuang et al. 2008; Lee 1999b). Therefore, its managers focus on financial statements analysis of container lines, which is an indispensable tool for evaluating a company's financial position and providing its stakeholders with financial information. Our literature survey over the last two decades shows that several methodologies have been applied for evaluating financial performance in association with financial ratios. Multi-criteria decision making (MCDM) approach has played an important role in solving multi-dimensional and complicated problems arising from business and real life. It has been widely applied in a variety of forms such as the weighted sum model (WSM), the weighted product model (WPM), analytic hierarchy process (AHP), Elimination and Choice Translating Reality (in French acronym, ELECTRE) (Benayoun et al. 1966), and technique for order preference by similarity of ideal solution (TOPSIS), which is an alternative to the ELECTRE method (Triantaphyllou et al. 1998). Since Zadeh (1965) proposed the fuzzy sets theory, the methods for solving fuzzy multi-criteria decision making problems have been developed by the extension and modification of MCDM (Atanassov 1986; Chen and Tan 1994; Hus and Chen 1994, 1996 and 1997; Cheng 1998; Lee 1999a; Chen 2000; Opricovic and Tzeng 2004; Atanassov et al. 2005; Tzeng et al. 2005; Tzeng et al. 2007; Wang et al. 2003; Shih 2008; Chou and Chang 2008; Zhang et al. 2005; Tung and Lee 2009).

In the process of these developments, the MCDM methods were applied to the financial performance evaluation of financial ratios in the transportation sector includes data envelopment analysis (DEA) (e.g. Bowlin 2004; Capobianco and Fernandes 2004; Halkos and Salamouris 2004; Scheraga 2004; Lin et al. 2005; Liang et al. 2006), fuzzy multi-criteria decision making (FMCDM) (Wang and Lee 2010), fuzzy multi-criteria group decision making (FMCGDM) (Wang 2008), TOPSIS (Feng and Wang 2000; Wang et al. (2003); Wang et al. 2004; Wang and Lee 2007), grey principal component analysis (Tung and Lee 2009), entropy and grey relation analysis (Lee et al. 2012), entropy and consistent fuzzy preference relation (CFPR) (Lee et al. 2014), fuzzy AHP (FAHP) and fuzzy axiomatic design (FAD) algorithms (Celik et al. 2009), clustering method with fuzzy relation (Wang and Lee 2008),Value at Risk (VaR) method and modified Sharp ratio (Chuang et al. 2008), and Tobit analysis (Scheraga 2004).

Feng and Wang (2000) applied TOPSIS method to evaluate the performances indicators consisting of marketing, productivity, and execution, and financial aspects of five domestic airlines in Taiwan and then to draw a ranking of the five companies. They also employed the grey relation analysis (GRA) to select the representative of performance indicators and to overcome the problems of small sample size and unknown distribution of the samples. After this process, in implementing TOPSIS method, developed by Hwang and Yoon (1981), to calculate the performance score and ranking, they took five steps; normalization of indicator values; determination of ideal and worst solution; calculation of the separation measure; calculation of the relative closeness to the ideal solution; and ranking of the preference order according to the descending order of the ideal solution. Wang et al. (2004) again applied TOPSIS method to evaluate the performance indicators of the highway bus companies in Taiwan, using mostly the same conceptual model, calculation process as their previous study (Feng and Wang 2000).

Wang (2009) evaluated financial performance of Taiwan container lines by combining GRA with FMCGDM. In the evaluating process, GRA was employed to partition financial ratios of three Taiwan container shipping lines for five periods equivalent two and half years and to find representative indices from the clusters and then FMCGDM method combining experts' opinions is applied to evaluate the financial performance of Taiwan container lines based on his previous work (Wang and Lee 2007). Linguistic weights of 15 criteria were given by four experts. In their paper 15 clusters are drawn with representative indices by arguing FMCGDM is workable even under small number of samples. Wang and Lee (2010) proposed FMCDM to evaluate financial performance of container shipping with utilization of GRA to cluster financial ratios and find representative indicators. Their paper, like Wang's papers (2008, 2009), employed 21 financial ratios of three Taiwan container shipping lines for five periods, from the 3rd season of 2003 to the 3rd season of 2004. Linguistic weights of 15 criteria are given by four financial experts. They claimed that FMCDM method based on extended fuzzy preference relation with strength and weakness indices can evaluate financial performance, by aggregating both indices into total indices and the three companies can be ranked easily.

Wang with his colleagues applied fuzzy TOPSIS, FMCDM, FMCGDM, and GRA to evaluate financial performance of airlines and container shipping companies in Taiwan. It seems that the above are based on the same conceptual model, utilizing GRA as the distance index in order to cluster the related financial ratios. (Wang and Lee 2008) partitioned the financial ratios and identify the representative indicators seem the main purpose of these researches, using clustering analysis, such as fuzzy relations and K-means. Grouping the financial ratios to find the representative indicators to evaluate the shipping companies will lose information of the data investigated, unlike entropy approach, which will be applied in our research. Wang's papers (2008 and 2009) employed a short data period with only five seasons equivalent to two and half years for 21 ~ 23 financial ratios as well as small number of experts (only four) even without a brief description of their characteristics. As long as the main aims of their works is, among others, to test applicability and feasibilities of several methods applied to evaluation of financial ratios of airport, airline and container shipping companies, their contribution should be recognized. But their data period is too short to draw more generalized conclusion from their analysis. Furthermore, neither implication nor discussion on the ranking result has been made in their papers. Our chapter contributes, among others, to filling such data gap with longer data period and more data available.

Although identifying the relative weights of the evaluation criteria is a very important part of evaluation problem, the researches mentioned above do not emphasize the weights identification. Ma et al. (1999) and Xu (2004) pointed out that the weight identification methods can be divided into subjective and objective approaches. The former, such as Analytic Hierarchy Process (AHP) introduced by Saaty (1980), always needs to collect the subjective preferences of the decision makers. The greater the number of evaluating objects, the more difficult the evaluation work becomes. The weights identified by this method are volatile due to the unreliability of the judgment of decision makers; in addition, the weights have nothing to do with the information given (Jessop 2004). The objective approach, such as Entropy introduced by Shannon (1948), is a well-known method to identify the weights of evaluation criteria. It does not need to collect the subjective perceptions of decision makers, but collect the performance of the evaluation objects, e.g. criteria. The weights identified by Entropy are the measurement of the disorder degree of the evaluation system. Useful information provided by the system can be measured in the form of weights (Zou et al. 2006). Of course, two types of weights identification method, objective and subjective approach, are practical when dealing with the evaluation problem. Considering the different characteristics of evaluation system, the task to choose an adequate approach is the researcher's own responsibility.

Based on our arguments above and our literature review, we have found that most of these researches utilize the experts' subjective opinions as the relative weights of financial ratios and combine them with GRA to rank the companies positions. In this chapter, a new concept is introduced to analyze impact of recent economic slowdown on shipping companies. Relative weights of the ratios are identified by an objective approach, Entropy, which deals with the input data of real values of the financial ratios of the four shipping companies in Korea and Taiwan in the period of 1999–2009, because it will not lose any information of the financial ratios data. GRA is then combined with the objective weights to find the trend of ranking of the four shipping companies discussed in this study.

3 Methodology and Calculation Example

According to the above literature review, we can make sure that the Entropy and GRA are workable when dealing with the evaluation problem of shipping companies. Having briefly reviewed GRA and entropy in this section, we propose their operation procedures of Entropy and GRA to evaluate the financial ratios of the four shipping companies, Evergreen (EG), Yang Ming (YM), Hanjin (HJ), and Hyundai Merchant Marine (HMM), under our study and to rank their financial performance.

3.1 Entropy

Step 1. Data: Performance/Evaluation Matrix.

First, input data of entropy should be collected in the form of Eq. (9.1), which means the performance of each alternative under considering of each evaluation criteria.

$$\begin{array}{c c} & \begin{array}{ccccc} C_1 & \cdots & C_j & \cdots & C_n \end{array} \\ \begin{array}{c} a_1 \\ \vdots \\ a_i \\ \vdots \\ a_m \end{array} & \begin{bmatrix} x_{11} & \cdots & x_{1j} & \cdots & x_{1n} \\ \vdots & & \vdots & & \vdots \\ x_{i1} & \cdots & x_{ij} & \cdots & x_{in} \\ \vdots & & \vdots & & \vdots \\ x_{m1} & \cdots & x_{mj} & \cdots & x_{mn} \end{bmatrix} \end{array} \tag{9.1}$$

Step 2. Find p_{ij}.

This step means normalizing the matrix above.

$$[p_{ij}]_{m\times n} = \left[x_{ij}/\sum_{i=1}^{m} x_{ij}\right]_{m\times n}$$

Step 3. Entropy *for all Criteria.*

Calculate the entropy of all criteria under Eq. (9.2).

$$e_j = -k\sum_{i=1}^{m} p_{ij}\ln p_{ij} \ \ \forall j \tag{9.2}$$

where k is Boltzman's constant, which equals $k = 1/\ln m$ which guarantees that $0 \leq e_j \leq 1$

Step 4. The degree of diversification $\bar{e}_j$ of the information provided by the alternative performance/evaluation value of criteria j can be defined as

$$\bar{e}_j = 1 - e_j \tag{9.3}$$

Step 5. Normalization for criteria

Make the relative weights within the range (0, 1) to satisfy the limitation of GRA, the value obtained in step 4 should be normalized by Eq. (9.4).

$$r_j = \bar{e}_j/\sum_{j=1}^{n} \bar{e}_j \ \ \forall j \tag{9.4}$$

where $\sum_{j=1}^{n} r_j = 1$

3.2 Grey Relation Analysis

Table 9.1 shows the data to be collected and investigated in advanced, including alternatives(x_i), evaluation criteria(c_j), relative weights(w_j), and alternative performance under each criteria ($x_{ij}(j)$) which should be normalized before in advance. After that, the aspired (x^*) and the worst (x^-) values of alternatives can be identified by the performance matrix, where $x^*(j) = \max_i (x_{ij}(j))$ and $x^-(j) = \min_i (x_{ij}(j))$.

On the basis of data collection and calculation of the aspired value and the worst value as shown in Table 9.1, related analysis by GRA can be accomplished by the following steps:

Step 1. Coefficients of grey relation for aspired values

$$|\,\gamma(x^*(j), x_i(j)) = \frac{\min_i \min_j |\, x^*(j) - x_i(j) \,| + \varsigma \max_i \max_j |\, x^*(j) - x_i(j) \,|}{|\, x^*(j) - x_i(j) \,| + \varsigma \max_i \max_j |\, x^*(j) - x_i(j) \,|}$$

Grade (degree) of grey relation (larger is better)

$$\gamma(x^*, x_i) = \sum_{j=1}^{n} w_j \gamma(x^*(j), x_i(j)) \tag{9.5}$$

where the weight w_j can be obtained by the objective or subjective approach, such as Entropy or AHP.

Step 2. Coefficients of grey relation for the worst values

$$\gamma(x^-(j), x_i(j)) = \frac{\min_i \min_j |\, x^-(j) - x_i(j) \,| + \varsigma \max_i \max_j |\, x^-(j) - x_i(j) \,|}{|\, x^-(j) - x_i(j) \,| + \varsigma \max_i \max_j |\, x^-(j) - x_i(j) \,|}$$

Grade (degree) of grey relation (larger is worse, the small is better)

Table 9.1 Data matrix of grey relation analysis

	Criteria				
	c_1	$\cdots$	c_j	$\cdots$	c_n
Alternatives	w_1	$\cdots$	w_j	$\cdots$	w_n
x_1	$x_1(1)$	$\cdots$	$x_1(j)$	$\cdots$	$x_1(n)$
$\vdots$	$\vdots$		$\vdots$		$\vdots$
x_i	$x_i(1)$	$\cdots$	$x_i(j)$	$\cdots$	$x_i(n)$
$\vdots$	$\vdots$		$\vdots$		$\vdots$
x_m	$x_m(1)$	$\cdots$	$x_m(j)$	$\cdots$	$x_m(n)$
Aspired value x^*	$x^*(1)$	$\cdots$	$x^*(j)$	$\cdots$	$x^*(n)$
The worst value x^-	$x^-(1)$	$\cdots$	$x^-(j)$	$\cdots$	$x^-(n)$

[Note] Data matrix: normalization

$$\gamma(x^{-}, x_i) = \sum_{j=1}^{n} w_j \gamma(x^{-}(j), x_i(j)) \tag{9.6}$$

Step 3. Relative grey relation scores.

Combine above Eq. (9.1) and (9.2) for ranking based on the relative grey relation of aspired and the worst values.

$$R_i = \frac{\gamma(x^{*}, x_i)}{\gamma(x^{-}, x_i)} \tag{9.7}$$

3.3 Analysis Concept Diagrams

Entropy and GRA as MCDM methods are applied in this chapter. In the classic MCDM process, criteria weights and alternatives performances are the major inputs for evaluation. The concept of this procedure is shown in Fig. 9.1.

In the classic MCDM procedure, researchers should investigate the criteria weights and alternatives performances. Criteria weights can be identified by subjective or objective weighting methods, such as AHP and entropy. Subjective weights can be identified by related methods (e.g., AHP, ANP) from experts' pairwise comparison results; on the other hand, objective weights can be derived from alternative performance by entropy. Then, evaluation methods combine these weights and performances to find the ranks of alternatives.

This chapter tries to apply entropy and GRA to find the trends of the rankings of four shipping companies in Taiwan and Korea in the period of 1999–2009. The revised operation procedure has been revised as shown in Fig. 9.2 where entropy is applied to find the objective weights of each financial ratio. GRA combines the objective weights of financial ratios with the performance of the container shipping companies to find the ranks of four companies.

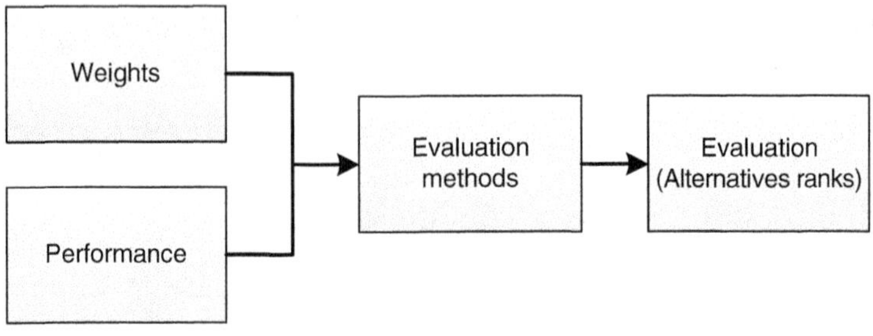

Fig. 9.1 Classic MCDM procedure

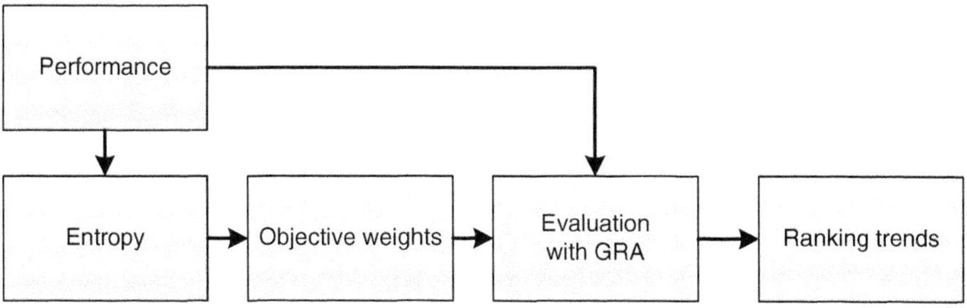

Fig. 9.2 Revised MCDM operation procedure

3.4 Data

The samples under our study consist of four container companies which belong to top 20 in the world: Evergreen Marine Corporations and Yang Ming Lines in Taiwan and Hanjin Shipping Company Ltd. and Hyundai Merchant Marine Company Ltd. in Korea. Related information can be found in Table 9.2.

At the initial stage of our study, total 45 financial ratios were drawn from their balance sheets, income statements, and cash flow statements for the period of 1999–2009. The financial ratios are categorized into six: (i) liquidity, (ii) profitability, (iii) return on investment, (iv) activity efficiency (turnover), (v) financial leverage, and (vi) cash flow. Considering our research purpose and efficient design of questionnaires, we reduced them by 25 ratios, of which four ratios, i.e., fixed asset to long-term liabilities (4), operating profit margin (7), operation cost ratio (10), and operating cash flow ratio (23) were excluded from our analysis because they are very closely duplicated with others such as (5), (6), (8), and (24) considering their formula and implications. Consequently, 21 ratios are used for our empirical test in this section. (see Table 9.3.)

3.5 Calculation Example

A calculation example for the revised MCDM operation procedure is demonstrated in the following sections, taking the 1999 data.

3.5.1 Entropy

The calculation process of entropy is described as follows:

Step 1. Data: Performance/evaluation matrix.

The purpose of this step is data collection, which is shown in Table 9.4. The numbers in this table is the financial performance of container shipping

Table 9.2 Selected top-20 container shipping companies in the world

		Total		Owned		Chartered			Orderbook		
Rank	Operator	TEU	Ships	TEU	Ships	TEU	Ships	% Chart	TEU	Ships	% existing
1	APM-Maersk	2,059,526	546	1,137,574	211	921,952	335	44.8%	411,090	67	20.0%
2	Mediterranean Shg. Co.	1,546,364	400	834,845	201	711,519	199	46.0%	586,360	49	37.9%
3	CMA CGM Group	1,040,161	360	354,924	87	685,237	273	65.9%	456,183	54	43.9%
4	**Evergreen Line**	**554,316**	**149**	**319,263**	**87**	**235,053**	**62**	**42.4%**			
5	APL	553,516	141	165,939	44	387,577	97	70.0%	128,362	16	23.2%
6	Hapag-Lloyd	515,276	124	269,685	58	245,591	66	47.7%	87,500	10	17.0%
7	COSCO Container L.	452,828	132	270,751	91	182,077	41	40.2%	409,826	53	90.5%
9	**Hanjin Shipping**	**427,878**	**96**	**95,488**	**18**	**332,390**	**78**	**77.7%**	**257,698**	**27**	**60.2%**
15	Zim	308,836	94	155,730	35	153,106	59	49.6%	195,386	20	63.3%
16	**Yang Ming Line**	**308,664**	**76**	**187,201**	**45**	**121,463**	**31**	**39.4%**	**141,402**	**22**	**45.8%**
18	**Hyundai M.M.**	**276,332**	**53**	**79,053**	**13**	**197,279**	**40**	**71.4%**	**71,810**	**6**	**26.0%**
20	PIL (Pacific Int. Line)	195,546	110	128,748	79	66,798	31	34.2%	60,824	14	31.1%

Source: Alphaliner

Table 9.3 Financial ratios with formula by group

Group	Value no.	Financial ratio	Formula
Liquidity	1	Current ratio	Current assets / Current liabilities
	2	Times interest earned ratio	EBIT / Interest
	3	Equity ratio	Shareholder's equity / Total assets
	4	***Fixed asset to long-term liabilities***	***Fixed asset / Long-term debt***
	5	Fixed assets to long-term fund ratio	Fixed assets / Long-term fund
Profitability	6	Gross profit margin	Gross income (sales - cost of goods) / Sales
	7	***Operating profit margin***	***Operating income / Sales***
	8	Net profit margin	Net income / Sales
	9	Earnings before tax ratio (EBT)	EBT / Operation revenue
	10	***Operation cost ratio***	***Sales cost/ Sales***
Return on investment	11	Return on long-term capital	Net income / Long-term fund
	12	Return on equity (ROE)	Net income / Shareholder's equity
	13	Return on total assets (ROA)	Net income / Total assets
Activity efficiency	14	Total asset turnover	Sales / Total assets
	15	Fixed asset turnover	Sales / Net fixed assets
	16	Stockholder's equity turnover	Sales / Shareholder's equity
	17	Total liabilities turnover	Sales / Total debt
	18	Working capital turnover	Sales / Operating capital (current assets - current liabilities)
Financial leverage	19	Debt ratio	Total debt / Total assets
	20	Debt to equity ratio	Total debt / Shareholder's equity
	21	Long-term debt to equity ratio	Long-term debt / Shareholder's equity
	22	Long-term debt to long-term capital ratio	Long-term debt / Long-term fund
Cash flow	***23***	***Operating cash flow ratio***	***Cash flow from operation / Current liabilities***
	24	Cash flow to net income ratio	Cash flow from operations / Net income
	25	Cash flow adequacy ratio	Cash flow from operation / Long-term debt

Source: Compiled by the authors
Note: The four ratios (4), (7), (10), and (23) in bold type were excluded from our analysis.

Table 9.4 Performance of container shipping companies in financial ratios

	Step1- performance	YM	EG	HMM	HJ	Sum
F1	Current ratio (%)	120.30	110.99	55.24	98.26	384.79
F2	Times interest coverage ratio	4.68	1.72	0.11	1.33	7.83
F6	Gross profit margin	5.28	20.04	10.73	5.10	41.14
F8	Net profit margin	10.71	11.27	1.02	7.41	30.40
F9	Income before tax ratio (EBT) (%)	13.16	13.62	1.17	10.31	38.27
F11	Return on long-term capital	14.43	10.91	1.77	9.43	36.54
F12	Return on equity (ROE)	54.18	49.93	1.16	70.46	175.73
F13	Return on total assets (ROA)	5.22	3.42	0.41	2.81	11.87
F14	Total Asser Turnover	1.17	0.36	0.70	0.82	3.06
F15	Fixed Asset Turnover	2.62	1.10	1.14	1.23	6.10
F16	Stockholder's equity turnover	1.85	0.68	7.66	5.82	16.01
F17	Total liabilities turnover	3.16	0.78	0.78	0.96	5.67
F18	Working Capital Turnover	251.40	237.25	216.23	1.24	706.12
F24	Cash flow to net income ratio	330.74	624.87	1.12	1035.17	1991.90
F25	Cash flow adequacy ratio	71.14	32.07	4.54	5.15	112.90
F19	Total debt to asset ratio (Debt ratio) (%)	36.99	46.67	90.81	85.88	260.36
F3	Equity to total assets ratio (Equity ratio)	0.63	0.53	0.09	0.14	1.40
F5	Fixed Assets to Long-term Fund Ratio	0.56	0.41	1.06	0.85	2.87
F20	Debt to Equity Ratio	0.59	0.88	988.50	6.08	996.04
F21	Long-term debt to Equity ratio	0.27	0.52	5.34	4.56	10.68
F22	Long-term debt to Long-term Capital ratio	0.21	0.34	0.84	0.82	2.21

Notes: *EG* Evergreen Shipping Company, *YM* Yang Ming Shipping Company, *HJ* Hanjin Shipping Company, and *HMM* Hyundai Merchant Marine Company

companies investigated in their official websites. The last column is the sum of each financial ratio performance, which will be used in the following step.

Step 2. Find p_{ij}.
This step normalizes the financial ratio performance using the sum values calculated in Table 9.4. Performances of normalized financial ratios are illustrated in Table 9.5.

Step 3 ~ *Step 5*. Entropy calculation of all criteria.
The results of entropy calculations are integrated in Table 9.6, including Step 3 to Step 5. The equations for calculation can be found in Eq. (9.2) to (9.4) (see Sect. 3.1 in this chapter). The results of step 5, *rj*, represent the objective weights of financial ratios.

Table 9.5 Normalized financial ratio performance

	Step 2-pij	YM	EG	HMM	HJ
F1	Current ratio (%)	0.313	0.288	0.144	0.255
F2	Times interest coverage ratio	0.597	0.219	0.014	0.170
F6	Gross profit margin	0.128	0.487	0.261	0.124
F8	Net profit margin	0.352	0.371	0.033	0.244
F9	Income before tax ratio (EBT) (%)	0.344	0.356	0.031	0.270
F11	Return on long-term capital	0.395	0.299	0.048	0.258
F12	Return on equity (ROE)	0.308	0.284	0.007	0.401
F13	Return on total assets (ROA)	0.440	0.288	0.035	0.237
F14	Total Asser Turnover	0.382	0.119	0.230	0.269
F15	Fixed Asset Turnover	0.429	0.181	0.187	0.202
F16	Stockholder's equity turnover	0.116	0.042	0.478	0.363
F17	Total liabilities turnover	0.557	0.137	0.137	0.169
F18	Working Capital Turnover	0.356	0.336	0.306	0.002
F24	Cash flow to net income ratio	0.166	0.314	0.001	0.520
F25	Cash flow adequacy ratio	0.630	0.284	0.040	0.046
F19	Total debt to asset ratio (Debt ratio) (%)	0.142	0.179	0.349	0.330
F3	Equity to total assets ratio	0.451	0.382	0.066	0.101
F5	Fixed Assets to Long-term Fund Ratio	0.195	0.141	0.368	0.296
F20	Debt to Equity Ratio	0.001	0.001	0.992	0.006
F21	Long-term debt to Equity ratio	0.025	0.048	0.500	0.427
F22	Long-term debt to Long-term Capital ratio	0.095	0.154	0.380	0.370

3.5.2 Grey Relation Analysis

Grey relation analysis (GRA) as an evaluation method in MCDM is applied to rank the alternatives in evaluation problems by their performances. In other words, the inputs of GRA in this chapter are the financial performances of container shipping companies, as shown in Table 9.7 which is the same as Table 9.4. The min and max values are the biggest and smallest financial ratios performances of the companies, which will be used for calculating the coefficients of grey relation. The last column in Table 9.7 shows the characteristics of financial ratios, which are categorized as larger-the-better (benefit), smaller-the-better (cost) or nominal-the-better (objective bound, OB).

On the basis of data collection and calculation of aspired and the worst values as shown in Table 9.8, the calculations follow the Step 1 to Step 3 which can also be found in Eq. (9.5) to (9.7) (see Sect. 3.2 in this chapter). The last column in Table 9.8 shows the objective weights identified by entropy.

The GRA calculations shown in the bottom row in Table 9.8 shows the ranks of four container shipping companies according to their performances and objective weights in 1999. Repeating the same steps above, we can find the ranks of container

Table 9.6 Entropy calculation for all criteria

	Steps	Step 3- lnpij					Step 4		Step 5
	Financial ratios	YM	EG	HMM	HJ	Σpij*lnpij	eij	1-eij	rj
F1	Current ratio (%)	−1.1627	−1.2433	−1.9410	−1.3651	−1.3494	0.9734	0.0266	0.0065
F2	Times interest coverage ratio	−0.5153	−1.5164	−4.2940	−1.7748	−1.0001	0.7214	0.2786	0.0680
F6	Gross profit margin	−2.0540	−0.7194	−1.3436	−2.0887	−1.2230	0.8822	0.1178	0.0288
F8	Net profit margin	−1.0435	−0.9927	−3.3973	−1.4116	−1.1932	0.8607	0.1393	0.0340
F9	Income before tax ratio (EBT) (%)	−1.0673	−1.0332	−3.4836	−1.3111	−1.1951	0.8621	0.1379	0.0337
F11	Return on long-term capital	−0.9293	−1.2085	−3.0278	−1.3544	−1.2240	0.8829	0.1171	0.0286
F12	Return on equity (ROE)	−1.1767	−1.2582	−5.0170	−0.9140	−1.1200	0.8079	0.1921	0.0469
F13	Return on total assets (ROA)	−0.8205	−1.2438	−3.3545	−1.4414	−1.1780	0.8497	0.1503	0.0367
F14	Total Asser Turnover	−0.9614	−2.1324	−1.4682	−1.3139	−1.3117	0.9462	0.0538	0.0131
F15	Fixed Asset Turnover	−0.8452	−1.7084	−1.6747	−1.5994	−1.3093	0.9445	0.0555	0.0136
F16	Stockholder's equity turnover	−2.1558	−3.1601	−0.7372	−1.0125	−1.1043	0.7966	0.2034	0.0497
F17	Total liabilities turnover	−0.5844	−1.9878	−1.9893	−1.7792	−1.1705	0.8443	0.1557	0.0380
F18	Working Capital Turnover	−1.0327	−1.0907	−1.1834	−6.3482	−1.1076	0.7990	0.2010	0.0491
F24	Cash flow to net income ratio	−1.7955	−1.1593	−7.4836	−0.6545	−1.0062	0.7258	0.2742	0.0670
F25	Cash flow adequacy ratio	−0.4618	−1.2585	−3.2136	−3.0880	−0.9185	0.6626	0.3374	0.0824
F19	Total debt to asset ratio (Debt ratio, %)	−1.9514	−1.7190	−1.0533	−1.1091	−1.3186	0.9512	0.0488	0.0119
F3	Equity to total assets ratio	−0.7958	−0.9626	−2.7213	−2.2917	−1.1374	0.8205	0.1795	0.0438
F5	Fixed Assets to Long-term Fund Ratio	−1.6373	−1.9572	−0.9988	−1.2177	−1.3231	0.9544	0.0456	0.0111
F20	Debt to Equity Ratio	−7.4364	−7.0372	−0.0076	−5.0982	−0.0493	0.0355	0.9645	0.2355
F21	Long-term debt to Equity ratio	−3.6868	−3.0303	−0.6933	−0.8517	−0.9487	0.6844	0.3156	0.0771
F22	Long-term debt to Long-term Capital ratio	−2.3502	−1.8726	−0.9663	−0.9930	−1.2475	0.8999	0.1001	0.0245

Table 9.7 Performance of the four container shipping companies

	YM	EG	HMM	HJ	Min.	Max.	category
F1	120.300	110.986	55.242	98.258	55.242	120.300	benefit
F2	4.679	1.719	0.107	1.328	0.107	4.679	benefit
F6	5.275	20.037	10.733	5.095	5.095	20.037	benefit
F8	10.708	11.266	1.017	7.411	1.017	11.266	benefit
F9	13.161	13.618	1.175	10.314	1.175	13.618	benefit
F11	14.426	10.912	1.769	9.430	1.769	14.426	benefit
F12	54.177	49.934	1.164	70.455	1.164	70.455	benefit
F13	5.225	3.422	0.415	2.808	0.415	5.225	benefit
F14	1.168	0.362	0.704	0.821	0.362	1.168	benefit
F15	2.620	1.105	1.143	1.232	1.105	2.620	benefit
F16	1.854	0.679	7.662	5.818	0.679	7.662	benefit
F17	3.159	0.776	0.775	0.956	0.775	3.159	benefit
F18	251.403	237.250	216.235	1.236	1.236	251.403	benefit
F24	330.743	624.870	1.120	1035.168	1.120	1035.168	benefit
F25	71.139	32.071	4.540	5.147	4.540	71.139	benefit
F19	36.991	46.670	90.813	85.883	36.991	90.813	cost
F3	0.630	0.533	0.092	0.141	0.092	0.630	OB
F5	0.558	0.406	1.058	0.850	0.406	1.058	OB
F20	0.587	0.875	988.497	6.084	0.587	988.497	OB
F21	0.268	0.516	5.339	4.557	0.268	5.339	OB
F22	0.211	0.340	0.842	0.820	0.211	0.842	OB

shipping companies from 2000 to 2009. The ranking trends of these companies are shown in Fig. 9.8.

4 Results

According to our proposed steps described in the previous section, the relative weights of financial ratios are identified by entropy as an objective approach which measures the degree of diversification of the financial ratios data and then the four companies is to be ranked by GRA.

4.1 Weight Variation Trend

The relative weights of financial ratios are identified by entropy as an objective approach. Table 9.9 shows relative objective weights of the 21 financial ratios of the four shipping companies every year, average value of objective weight of each financial ratio, and coefficient variance (CV). Average values range from 0.009 to

Table 9.8 Coefficients of grey relation for aspired and worse level

	Normalization								Step 1-γ(x*,xi)				Step 2-γ(x-,xi)				
	YM	EG	HMM	HJ	YM	EG	HMM	HJ	YM	EG	HMM	HJ	YM	EG	HMM	HJ	weights
F1	1.000	0.857	0.000	0.661	0.000	0.143	1.000	0.339	1.000	0.777	0.333	0.596	0.333	0.369	1.000	0.431	0.007
F2	1.000	0.353	0.000	0.267	0.000	0.647	1.000	0.733	1.000	0.436	0.333	0.406	0.333	0.586	1.000	0.652	0.068
F6	0.012	1.000	0.377	0.000	0.988	0.000	0.623	1.000	0.336	1.000	0.445	0.333	0.976	0.333	0.570	1.000	0.029
F8	0.946	1.000	0.000	0.624	0.054	0.000	1.000	0.376	0.902	1.000	0.333	0.571	0.346	0.333	1.000	0.445	0.034
F9	0.963	1.000	0.000	0.735	0.037	0.000	1.000	0.265	0.932	1.000	0.333	0.653	0.342	0.333	1.000	0.405	0.034
F11	1.000	0.722	0.000	0.605	0.000	0.278	1.000	0.395	1.000	0.643	0.333	0.559	0.333	0.409	1.000	0.452	0.029
F12	0.765	0.704	0.000	1.000	0.235	0.296	1.000	0.000	0.680	0.628	0.333	1.000	0.395	0.415	1.000	0.333	0.047
F13	1.000	0.625	0.000	0.498	0.000	0.375	1.000	0.502	1.000	0.572	0.333	0.499	0.333	0.444	1.000	0.501	0.037
F14	1.000	0.000	0.424	0.569	0.000	1.000	0.576	0.431	1.000	0.333	0.465	0.537	0.333	1.000	0.541	0.468	0.013
F15	1.000	0.000	0.025	0.084	0.000	1.000	0.975	0.916	1.000	0.333	0.339	0.353	0.333	1.000	0.952	0.856	0.014
F16	0.168	0.000	1.000	0.736	0.832	1.000	0.000	0.264	0.375	0.333	1.000	0.654	0.748	1.000	0.333	0.405	0.050
F17	1.000	0.000	0.000	0.076	0.000	1.000	1.000	0.924	1.000	0.333	0.333	0.351	0.333	0.999	1.000	0.868	0.038
F18	1.000	0.943	0.859	0.000	0.000	0.057	0.141	1.000	1.000	0.898	0.781	0.333	0.333	0.346	0.368	1.000	0.049
F24	0.319	0.603	0.000	1.000	0.681	0.397	1.000	0.000	0.423	0.558	0.333	1.000	0.611	0.453	1.000	0.333	0.067
F25	1.000	0.413	0.000	0.009	0.000	0.587	1.000	0.991	1.000	0.460	0.333	0.335	0.333	0.547	1.000	0.982	0.082
F19	1.000	0.820	0.000	0.092	0.000	0.180	1.000	0.908	1.000	0.735	0.333	0.355	0.333	0.379	1.000	0.845	0.012
F3	0.000	0.000	0.001	0.000	0.001	0.000	0.000	0.000	0.999	0.999	1.000	1.000	1.000	1.000	0.999	0.999	0.044
F5	0.001	0.001	0.000	0.000	0.000	0.000	0.001	0.000	1.000	1.000	0.999	0.999	0.999	0.999	1.000	1.000	0.011
F20	1.000	1.000	0.000	0.994	0.000	0.000	1.000	0.006	1.000	0.999	0.333	0.989	0.333	0.333	1.000	0.335	0.236
F21	0.005	0.005	0.000	0.001	0.000	0.000	0.005	0.004	1.000	0.999	0.990	0.991	0.990	0.990	1.000	0.998	0.077
F22	0.001	0.001	0.000	0.000	0.000	0.000	0.001	0.001	1.000	1.000	0.999	0.999	0.999	0.999	1.000	1.000	0.024
									0.8906	0.7581	0.4969	0.7253	0.498	0.568	0.917	0.620	
											Step 3-	R	1.788	1.336	0.542	1.170	
												Rank	1	2	4	3	

Table 9.9 Objective weights of financial ratios of the four shipping companies

Weights	F1	F2	F3	F5	F6	F8	F9	F11	F12	F13	F14	F15	F16	F17	F18	F19	F20	F21	F22	F24	F25
1999	0.007	0.068	0.044	0.011	0.029	0.034	0.034	0.029	0.047	0.037	0.013	0.014	0.050	0.038	0.049	0.012	0.236	0.077	0.024	0.067	0.082
2000	0.017	0.065	0.040	0.017	0.029	0.036	0.034	0.027	0.080	0.037	0.015	0.022	0.051	0.029	0.059	0.010	0.240	0.074	0.020	0.066	0.033
2001	0.011	0.063	0.047	0.022	0.061	0.045	0.049	0.032	0.049	0.024	0.013	0.007	0.068	0.018	0.050	0.011	0.225	0.075	0.021	0.066	0.043
2002	0.011	0.008	0.027	0.021	0.049	0.046	0.029	0.042	0.060	0.020	0.013	0.001	0.052	0.006	0.060	0.006	0.278	0.064	0.014	0.135	0.055
2003	0.050	0.056	0.030	0.040	0.018	0.034	0.046	0.039	0.090	0.022	0.009	0.022	0.045	0.012	0.049	0.008	0.247	0.062	0.017	0.053	0.050
2004	0.059	0.042	0.018	0.064	0.007	0.031	0.021	0.002	0.024	0.000	0.015	0.024	0.053	0.009	0.086	0.009	0.320	0.064	0.017	0.074	0.059
2005	0.020	0.071	0.012	0.060	0.024	0.040	0.043	0.005	0.006	0.001	0.012	0.039	0.031	0.006	0.064	0.012	0.306	0.070	0.034	0.012	0.134
2006	0.011	0.017	0.005	0.033	0.058	0.054	0.054	0.077	0.041	0.061	0.011	0.027	0.018	0.006	0.046	0.005	0.228	0.036	0.019	0.114	0.082
2007	0.005	0.134	0.007	0.030	0.022	0.103	0.105	0.030	0.008	0.012	0.017	0.030	0.024	0.010	0.047	0.010	0.217	0.051	0.031	0.016	0.090
2008	0.001	0.007	0.006	0.040	0.034	0.055	0.044	0.077	0.097	0.072	0.023	0.053	0.029	0.019	0.050	0.006	0.235	0.031	0.013	0.067	0.039
2009	0.004	0.043	0.014	0.033	0.044	0.047	0.042	0.048	0.055	0.037	0.052	0.069	0.062	0.041	0.047	0.007	0.217	0.029	0.011	0.051	0.046
average	0.018	0.052	0.023	0.034	0.034	0.048	0.046	0.037	0.051	0.029	0.017	0.028	0.044	0.018	0.055	0.009	0.250	0.058	0.020	0.066	0.065
CV	0.975	0.635	0.629	0.450	0.462	0.384	0.439	0.597	0.551	0.704	0.624	0.641	0.335	0.667	0.199	0.258	0.130	0.285	0.325	0.500	0.418

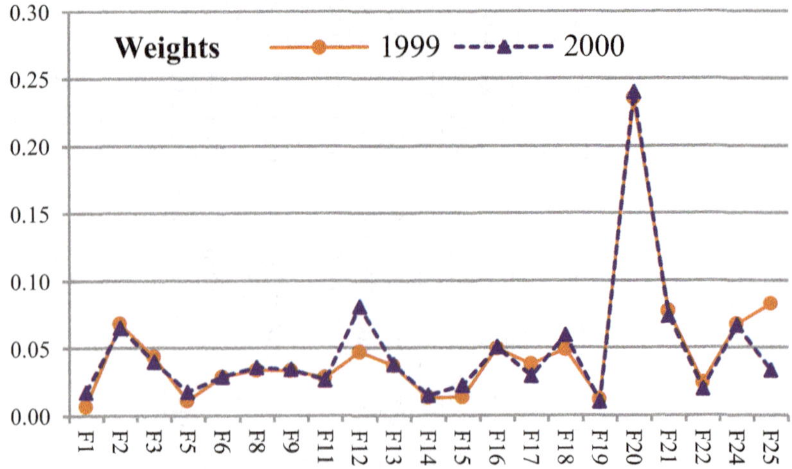

Fig. 9.3 Financial weights variations (1999–2000)

0.250, while individual weight for 1999–2009 from 0.0003 to 0.320. CV represents an index to measure the variance of weights, ranging from 0.13 to 0.975 over the period.

According to the results of entropy analysis, we can find that out of 21 ratios, the most important ratios during 1999–2009 are times interest earned ratio (F2), return on equity (F12), working capital turnover (F18), long-term debt to equity ratio (F21), cash flow to net income ratio (F24) and cash flow adequacy ratio (F25) (their average weight value is no less than 0.5). Out of the top six ratios in terms of average weight value, the ratios of the cash flow category (F24 and F25) are the most important. In contrast, current ratio (F1), total asset turnover (F14), total liabilities turnover (F17) and debt ratio (F19) (their average weight value is no more than 0.020) is relatively less important than the other ratios. Out of the bottom four ratios in terms of average weight value, the importance of the debt ratio (F19) is the lowest. It means that the higher the weight value of a financial ratio is, the larger its diversification is. In other words, the entropy enables us to find that a ratio with higher weight value is relatively more important than the other ratios with lower weight value, when their importance is evaluated among them. The ratio with the biggest weight is Debt to equity ratio (F20) with 0.320 in 2004, which says it is the most important ratio of the 21 ratios during the period of 1999–2009. This can be easily confirmed visually from Figs. 9.3, 9.4, 9.5, 9.6 and 9.7.

4.2 Ranking Variation Trend of Shipping Companies

The previous section describes the relative weights of financial ratios identified by entropy. In this section, the average weights and the performances of financial ratios

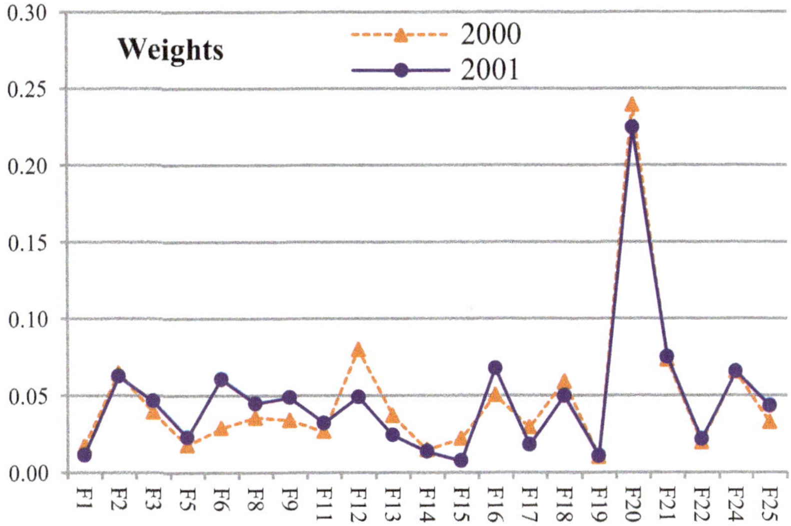

Fig. 9.4 Financial weights variations (2000–2001)

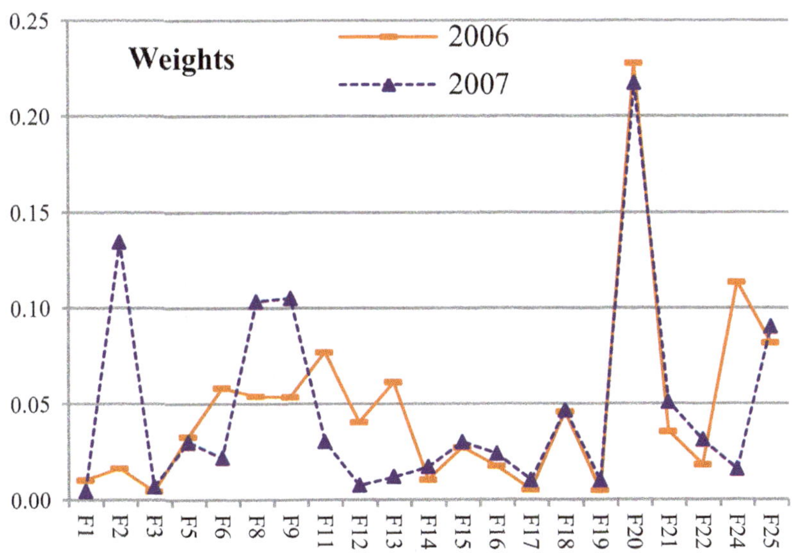

Fig. 9.5 Financial weights variations (2006–2007)

are combined in GRA to rank the four shipping companies. Table 9.10 shows rank score of each company and the relative grey relation score acquired by Eq. (9.7). Over the last 11 years, the two Taiwanese shipping companies, EG and YM, show good performances compared the two Korean companies, HJ and HMM. HMM's ranking was bottom from 1999 to 2007, but shows better results since 2008, while

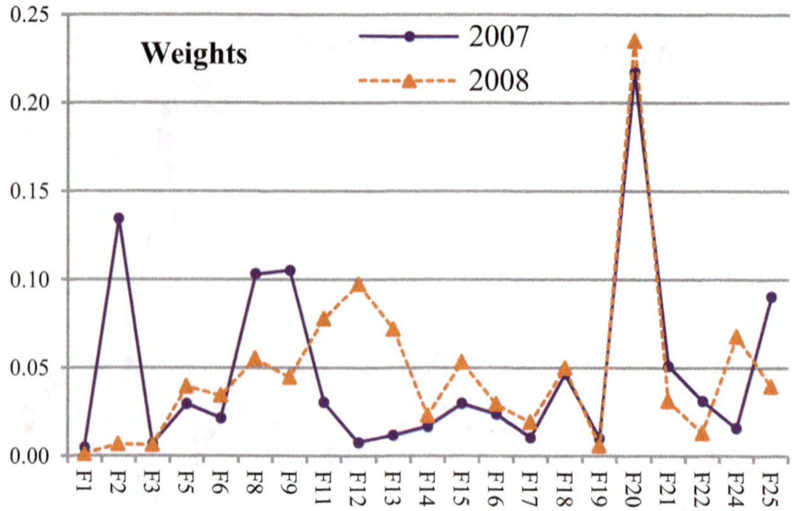

Fig. 9.6 Financial weights variations (2007–2008)

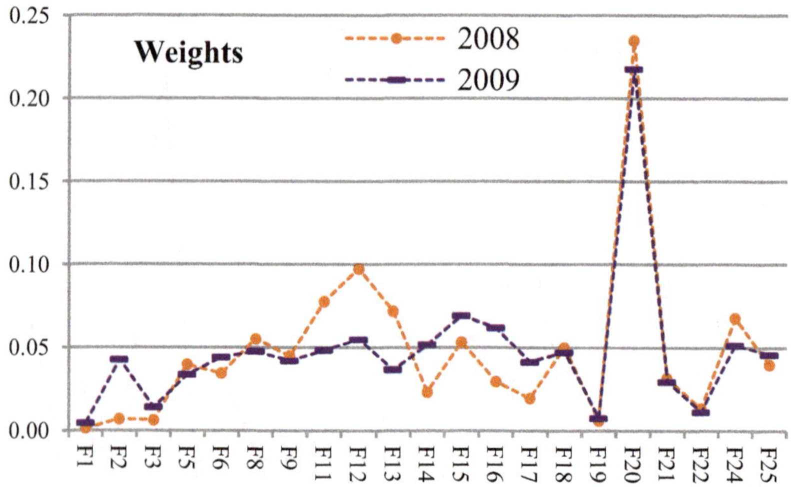

Fig. 9.7 Financial weights variations (2008–2009)

HJ downed from ranking top to the bottom. The rank variation trend of each company is plotted by the synthesis scores of the companies in line chart in Fig. 9.8.

Table 9.10 Rank variation trend of the four shipping companies

	1999	R	2000	R	2001	R	2002	R	2003	R	2004	R	2005	R	2006	R	2007	R	2008	R	2009	R
YM	1.787	1	1.550	1	1.0972	3	1.321	2	1.898	1	1.606	1	1.117	2	0.870	3	1.013	2	1.007	3	1.276	1
EG	1.335	2	1.322	3	1.4241	1	1.131	3	1.072	3	1.320	2	1.752	1	1.128	2	1.798	1	0.826	4	1.220	2
HMM	0.542	4	0.511	4	0.6200	4	0.745	4	0.52	4	0.685	4	0.576	4	0.618	4	0.509	4	1.208	2	1.102	3
HJ	1.169	3	1.385	2	1.3733	2	1.331	1	1.392	2	1.240	3	0.975	3	1.439	1	0.769	3	1.381	1	1.006	4

Notes: *EG* Evergreen, *YM* Yang Ming, *HMM* Hyundai Merchant Marine, *HJ* Hanjin

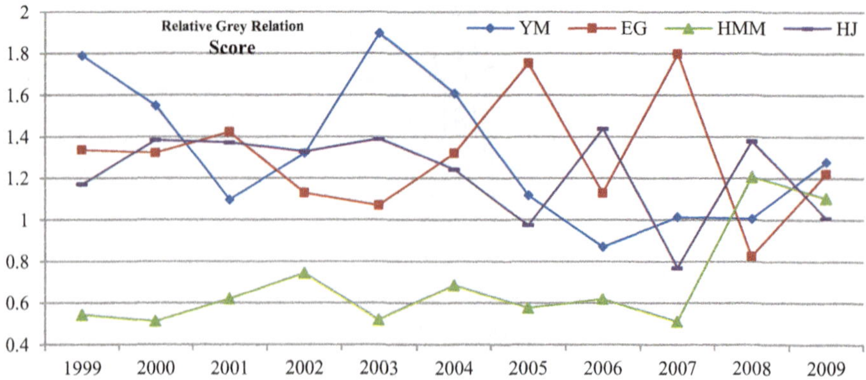

Fig. 9.8 Rank variation of the four shipping companies

5 Discussion and Business Implications

Our observations from the analysis and outputs (see Tables 9.4 and 9.5 and Figs. 9.3, 9.4, 9.5, 9.6 and 9.7) are summarized, if applicable, with implications as follows:

(1) *The overall financial performance of the two Taiwanese shipping companies, i.e. EG and YM, are better than the two Korean ones, i.e. HJ and HMM, between 1999 and 2009, except 2008.*

Korea had the financial crisis in 1997 of which impacts was extensively made on her national economy in many aspects. According to the International Monetary Fund (IMF) bailout programmes, the *Chaebols* (i.e. family-controlled conglomerates agreed to sign financial pacts with their respective creditor banks that required those companies to reduce their debt-to-equity ratios to less than 200%. Under these circumstances, Hanjin Shipping Co. Ltd. sold 31 ships including 29 container ships and raised US$720 million from the sales, to improve their debt-to-equity ratios (Lee 1999b). HMM later followed a similar route to improve its financial ratios by selling its car carrier division to foreign investors. Nevertheless, owing to their legacy of highly geared financial structure, both have shown worse financial performance compared to the two Taiwanese companies up to financial tsunami occurred in 2008.

(2) *EG and HJ show higher variation of rank than the other two companies since 2005, while the four companies' profitability sharply are being deteriorated since 2007.*

As shown in Fig. 9.8, the ranks of EG and HG indicate volatile movements since 2005, while the other rather stable. But Table 9.11 supports our findings that the financial position of the 4 container shipping companies has been seriously deteriorating in terms of profitability since 2008.

Table 9.11 Financial results of selected carriers for full-year 2008 and the first half 2009 (Unit: US$ million)

	Revenue		Operating profit				Operating profit margin		Net profit		
Carrier /Group	2008	2009	% Change	2008	2009	% Change	2008	2009	2008	2009	% Change
AP Moller-Maersk(container shipping)	30,436	9817	−30.1%	430	−879	−304.4%	3.1%	−9.0%	278	−961	−445.7%
China shipping container lines	2688	1307	−51.4%	142	−467	−428.9%	5.3%	−35.7%	103	−500	−585.4%
Evergreen marine corp.	2108	1208	−42.7%	87	−169	−294.3%	4.1%	−14.0%	49	−175	−457.1%
Hanjin shipping (all activities)	4051	2710	−33.1%	177	−419	−336.7%	4.4%	−15.5%	220	−542	−346.4%

After six months of 2009, the four companies recorded the combine net losses: HJ (−542 million US$ for consolidated data of all activities; −346.4% changes to the previous year), HMM (−270 million US$; −177.1% change), EG (−175 million US$ for consolidated data; −457.1% change), and YM (−210 US$ million for consolidated data; −588.4% change). APM (container shipping), Hapag-Lloyd, and Zim have been also unprofitable companies owing to this global economic recession, each recording red net profit −961, −704, and −305 million US$ in the first half of 2009, respectively. Hapag-Lloyd and Zim both faced bankruptcy before the rescue package provided with the mercy of their shareholders, creditors and governments. Both governments indicated to help their container shipping companies on the ground that ZIM as a going concern is "extremely important both from national, security and the economy as a whole", while "Hapag-Lloyd is not only of great significance for Hamburg but also a vital cornerstone and advertisement for Germany's entire maritime sector" (Drewry, 2009). This argument could be developed for Taiwanese and Korean container shipping companies on the basis of similar rationale for their economic security.

(3) *HJ is forerunner of the rank variation of the two Korean shipping companies during the period, except 2009.*

As discussed before, according to the big five Chaebols' pledge to conform to the government's debt-reduction guidelines, family-controlled business groups had to lower their debt-to-equity ratios to less than 200% by the end of 1999. The Financial Supervisory Committee (FSC), the nation's top financial watchdog, was in charge of monitoring the reduction progress of their debts through asset re-evaluation, sell-off, and attraction of foreign capital. This progress was monitored by the FSC in association with Chaebols' main creditor banks and reported to President of Korea. It can be said that the restructuring scope of HJ was more extensive in the early period of the crisis compared to HMM's to improve their debt-to-equity ratios (Lee 1999b). It seems that HJ's earlier restructuring efforts achieved better financial performances subsequent years compared to HMM's, as can be seen from Fig. 9.9.

(4) *The variation trend of rank of the four companies in* Fig. 9.8 *raises an interesting forecasting of the two Taiwanese companies' rank in 2010 onwards.*

The financial performances are interrelated to overall business strategy in container shipping sector in terms of service area, ship size, market strategy, ship operation strategy with sea speed adjustment to save fuel cost. The above factors are mostly same among the four companies so that there is little room for the competitors to make their performance different. In this circumstance, one of the factors to improve their financial performances can be drawn from the government policy on the shipping industry.

A tonnage tax system is a good example. The system had been introduced into Korea in 2005. The shipowners in high tax countries in developed maritime countries are in a less competitive position in terms of the crewing and operating costs compared to those under open registers that allow them to be manned by

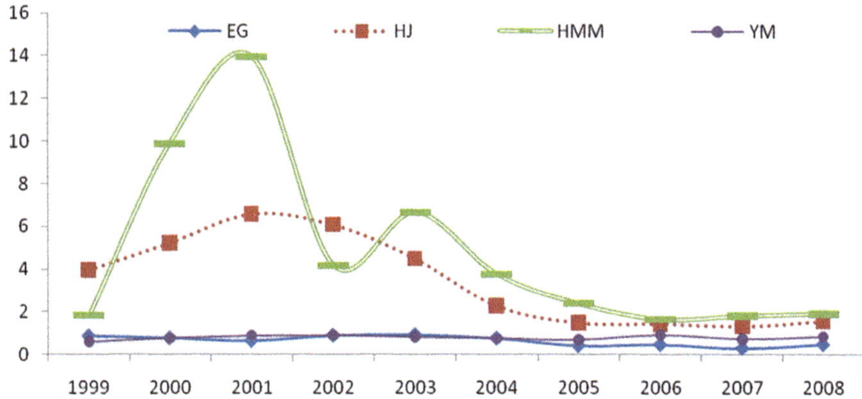

Fig. 9.9 Debt to equity ratio

non-national crew and also potentially enjoy a more moderate tax position. As a result, flagging-out was a common trend until the maritime developed countries introduced international ship registration system and tonnage tax system. Tonnage tax entails taxing shipping companies on the basis of the net tonnage owned and operated by them irrespective of whether they make profits. (Lee 1996; Bergantino and Marlow 1998; Brownrigg et al. 2001; Cullinane and Robertshaw 1996; Goulielmos 1998; Llácer 2003; Leggate and McConville 2005). Most European counties, including Germany, Denmark, Belgium, Finland, Norway, Spain and UK, and Asian countries of Korea and India are all under this tax system. Such introduction of new ship registration has generated multi-dimensional impacts on shipping business as well as the national economies not to mention maritime clusters. An internal report and the author's interview confirmed that Korean tonnage tax boosted bottom line profit and the Korean Shipowners' Association also recognized that it is very positive factor for Korean shipping companies' bottom line earnings. This is because the new tax regime has contributed to significantly reduce their tax burden over the last four years. Fortunately, a very recently Taiwanese government has announced to introduce the tonnage tax system. In association with several liberalization policies in shipping (Chang et al. 2006; Chiu 2007), the introduction of tonnage tax system may help improve financial position of the shipping companies in Taiwan.

6 Conclusions

This chapter applied Entropy and Grey Relation Analysis (GRA) to analyze the relative weights of financial ratios and the major four shipping companies in Korea and Taiwan: Evergreen, Yang Ming, Hanjin and Hyundai Merchant Marine, which are ranked according to the real value of their financial ratios taken and calculated from the financial statements audited by certified public accountants in the period of 1999–2009. Our thorough literature review showed that not only did most of the previous studies apply multi-criteria decision making method (MCDM) to measure the financial performance of shipping companies. Such methodology was based on the subjective perceptions of experts, while also displaying the shortcoming that financial ratios are difficult partition into groups because of the complex interrelationship among the ratios. Therefore, in this chapter, an objective weight identification method, i.e. Entropy has been applied to find the relative weights and variation trends of financial ratios, because it will preserve the integrity of the financial ratios data. GRA has been then combined with the objective weights to find the trend of ranking of the four shipping companies during the period of 1999–2009. Having said that, this chapter claims that a research gap has been filled with our approach from the viewpoint of methodology applied for evaluation of financial performance of shipping companies. Of course, the basic presuppositions of objective and subjective weight identification methods are different from each other and ultimately a reflection of decisions made by research teams. As a result, the results based on the identified weights are also variable. A related comparison of subjective and objective weight identification in shipping area is important, which should be conducted in the future to find advanced evidence and logic strategies to prepare management and stakeholders for unexpected shocks and severe financial conditions in the wake of a financial tsunami.

The higher the weight value of a financial ratio is, the larger its diversification is. In other words, the entropy output implies that the higher value ratio is relatively more important than other ratios with lower weight value, based on a comparative evaluation. According to the results of entropy analysis, times interest earned ratio, working capital turnover, long-term debt to equity ratio, cash flow to net income ratio and cash flow adequacy ratio were relatively more important than the other ratios during 1999–2009. GRA output says that the general financial performance of the two Taiwanese shipping companies, i.e. EG and YM, are better than the two Korean ones, i.e. HJ and HMM, between 1999 and 2007, except 2003. But in 2008, such situation was completely reversed. In Taiwan, YM's ranks are higher than EG's during the period of 1999–2009, except 2007. But both companies' profitability is sharply being deteriorated since 2007. In this respect, as we have seen the arguments and rescue programmes for Hapag-Lloyd and Zim in the previous section, policy instruments are required to be designed for improving financial performance of the container shipping companies in Taiwan with their own business strategy development in collaboration with tonnage tax system recently introduced in to the country. A comparison with Korean competitors suggests that a disciplined financial target borne out of a severe period of distress might result in better future financial performance. It is also important that Taiwanese shipping lines are given the necessary policy support. The above research should also be combined with pragmatic understanding of the Taiwanese operating environment for shipping lines, particularly with regard to access to capital. Previous research by the authors on changes to financial structures in Korea, suggest that specific financial environments may produce alternative behavior. The container shipping and financial markets are in a period of severe financial disequilibrium. Investors, shipping lines, stakeholders, financiers and maritime policy makers should potentially use the above research for well-informed decision making for both immediate problem solving and longer-term strengthening of financial structures.

References

Atanassov K (1986) Intuitionistic Fuzzy Set. Fuzzy Sets Syst 20:87–96

Atanassov K, Pasi G, Yager R (2005) Intuitionistic fuzzy interpretations of multi-criteria multi-person and multi-measurement tool decision making. Int J Syst Sci 36:859–868

Benayoun R, Roy B, Sussmann B (1966) Manuel de réference du programme electre. Note de synthèse, formation n.25, Direction scientifique. SEMA, France

Bergantino A, Marlow P (1998) Factors influencing the choice of flag: empirical evidence. Marit Policy Manag 25:157–174

Bowlin WF (2004) Financial analysis of civil reserve air fleet participants using data envelopment analysis. Eur J Oper Res 154:691–709

Brownrigg M, Dawe G, Mann M, Weston P (2001) Developments in UK shipping: the tonnage tax. Marit Policy Manag 28:213–223

Capobianco HMP, Fernandes E (2004) Capital structure in the world airline industry. Transp Res Part A Policy Pract 38:421–434

Celik M, Deha Er I, Ozok AF (2009) Application of fuzzy extended AHP methodology on shipping registry selection: the case of Turkish maritime industry. Expert Syst Appl 36:190–198

Chang CC, Lin K, J yuan B (2006) Developing direct shipping across the Taiwan Straits. Mar Policy 30:392–399

Chen CT (2000) Extensions of the TOPSIS for group decision-making under fuzzy environment. Fuzzy Sets Syst 114:1–9

Chen S, Tan J (1994) Handling multicriteria fuzzy decision-making problems based on vague set theory. Fuzzy Sets Syst 67:163–172

Cheng CH (1998) A new approach for ranking fuzzy numbers by distance method. Fuzzy Sets Syst 95:307–317

Chiu RH (2007) The liberalization of shipping in Taiwan. Mar Policy 31:258–265

Chou SY, Chang YH (2008) A decision support system for supplier selection based on a strategy-aligned fuzzy SMART approach. Expert Syst Appl 34:2241–2253

Chuang IY, Chiu YC, Edward Wang C (2008) The performance of Asian airlines in the recent financial turmoil based on VaR and modified Sharpe ratio. J Air Transp Manag 14:257–262

Cullinane K, Robertshaw M (1996) The influence of qualitative factors in Isle of Man ship registration decisions. Marit Policy Manag 23:321–336

Deng J (1982) Control problems of grey systems. Syst Control Lett 5:288–294

Drewry Publishing (2009) Container market review and forecast. Annual Report

Feng CM, Wang RT (2000) Performance evaluation for airlines including the consideration of financial ratios. J Air Transp Manag 6:133–142

Goulielmos AM (1998) Flagging out and the need for a new Greek maritime policy. Transp Policy 5:115–125

Guzhva VS, Pagiavlas N (2003) Corporate capital structure in turbulent times: a case study of the US airline industry. J Air Transp Manag 9:371–379

Halkos GE, Salamouris DS (2004) Efficiency measurement of the Greek commercial banks with the use of financial ratios: a data envelopment analysis approach. Manag Account Res 15:201–224

Heaver TD (2002) The Evolving Roles of Shipping Lines in International Logistics. Marit Econ Logist 4(3):210–230
Hsu HM, Chen CT (1997) Fuzzy credibility relation method for multiple criteria decision-making problems. Inf Sci (Ny) 96:79–91
Hus HM, Chen CT (1994) Fuzzy hierarchical weight analysis model for multicriteria decision problem. J Chinese Inst Industrial Eng 11(3):129–136
Hus HM, Chen CT (1996) Aggregation of fuzzy opinions under group decision making. Fuzzy Sets Syst 79:279–285
Hwang CL, Yoon K (1981) Multiple attributes decision making methods and applications. Springer, Berlin Heidelberg
Jessop A (2004) Minimally biased weight determination in personnel selection. Eur J Oper Res 153:433–444
Jung A, Schulz T, Wangner W (2009) Shipping industry drowning in financial woes, Der Spiegel 14
Lee TW (1996) Flagging options for the future: a turning point in Korean shipping policy? Marit Policy Manag 23:177–186
Lee HS (1999a) Automatic clustering of business processes in business systems planning. Eur J Oper Res 114:354–362. doi:10.1016/S0377-2217(98)00125-8
Lee TW (1999b) Restructuring of the economy and its impacts on the Korean maritime industry. Marit Policy Manag 26:311–325
Lee TW, Lin CW, Shin SH (2012) A comparative study on financial positions of shipping companies in Taiwan and Korea using entropy and grey relation analysis. Expert Syst Appl 39(5):5649–5657
Lee TW, Lin CW, Chung YS (2014) Comparison analysis for subjective and objective weights of financial position of container shipping companies. Marit Policy Manag 41(3):241–250
Leggate H, McConville J (2005) Tonnage tax: is it working? Marit Policy Manag 32:177–186
Levinson M (2006) The box: how the shipping container made the world smaller and the world economy bigger. Princeton University Press, Princeton
Liang GS, Liu CF, Lin WC, Yeh CH (2006) A data envelopment analysis of shipping industry bond ratings. Tamkang J Sci Eng 9:403–408
Lin WC, Liu CF, Chu CW (2005) Performance efficiency evaluation of the Taiwan's shipping industry : an application of data envelopment analysis. Proc East Asia Soc Transp Stud 5:467–476
Llácer FJM (2003) Open registers: past, present and future. Mar Policy 27:513–523
Ma J, Fan ZP, Huang LH (1999) A subjective and objective integrated approach to determine attribute weights. Eur J Oper Res 112:397–404
Notteboom T, Rodrigue J-P (2008) Containerisation, box logistics and global supply chains: the integration of ports and liner shipping networks. Marit Econ Logist 10:152–174. doi:10.1057/palgrave.mel.9100196
Opricovic S, Tzeng GH (2004) Compromise solution by MCDM methods: a comparative analysis of VIKOR and TOPSIS. Eur J Oper Res 156:445–455
Panayides PM (2006) Maritime logistics and global supply chains: towards a research agenda. Marit Econ Logist 8:3–18
Saaty TL (1980) The analytic hierarchy process. McGraw-Hill, New York
Scheraga CA (2004) Operational efficiency versus financial mobility in the global airline industry: a data envelopment and Tobit analysis. Transp Res Part A Policy Pract 38:383–404
Shannon CE (1948) A mathematical theory of communication. Bell Syst Tech J 27:379–423
Shih HS (2008) Incremental analysis for MCDM with an application to group TOPSIS. Eur J Oper Res 186:720–734
Song DW, Lee PT-W (2009) Maritime logistics in the global supply chain. Int J Logist Res Appl 12:83–84

Triantaphyllou E, Shu B, Sanchez SN, Ray T (1998) Multi-criteria decision Making: an operations research approach. In: Webster JG (ed) Encyclopedia of electrical and electronics engineering. John Wiley & Sons, New York, pp 175–186

Tung CT, Lee Y-J (2009) A novel approach to construct grey principal component analysis evaluation model. Expert Syst Appl 36:5916–5920

Tzeng GH, Lin CW, Opricovic S (2005) Multi-criteria analysis of alternative-fuel buses for public transportation. Energy Policy 33:1373–1383

Tzeng GH, Chiang CH, Li CW (2007) Evaluating intertwined effects in e-learning programs: a novel hybrid MCDM model based on factor analysis and DEMATEL. Expert Syst Appl 32:1028–1044

Wang YJ (2008) Applying FMCDM to evaluate financial performance of domestic airlines in Taiwan. Expert Syst Appl 34:1837–1845

Wang YJ (2009) Combining grey relation analysis with FMCGDM to evaluate financial performance of Taiwan container lines. Expert Syst Appl 36:2424–2432

Wang YJ, Lee HS (2007) Generalizing TOPSIS for fuzzy multiple-criteria group decision-making. Comput Math with Appl 53:1762–1772

Wang YJ, Lee HS (2008) A clustering method to identify representative financial ratios. Inf Sci (Ny) 178:1087–1097

Wang YJ, Lee HS (2010) Evaluating financial performance of Taiwan container shipping companies by strength and weakness indices. Int J Comput Math 87:38–52

Wang YJ, Lee HS, Lin K (2003) Fuzzy TOPSIS for multi-criteria decision-making. Int Math J 3:367–379

Wang RT, Ho CT, Feng CM, Yang YK (2004) A comparative analysis of the operational performance of Taiwan's major airports. J Air Transp Manag 10:353–360

Wang YJ (2008) Applying FMCDM to evaluate financial performance of domestic airlines in 572 Taiwan. Expert Syst Appl 34:1837–1845 573

Wang YJ (2009) Combining grey relation analysis with FMCGDM to evaluate financial performance of Taiwan container lines. Expert Syst Appl 36:2424–2432

Xu X (2004) A note on the subjective and objective integrated approach to determine attribute weights. Eur J Oper Res 156:530–532

Zadeh LA (1965) Fuzzy sets. Inf Control 8(2):338–353

Zhang G, Zhang N, Wang Q (2005) Container ports development and regional economic growth: an empirical research on the pearl river delta region of China. Proc East Asia Soc Transp Stud 5:2136–2150

Zou ZH, Yun Y, Sun JN (2006) Entropy method for determination of weight of evaluating indicators in fuzzy synthetic evaluation for water quality assessment. J Environ Sci (China) 18:1020–1023

Chapter 10
The Use of the Hybrid Fuzzy-Delphi-TOPSIS Approach in Identifying Optimal Bunkering Ports for Shipping Lines

Ying Wang, Gi-Tae Yeo, and Adolf K.Y. Ng

Abstract With sustained high bunker prices, new methods for choosing optimal bunkering ports to save total operating costs have appeared in research involving liner shipping companies. Generally speaking, the bunkering port selection problem is solved by utilizing ship planning software. However, this can only work optimally when ship arrivals can be forecasted rather accurately, and its primary limitation is that it ignores unforeseen circumstances in actual operations. Hitherto, there are no fixed rules for bunkering port selection. To address this problem, this chapter develops a benchmarking framework that evaluates bunkering ports' performances within regular liner routes in order to choose optimal ones. Bunkering port selection is typically a multi-criteria group decision problem, and in many practical situations, decision makers cannot form proper judgments using incomplete and uncertain information in an environment with exact and crisp values; thus, fuzzy numbers are proposed in this chapter. A hybrid Fuzzy-Delphi-TOPSIS based methodology that divides the benchmarking into three stages is employed to support the entire framework. Additionally, a sensitivity analysis is performed. The proposed framework can enable decision makers to better understand the complex relationships of the relevant key performance factors and assist managers in comprehending the present strengths and weaknesses of their strategies.

Keywords Bunkering port • Key performance factor (KPF) • Liner shipping companies • Fuzzy-Delphi-TOPSIS

Y. Wang
School of Economics and Management, Yantai University, Yantai, China

G.-T. Yeo (✉)
Graduate School of Logistics, Incheon National University, Incheon, South Korea
e-mail: ktyeo@inu.ac.kr

A.K.Y. Ng
Department of Supply Chain Management, Asper School of Business, University of Manitoba, Winnipeg, MB, Canada

Transport Institute, Asper School of Business, University of Manitoba, Winnipeg, MB, Canada

P.T.-W. Lee, Z. Yang (eds.), *Multi-Criteria Decision Making in Maritime Studies and Logistics*, International Series in Operations Research & Management Science 260,
DOI 10.1007/978-3-319-62338-2_10

1 Introduction

Bunker fuel is used by international seagoing ships. Given that 90% of world trade is carried out by sea (IMO 2009) and must work within evolving international marine transportation requirements, the world bunker demand is increasing. Bunker prices have risen considerably in recent years, and fuel costs form more than half of a liner shipping company's total operating costs (Yao et al. 2012). In addition, environmental policies are another difficulty for the operation of liner shipping companies. With increasing bunker prices, many liner shipping companies try to save fuel by making some operational adjustments, including: (1) redeployment of ships, (2) consolidation of services, (3) speed adjustment, (4) reduction of resistance, and (5) bunkering port selection (Mazraati 2011). Among them, optimizing bunkering port selection is crucial (Besbes and Savin 2009) to save on total running costs.

Generally speaking, vessels call a port for various purposes, such as taking bunkers, going to shipyards for repair, stevedoring cargoes at a terminal, or a combination of the above (Huang et al. 2011). In tramp routes, bunkering service is required only when there is insufficient fuel or when the bunker prices are attractive; in such instances, bunkering port selections are simple. However, with regard to regular liner routes, the bunkering port selection processes are more complicated because liner shipping companies prefer a combination purpose of obtaining bunkering services at the ports and berthing for stevedoring cargoes; hence, there are more factors under consideration (Hu 2005). It is important to study liner routes so that liner shipping companies can maintain their shipping schedules at each port and reduce their operating costs.

This chapter aims to develop a benchmarking framework for choosing optimal bunkering ports for liner shipping companies along a regular liner route by evaluating the bunkering ports' performances. Therefore, the key performance factors (KPFs) of bunkering ports are identified; further, via a case study analysis, the strengths and weaknesses of these alternative bunkering ports can be understood, and the developed benchmarking rule can be determined to be applicable or not. Additionally, a sensitivity analysis is performed to determine the changes in the final alternative selection with variation taken from expert opinions. Finally, some implications for continuous improvement of the bunkering ports are given. Bunkering port selection is both a fundamental issue and a challenging problem for liner shipping companies, since bunkering port selection is typically a multi-criteria group decision problem that has two types of uncertainties: (1) weighting values to proxy KPFs and (2) crisp input data. The first type of uncertainty may arise during the decision making process and may include stakeholders with different interests. The second type can result from transferring data to crisp values (Jun et al. 2012). This chapter develops a hybrid Fuzzy-Delphi-TOPSIS model to overcome these uncertainties. First, decision makers were selected from liner shipping companies as questionnaire respondents. Therefore, the worry about different interests among stakeholders is unnecessary. Second, fuzzy numbers are transferred to crisp values at the end of the assessment on order to maintain the original opinions of decision makers.

2 Literature Review

In previous research, much interest focuses on the factors that affect the performance of bunkering ports when liner shipping companies obtain bunker supplies in these ports. Acosta et al. (2011) found that the factors affecting bunkering competitiveness include low bunkering price, few legal restrictions, quick bunkering and geographical advantage. Yao et al. (2012) revealed through a case study that bunkering port decisions are mainly affected by the evolution of bunker fuel prices along the service route, as well as the speed of the ship. Liner shipping companies are likely to choose the first port after a long voyage as a bunkering port due to its geographical advantage. Bunkering associated conditions, including weather, access to port, security, bunker fuel capacities, bunkering facilities, and fueling speed, are also important because of the short port time (Notteboom and Vernimmen 2009; Wang and Meng 2012a). Chang and Chen (2006) developed a knowledge-based simulation model to improve the bunkering service capacity in the port of Kaohsiung. The average bunkering service time, mean waiting time, bunker barge usage, and berth utilization efficiency are important indicators. The type of fuel, the amount needed, and the schedule of arrival and departure are also vital information needed before obtaining bunkering service. Dinwoodie et al. (2012) proposed that reciprocal information sharing among stakeholders would enhance efficiency and reduce the risk of flawed bunkering operations. Also, the quality of marine fuel oils affects factors such as ship handling, engine operation, bunker consumption, and the environmental impact of emissions (Fu 2009; Wang 2002; Yuan 2012; Anfindsen et al. 2012).

Oil spills and leakages during bunkering operations may increase pollution damage to the environment and is costly to clean. Safe bunkering to prevent pollution, fire, and other potential risks is essential for both vessels and ports (You 2008; Dong et al. 2011; Talley et al. 2012). Hu (2005) noted that bunkering port selection depended on fuel quality, port charges, and the efficiency of the bunkering supply. Also, the bunkering suppliers in the ports are significant. Liner shipping companies prefer to obtain domestic rather than foreign bunkering services. In addition, utilizing more advanced bunkering equipment may result in more efficient bunkering, faster refueling speed, and higher levels of security (Wu 2011).

Recent research related to solving bunkering port selection problems has mostly employed ship planning software with real data inputs, such as shipping routes (Wang and Meng 2012b), the distance between ports of call (Wang and Meng 2012c), bunker prices (Notteboom and Cariou 2011), emission taxes (Kim et al. 2012; Corbett et al. 2009), ship time costs (Kim et al. 2012), inventory carrying costs (Psaraftis and Kontovas 2010), fixed ordering costs, port time (Du et al. 2011; Qi and Song 2012; Wang and Meng 2012d), to name but a few. However, bunkering operations involve a combination of transport and fueling services with a dynamic data element (Chang and Chen 2006). There are many hypotheses (for instance, all ships deployed in the service are homogenous, a penalty is not incurred, the terminal handling costs per container do not alter with vessel size or route length, etc.) and limiting conditions (for instance, known port time at each port along the service route; the same load cost, discharge cost, and transshipment cost; the container handling efficiency of each port of call) involved in software analyses, and these analyses also cannot take liner shipping companies' preferences into account. There are no fixed rules for bunkering port selection among several alternatives as of yet; decision making may simply be based on liner shipping companies' preferences, but such preferences are often unknown to ports. Indeed, it is not easy to compile a comprehensive set of rules that reflects the preference of liner shipping companies, which often make choices among the alternative bunkering ports due to unforeseen circumstances in actual operations.

3 Methodology

3.1 A Hybrid Fuzzy-Delphi-TOPSIS Approach

Bunkering port selection is a decision making problem based on liner shipping companies' preferences. In many cases, the preferential model of decision making is uncertain, and it is difficult for decision makers to provide exact numerical values for comparative ratios (Tsai et al. 2010). This chapter hence proposes using fuzzy theory to resolve the uncertainty and imprecision of performance evaluations, in which the comparative judgments of a decision maker are represented as fuzzy triangular numbers. To more accurately reflect the original opinions of decision makers, a Fuzzy-Delphi-TOPSIS methodology, which can handle both the quantitative and qualitative elements of a problem, is used.

Fuzzy-Delphi-TOPSIS is a methodology combining the Fuzzy Delphi and Fuzzy TOPSIS methods for optimal decision making strategies. The Fuzzy Delphi method can resolve uncertainty regarding decision space and combine the advantages of statistical methods (Prusty et al. 2010; Ma et al. 2011; Duru et al. 2012). It has four advantages: (1) to decrease the time of questionnaire surveys, (2) to avoid distorting individual expert opinions, (3) to clearly express the semantic structure of predicted items, and (4) to consider the fuzzy nature during the interview process (Chang and Wang 2006). Fuzzy TOPSIS, one of the MCDM Multiple Criteria Decision Making (MCDM) techniques, is widely used to quantify the performance measures of the alternatives by through extensive research (Kim et al. 2013). It can embody the fuzzy nature of the comparison or evaluation process and strengthen the comprehensiveness and rationality of the decision making process (Bao et al. 2012).

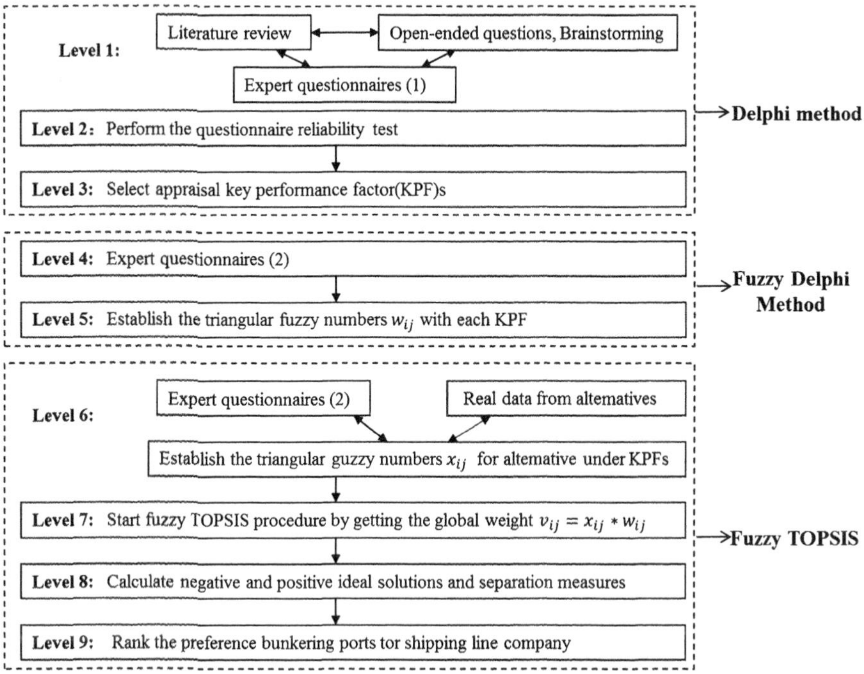

Fig. 10.1 Generalized framework through Fuzzy-Delphi-TOPSIS based approach

3.2 *Research Process*

The research process can be found in Fig.10.1. A hybrid Fuzzy-Delphi-TOPSIS based methodology divides the whole benchmarking process into three stages. The first stage includes identification, synthesis and prioritization of the KPFs that may affect the bunkering port selection of liner shipping companies via the Delphi method. The second stage sets up the fuzzy matrix and compute the weights of each KPF using the Fuzzy Delphi Method. The third stage undertakes a Fuzzy TOPSIS based assessment of possible alternative bunkering ports so as to save operation costs and time.

3.2.1 Delphi Method

The procedure of the Delphi Method is shown as follows:

1. Level 1: On the basis of extensive literature reviews and consultations with experts, collect all possible KPFs that may affect bunkering service when the vessels are in port. Ask open-ended questions and brainstorm about whether any of the possible KPFs have been double-counted or whether any KPFs should be considered that are not on the list. Next, ask domain experts to rank the

importance of each KPF by assigning a range from 1 to 7 on a questionnaire survey (1), where 1 indicates the least important and 7 the most important.

2. Level 2: Questionnaire reliability test. When the survey is completed, the degree of importance of each KPF can be obtained. Also, the reliability of the responses needs to be tested. Cronbach's Alpha is applied to test reliability (Chong et al. 2012) in this chapter. If Cronbach's Alpha is less than 0.35, the corresponding datum is not reliable and must be deleted. Those are greater than 0.35 are viewed as reliable (Wang and Lin 2008).
3. Level 3: Obtain the importance of the KPFs by the geometric mean so that the impact of extreme values can be avoided; a higher value indicates a higher degree of importance, and a lower value indicates a lower degree of importance. The threshold value r is also determined. If the geometric mean value of the KPF is no less than r, the KPF is accepted, and vice versa. In this chapter, the threshold value r was set at 4; therefore, the selection of the KPFs is as follows:

   If the geometric mean $\geq r = 4$, the appraisal KPF is accepted.
   If the geometric mean $< r = 4$, the appraisal KPF is rejected.

3.2.2 Fuzzy Delphi Method

The process of Fuzzy Delphi is shown as follows:

1. Level 4: Determine the importance of each KPF. After the crucial KPFs are selected in Level 3, expert questionnaires (2) are applied to the 15 KPFs to evaluate their importance based on a 10-point scale (Kuo and Chen 2008); a higher point value indicates higher importance, and a lower point value indicates lower importance.
2. Level 5: Set up the triangular fuzzy numbers (TFNs) $\tilde{w}_{ij}$ as defined in Eq. (10.1). In this work, the TFNs representing the pessimistic, moderate, and optimistic estimate are used to represent the opinions of experts for each activity time (Hsu and Yang 2000).

$$\tilde{w}_{ij} = \left(a_{ij}, b_{ij}, c_{ij}\right), a_{ij} \leq b_{ij} \leq c_{ij} \tag{10.1}$$

$$a_{ij} = \min\left(M_{ijk}\right) \tag{10.2}$$

$$b_{ij} = \sqrt[n]{\prod_{k=1}^{n} M_{ijk}}, k = 1, 2, \ldots, n \tag{10.3}$$

$$c_{ij} = \max\left(M_{ijk}\right) \tag{10.4}$$

where M_{ijk} indicates the appraisal value of the kth expert for the KPF, a_{ij} indicates the bottom threshold of all the experts' appraisal value; b_{ij} indicates the geometric mean of all the experts' appraisal value, c_{ij} indicates the ceiling of all the experts' appraisal value, and n is the number of experts in a group.

Table 10.1 Linguistic variables for the preference of each alternative

Linguistic scale	Fuzzy score
Poor (P)	(1, 1, 2)
Medium poor (MP)	(2, 3, 4)
Fair (F)	(4, 5, 6)
Medium good (MG)	(6, 7, 8)
Good (G)	(8, 9,10)

3.2.3 Fuzzy TOPSIS

The Fuzzy TOPSIS steps are as follows:

Level 6: Using expert questionnaires (2), the fuzzy linguistic values ($\tilde{x}_{ij}, i = 1, 2, \ldots, n, j = 1, 2, \ldots, m$) of the alternatives concerning the KPFs are chosen to determine the importance of each alternatives using the fuzzy linguistic rating ($\tilde{x}_{ij}$), which keeps the ranges of normalized triangular fuzzy numbers that belong to [0, 10] shown in Table 10.1 (Buyukozkan and Cifci 2012).

When the KPF has a crisp quantity value for each alternative, we should also transform the crisp quantities for the KPFs into fuzzy numbers. Let $R_{ij}^{o} = \left(c_{ij}^{o},\ a_{ij}^{o},\ b_{ij}^{o}\right)$ be the crisp quantity from the real data to the alternative An, where $c_{ij}^{o} = a_{ij}^{o} = b_{ij}^{o}$. The KPFs are determined in various units and must be transformed into dimensionless indices (or numbers) to ensure compatibility with the linguistic numbers of the KPFs. The alternative with the maximum benefit (or the minimum cost) should have the highest number. In this chapter, we adopt the transform method provided by Chou (2010) as follows:

$$\bar{R}_{ij} = \left\{R_{ij}^{o} / \max_{i}\left\{b_{ij}^{o}\right\}\right\} \times 10 \tag{10.5}$$

Where $\max_{i}\left\{b_{ij}^{o}\right\} > 0$, $\bar{R}_{ij}$ denotes the transformed fuzzy number of objective benefit R_{ij}^{o}, $\bar{R}_{ij}$ becomes larger when the objective benefit R_{ij}^{o} is larger.

$$\bar{R}_{ij} = \left\{\min_{i}\left\{c_{ij}^{o}\right\} / R_{ij}^{o}\right\} \times 10 \tag{10.6}$$

Where $\min_{i}\left\{c_{ij}^{o}\right\} > 0$, $\bar{R}_{ij}$ denotes the transformed fuzzy number of objective benefit R_{ij}^{o}, $\bar{R}_{ij}$ becomes smaller when the objective cost R_{ij}^{o} is larger.

2. Level 7: Compute the weighted normalized fuzzy-decision matrix by

$$\tilde{v} = \left[\tilde{v}_{ij}\right]_{n*j}, i = 1, 2, \ldots, n, j = 1, 2, \ldots, m \quad (10.7)$$

$$\tilde{v}_{ij} = \tilde{x}_{ij} {}^{*} \tilde{w}_{j} \quad (10.8)$$

$\tilde{w}_j$ is obtained from the fuzzy Delphi Method via expert questionnaires (2) shown above.

3. Level 8: Determine the positive-ideal (FPIS, A^*) and negative-ideal (FNIS, A^-) solutions from Eqs. (10.9) and (10.10), and then calculate the distances of each alternative from the ideal solution and the negative-ideal solution:

$$A^* = \{v_1^*, \ldots, v_i^*\} = \left\{\left(\max_j v_{ij}, i \in \Omega_b\right), \left(\min_j v_{ij}, i \in \Omega_c\right)\right\} \quad (10.9)$$

$$A^- = \{v_1^-, \ldots, v_i^-\} = \left\{\left(\min_j v_{ij}, i \in \Omega_b\right), \left(\max_j v_{ij}, i \in \Omega_c\right)\right\} \quad (10.10)$$

Ω_b are the sets of benefit criteria and Ω_c are the sets of cost criteria

$$D_i^* = \sum_{j=1}^{m} d(\tilde{V}_{ij}, \tilde{V}_i), i = 1, 2, \ldots, n \quad (10.11)$$

$$D_i^- = \sum_{j=1}^{m} d(\tilde{V}_{ij}, \tilde{V}_i), j = 1, 2, \ldots, m \quad (10.12)$$

$$d(\tilde{a}, \tilde{b}) = \sqrt{(1/3)\left[(a_1 - b_1)^2 + (a_2 - b_2)^2 + (a_3 - b_3)^2\right]} \quad (10.13)$$

$\tilde{a}$ and $\tilde{b}$ are two triangular fuzzy numbers, which is shown by the triplets (a_1 , a_2 , a_3) and (b_1 , b_2 , b_3).

4. Level 9: Determine the relative closeness of each alternative to the ideal solution. The relative closeness of the alternative A_i in relation to A^* is characterized as below:

$$FC_i = D_i^- / (D_i^* + D_i^-), i = 1, 2, \ldots, n \quad (10.14)$$

4 A Case Study

In this section, a case study on the regional route in East Asia is performed to determine whether the developed benchmarking rule is appropriate or not. There are two reasons why the regional shipping route in East Asia was selected rather than an inter-continental route:

1. The substantial growth of marine traffic in East Asia has increased the bunkering service requests at the calling ports in this region. However, hitherto, there is scant research on the analysis of bunkering ports in this region.
2. Given that the distance between calling ports is shorter than those on an inter-continental route, the competition among calling ports for bunkering service is intense. Moreover, the limitations from environmental policies in East Asia are fewer than the Emission Control Areas[1] in western countries. This leads liner shipping companies to prefer to obtain bunkering services in this region in order to circumvent tedious and strict regulations from environmental policies during the bunkering process.

Therefore, a representative regular shipping route, China-Korea-Japan is chosen as the evaluation object shown in Fig. 10.2. According to the port time and the container volumes in the ports of call, four alternative bunkering ports (Xingang, Dalian, Busan and Niigata) are selected for further evaluation.

4.1 Selection of Appraisal KPFs-Delphi Method

In the first stage, previous research related to bunkering port selection with 21 factors as bunkering port performance evaluation criteria was circulated among experts to obtain better insight into the problem by expert questionnaire as mentioned in Fig. 10.1 to select the appraisal KPFs for bunkering port evaluation shown in Table 10.2. To determine the crucial factors among all of the factors obtained from the experts' opinions more objectively, a 7-point scale was simultaneously employed to select the appraisal KPFs from the opinions of experts. Additionally, Cronbach's Alpha was applied to test the reliability of the questionnaire before the selection of appraisal KPFs. Taking factor "bunker price" for example, the degree of 7-point scale from 1 to 7 is considered from very low to very high. Totally, 12 respondents[2] from liner shipping companies, consisting of CEOs, general managers, and operations managers who possessed more than 15 years of professional experience, were interviewed during a brainstorming session to identify the

[1]The Emission Control Areas (ECAs) have been established in the Baltic Sea area, the North Sea area, and the North American area, but not in East Asia (Schinas and Stefanakos 2012).

[2]The suitable sample size of the expert questionnaire used in the Delphi method should not be too large (about 10–15 experts) (Adler and Ziglio 1996; Ng 2006; Ma et al. 2011).

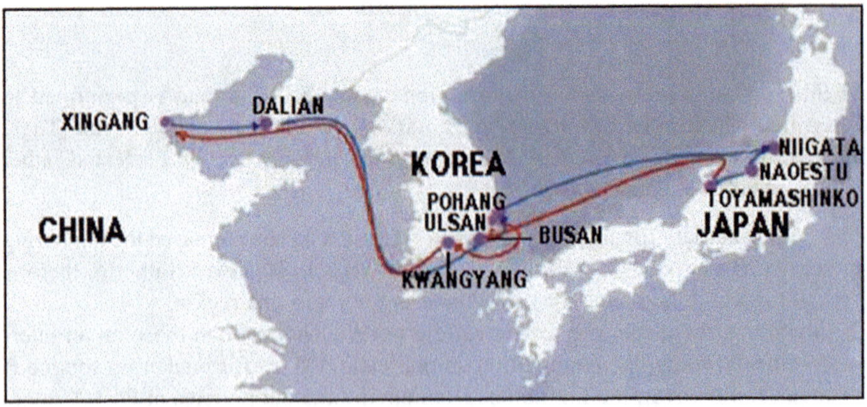

Fig. 10.2 A The representative China-Korea-Japan liner routes (A sample China-Korea-Japan regular liner shipping route taken from the CK Line (http://www.ckline.co.kr/) is applied, with the sequence of the voyage being Xingang-Dalian-Busan-Pohang-Niigata-Naoestu-Toyamashinko-Busan-Ulsan-Kwangyang-Xingang-Dalian)

Table 10.2 Expert questionnaire 1: sample question asking for opinions on "bunker price" factor

	Very low	Low	Medium low	Fair	Medium high	High	Very high
Factors	1	2	3	4	5	6	7
Bunker price							O
Port tariffs					O		
Port time							O

KPFs. In all, the expert questionnaire 1 was conducted within a period of 32 days (from 2 Nov 2012 to 3 Dec 2012).

After gathering the opinions from these 12 experts, Cronbach's Alpha of the questionnaire can be calculated by software SPSS. The value of 0.687 that was obtained is greater than 0.35 and is therefore viewed as the reliable.[3] After confirming the reliability of the questionnaire, according to the procedure of the Delphi Method mentioned in Sect. 3.2.1, the geometric mean value should be calculated to decide the final appraisal KPFs for bunkering port evaluation. The rule is if the geometric mean ≥ 4, the factor is accepted as appraisal KPF, if not, the factor is rejected. The factor scores of each of the 12 experts were obtained from the expert questionnaire 1. As the factor "bunker price" as an example, 12 experts have 12 opinions about the importance degree of this factor. Therefore, the geometric mean value of this factor is shown as follows:

[3]If any Cronbach's Alpha is less than 0.35, the corresponding datum is not reliable and will be deleted. Those more than 0.35 are viewed as reliable (Wang and Lin 2008).

Table 10.3 KPFs before/after analysis

No.	KPFs (before: 21)	Geometric value	KPFs (after: 15)
1	Bunker price	6.74	Bunker price
2	Geographical advantage	5.97	Geographical advantage
3	Port bunker fuel capacity	5.85	Port bunker fuel capacity
4	Supply waiting time	5.57	Supply waiting time
5	Bunker quality	6.38	Bunker quality
6	Safety of bunkering	6.22	Safety of bunkering
7	Port tariffs	5.84	Port tariffs
8	Information sharing among stakeholders	5.29	Information sharing among stakeholders
9	Port weather conditions	5.17	Port weather condition
10	Environmental restrictions effects	5.32	Environmental restrictions effects
11	Port time	6.23	Port time
12	Volume of containers	6.14	Volume of containers
13	Port bunker suppliers	4.25	Port bunker suppliers
14	Port bunkering supply regulations	4.08	Port bunkering supply regulations
15	Efficiency of bunker supply	5.98	Efficiency of bunker supply
16	*Experienced human resources*	*3.98*	
17	*Port congestion condition*	*3.62*	
18	*Bunkering service at night*	*3.22*	
19	*Small order bunkering service*	*2.99*	
20	*Bunkering risk management*	*3.45*	
21	*Available bunkering barge*	*3.76*	

Geometric mean value of "bunker price" $= \sqrt[12]{7*6*7*7*7*6*6*7*7*7*7*7} = 6.74$

Fifteen crucial KPFs with a value greater than 4 were selected as bunkering port performance evaluation criteria after the geometric mean value calculation, as shown in Table 10.3, they are bunker price, geographical advantage, port bunker fuel capacity, supply waiting time, bunker quality, safety of bunkering, port tariffs, information sharing among stakeholders, port weather condition, environmental restrictions effects, port time, volume of containers, port bunker suppliers, port bunkering supply regulations and efficiency of bunker supply; also factor experienced human resources, port congestion condition, bunkering service at night, small order bunkering service, bunkering risk management and available bunkering barge with a value less than four were rejected.

4.2 Determination of KPF Weights: Fuzzy Delphi Method

Since the decision making of bunkering port selection is based on liner shipping companies' preferences but not service suppliers, the perspectives from liner shipping companies are significant. In the second questionnaire survey, because the target is a China-Korea-Japan route, three experts in the first questionnaire survey who are working in companies belonging to China, Korea, and Japan were chosen as respondents to determine the weights of KPFs using a 10 point performance scale. The second questionnaire includes two parts as shown in Fig. 10.1: part 1 is to establish the triangular fuzzy numbers of KPFs by Fuzzy Delphi Method and part 2 is to evaluate these KPFs and alternative bunkering ports by Fuzzy-TOPSIS method. Firstly, in this section, the triangular fuzzy numbers of KPFs can be established. Under the Fuzzy Delphi Method, the triangular fuzzy numbers of KPFs can be collected by Eqs. (10.1) to (10.4) mentioned in Sect. 3.2.2 as shown below.

Using Eqs. (10.2), (10.3) and (10.4), the triangular fuzzy numbers of KPF geographical advantage are calculated as:

$$a_{ij} = \min(M_{ijk}) = \min(9.0, 8.0, 8.0) = 8.0$$

$$b_{ij} = \sqrt[n]{\prod_{k=1}^{n} M_{ijk}} = \sqrt[3]{9.0 * 8.0 * 8.0} = 8.3$$

$$c_{ij} = \max(M_{ijk}) = \max(9.0, 8.0, 8.0) = 9.0$$

Therefore, the triangular fuzzy numbers of the KPF geographical advantage can be determined as (8.0, 8.3, 9.0). With the same calculation process, all the triangular fuzzy numbers of the KPFs can be obtained as shown in Table 10.4 with the name fuzzy weight. It indicates that respondents regard KPF bunker price, bunker quality, safety of bunkering, and port tariffs as important factors in selecting bunkering ports with the fuzzy weights as (10.0, 10.0, 10.0).

4.3 Assessment of Alternatives: Fuzzy TOPSIS Method

The selected 15 KPFs are used to assess alternatives containing both subjective and objective factors. To confirm the accuracy and objectivity of the evaluation results, objective factors that have crisp quantity values should be used in analysis as much as possible. However, most factors do not have crisp quantity value and need to be judged using subjective ideas. In this section, an integrative quantitative and qualitative analysis is employed to evaluate the performance of alternative bunkering ports in order to avoid overly subjective results. The evaluation of subjective KPFs performed by the judgment of decision makers in the second questionnaire survey via fuzzy theory and objective KPFs are evaluated using crisp quantity values. The fuzzy ratings under each subjective KPF can be given by the decision

Table 10.4 The fuzzy weights of KPFs and global fuzzy ratings of alternatives

KPFs	Fuzzy weights	Alternatives	Fuzzy ratings	Global ratings
Bunker price	(10.0,10.0,10.0)	Xingang	(9.7,9.7,9.7)	(97.0,97.0,97.0)
		Dalian	(9.8,9.8,9.8)	(98.0,98.0,98.0)
		Busan	(10.0,10.0,10.0)	(100,100,100)
		Niigata	(9.9.9.9,9.9)	(99.0,99.0,99.0)
Geographical advantage	(8.0,8.3,9.0)	Xingang	(4.6,5.6,6.6)	(36.6,46.4,59.4)
		Dalian	(3.6,4.7,5.8)	(29.1,39.2,51.9)
		Busan	(6.6,7.6,8.6)	(52.8,63.2,77.6)
		Niigata	(2.0,3.0,4.0)	(16.0,24.9,36.0)
Port bunker fuel capacity	(7.0,7.7,8.0)	Xingang	(3.2,4.2,5.2)	(34.5,40.0,52.8)
		Dalian	(4.9,5.2,6.6)	(36.7,48.2,58.1)
		Busan	(5.2,6.3,7.3)	(42.0,53.9,64.0)
		Niigata	(2.0,2.5,3.6)	(14.0,19.0,29.1)
Supply waiting time	(8.0,9.3,10.0)	Xingang	(3.2,4.2,5.2)	(25.4,39.2,52.4)
		Dalian	(4.6,5.6,6.6)	(36.6,52.0,66.0)
		Busan	(4.6,5.6,6.6)	(36.6,52.0,66.0)
		Niigata	(6.0,7.0,8.0)	(48.0,65.1,80.0)
Bunker quality	(10.0.10.0,10.0)	Xingang	(4.0,5.0,6.0)	(40.0,50.0,60.0)
		Dalian	(4.0,5.0,6.0)	(40.0,50.0,60.0)
		Busan	(6.6,7.6,8.6)	(66,76.1,86.2)
		Niigata	(6.0,7.0,8.0)	(60.0,70.0,80.0)
Volume of containers	(8.0,8.0,8.0)	Xingang	(5.3,5.3,5.3)	(42.4,42.4,42.4)
		Dalian	(2.7,2.7,2.7)	(21.6,21.6,21.6)
		Busan	(10.0,10.0,10.0)	(80.0,80.0,80.0)
		Niigata	(1.3,1.3,1.3)	(10.4,10.4,10.4)
Safety of bunkering	(10.0,10.0,10.0)	Xingang	(6.0,7.0,8.0)	(60.0,70.0,80.0)
		Dalian	(6.0,7.0,8.0)	(60.0,70.0,80.0)
		Busan	(6.0,7.0,8.0)	(60.0,70.0,80.0)
		Niigata	(7.3,8.3,9.3)	(72.7,82.8,92.8)
Port bunker suppliers	(8.0,8.3,9.0)	Xingang	(2.5,2.5,2.5)	(20.0,20.8,22.5)
		Dalian	(5.8,5.8,5.8)	(46.4,48.1,52.2)
		Busan	(10.0,10.0,10.0)	(80.0,83.0,90.0)
		Niigata	(5.5.5.0,5.0)	(40.0,41.5,45.0)
Port bunkering supply regulations	(6.0,7.3,8.0)	Xingang	(4.0,5.0,6.0)	(24.0,36.5,48.0)
		Dalian	(4.0,5.0,6.0)	(24.0,36.5,48.0)
		Busan	(6.0,7.0,8.0)	(36.0,51.1,64.0)
		Niigata	(1.6,2.1,3.2)	(9.5,15.2,25.4)
Port tariffs	(10.0,10.0,10.0)	Xingang	(2.5,3.6,4.6)	(25.2,35.6,45.8)
		Dalian	(2.5,3.6,4.6)	(25.2,35.6,45.8)
		Busan	(3.2,4.2,5.2)	(31.7,42.2,52.4)
		Niigata	(2.0,2.5,3.6)	(20.0,24.7,36.3)
Information sharing among stakeholders	(9.0,9.0,9.0)	Xingang	(2.5,3.6,4.6)	(22.7,32.0,41.2)
		Dalian	(2.5,3.6,4.6)	(22.7,32.0,41.2)
		Busan	(6.0,7.0,8.0)	(54.0,63.0,72.0)
		Niigata	(2.5,3.6,4.6)	(22.7,32.0,41.2)

(continued)

Table 10.4 (continued)

KPFs	Fuzzy weights	Alternatives	Fuzzy ratings	Global ratings
Port weather conditions	(6.0,7.9,9.0)	Xingang	(6.0,7.0,8.0)	(36.0,55.3,72.0)
		Dalian	(6.0,7.0,8.0)	(36.0,55.3,72.0)
		Busan	(5.2,6.3,7.3)	(31.4,49.4,65.4)
		Niigata	(5.2,6.3,7.3)	(31.4,49.4,65.4)
Efficiency of bunker supply	(9.0,9.3,10.0)	Xingang	(7.3,8.3,9.3)	(65.4,77.0,92.8)
		Dalian	(7.3,8.3,9.3)	(65.4,77.0,92.8)
		Busan	(6.0,7.0,8.0)	(54.0,65.1,80.0)
		Niigata	(4.0,5.0,6.0)	(36.0,46.5,60.0)
Environmental restrictions effects	(6.0,7.6,9.0)	Xingang	(7.3,8.3,9.3)	(43.6,62.9,83.5)
		Dalian	(7.3,8.3,9.3)	(43.6,62.9,83.5)
		Busan	(6.0,7.0,8.0)	(36.0,53.2,72.0)
		Niigata	(2.5,3.6,4.6)	(15.1,27.0,41.2)
Port time	(9.0,9.3,10.0)	Xingang	(10.0,10.0,10.0)	(90.0,93.0,100)
		Dalian	(3.9,3.9,3.9)	(35.1,36.3,39.0)
		Busan	(6.8,6.8,6.8)	(61.2,63.2,68.0)
		Niigata	(8.0,8.0,8.0)	(72.0,74.4,80.0)

makers according to the linguistic variables, i.e., poor, medium poor, fair, medium good, and good. Taking subjective KPF “geographical advantage”(see Table 10.5), the three experts’ opinions of the performance of four alternative bunkering ports under the subjective KPF “geographical advantage” indicate that expert 1 thinks that the performance of four alternative bunkering ports under “geographical advantage” as ‘Medium good’, ‘Medium good’, ‘Good’, and ‘Poor’, respectively. By transforming these opinions to linguistic variables, the triangular fuzzy numbers are (6,7,8), (6,7,8), (8,9,10), and (1,1,2), respectively. In a similar way, the opinion of expert 2 can be transformed as (4,5,6), (2,3,4), (8,9,10), (2,3,4) and expert 3 is as (4,5,6), (2,3,4), (8,9,10) and (8,9,10), respectively. After obtaining the triangular fuzzy numbers of each expert, the fuzzy ratings of each alternative bunkering port under the subjective KPF “geographical advantage” can be calculated by Eq. (10.3), i.e., the geometric mean of these three triangular fuzzy numbers gathered from three experts. For example, the fuzzy ratings of alternative bunkering port Xingang under “geographical advantage” can be calculated as:

$$\begin{aligned}\text{Fuzzy ratings of alternative bunkering port Xingang} &= \sqrt[3]{6*4*4} = 4.6(\text{a})\\ &= \sqrt[3]{7*5*5} = 5.6(\text{b})\\ &= \sqrt[3]{8*6*6} = 6.6(\text{c})\end{aligned}$$

Therefore, the fuzzy rating of alternative bunkering port of Xingang is (4.6, 5.6, 6.6). In a similar way, all the fuzzy ratings of four alternative bunkering ports under the subjective KPFs can be obtained shown in Table 10.5.

Table 10.5 Expert questionnaire 2: Part 2: sample question asking for opinions on the subjective KPF "geographical advantage"

Alternatives (expert 1)	What are your thoughts on the bunkering port performance in respect of factor geographical advantage? (tick one box on each row)				
	Poor (P)	Medium poor (MP)	Fair (F)	Medium good (MG)	Good (G)
Xingang				O	
Dalian				O	
Busan					O
Niigata	O				
Alternatives (expert 2)	What are your thoughts on the bunkering port performance in respect of factor geographical advantage? (tick one box on each row)				
	Poor (P)	Medium poor (MP)	Fair (F)	Medium good (MG)	Good (G)
Xingang			O		
Dalian		O			
Busan					O
Niigata		O			
Alternatives (expert 3)	What are your thoughts on the bunkering port performance in respect of factor geographical advantage? (tick one box on each row)				
	Poor (P)	Medium poor (MP)	Fair (F)	Medium good (MG)	Good (G)
Xingang			O		
Dalian		O			
Busan					O
Niigata					O

When the KPFs have crisp quantity value for each alternative, the crisp quantities for the KPFs should be transformed into fuzzy numbers to ensure compatibility with the linguistic numbers of the KPFs. The crisp quantities for objective KPFs transformed into fuzzy ratings can be calculated by Eqs. (10.5) and (10.6). The crisp quantity value "bunker price" for each alternative can be obtained by Bunker Ports News World- wide (2013) and Bunker World (2013), "volume of containers" can be obtained in Containerization Year Book 2012, "port bunker suppliers" and "port time" can be obtained by CK Line shipping company in 2013.

In the situation of benefit KPF "volume of containers (TEU)", Eq. (10.5) is employed as:

$$\text{Fuzzy ratings of alternative bunkering port of Xingang} = \left\{ R^o_{ij} / \max_i \left\{ b^o_{ij} \right\} \right\} \times 10$$

$$= \left(\frac{8562600}{16175000} \right) \times 10 = 5.3$$

Therefore, the fuzzy ratings of alternative bunkering port of Xingang under objective KPF "volume of containers (TEU)" is (5.3,5.3,5.3).

Table 10.6 Fuzzy ratings for alternatives under objective KPFs

Objective KPFs	Quantity	Fuzzy rating
Bunker price ($USD)		
Xingang	690	(9.7,9.7,9.7)
Dalian	687	(9.8,9.8,9.8)
Busan	670	(10.0,10.0,10.0)
Niigata	677	(9.9,9.9,9.9)
Volume of containers (TEU)		
Xingang	8,562,600	(5.3,5.3,5.3)
Dalian	4,628,200	(2.7,2.7,2.7)
Busan	16,175,000	(10.0,10.0,10.0)
Niigata	204,960	(1.3,1.3,1.3)
Port bunker suppliers (bunker companies)		
Xingang	3	(2.5,2.5,2.5)
Dalian	7	(5.8,5.8,5.8)
Busan	12	(10.0,10.0,10.0)
Niigata	6	(5.0,5.0,5.0)
Port time (h)		
Xingang	44	(10.0,10.0,10.0)
Dalian	17	(3.9,3.9,3.9)
Busan	30	(6.8,6.8,6.8)
Niigata	35	(8.0,8.0,8.0)

Source: Bunker Ports News World- wide. Available from: http://www.bunkerportsnews.com/
Bunker World. Available from: http://www.bunkerworld.com/
CK Line. Available from: http://www.ckline.co.kr/
Containerization Year Book 2012

In the situation of cost KPF "bunker price ($USD)", Eq. (10.6) is employed as:

$$\text{Fuzzy ratings of alternative bunkering port Xingang} = \left\{ \min_{i} \left\{ c_{ij}^{o} \right\} / R_{ij}^{o} \right\} \times 10$$

$$= (670/690) \times 10 = 9.7$$

Therefore, the fuzzy ratings of alternative bunkering port of Xingang under objective KPF "bunker price (in USD)" is (9.7,9.7,9.7). In a similar way, all the fuzzy ratings of four alternative bunkering ports under the objective KPFs are shown in Table 10.6.

The global fuzzy rating based on the fuzzy ratings multiplied by fuzzy weights can be obtained using Eq. (10.8). For example, to obtained the global rating of Xingang alternative bunkering port under KPF "bunker price", firstly, the fuzzy weights of KPF "bunker price" and the fuzzy ratings of Xingang alternative bunkering port under KPF "bunker price" should be obtained as (10.0,10.0,10.0) and (9.7,9.7,9.7) respectively.

The global rating of Xingang as an alternative bunkering port under KPF "bunker price" $= (10.0*9.7, 10.0*9.7, 10.0*9.7) = (97.0, 97.0, 97.0)$.

In a similar way, all the global fuzzy ratings of four alternative bunkering ports under each KPF can be obtained.

To evaluate the four alternative bunkering ports, the distances of each alternative from the positive-ideal solution $\left(D_i^*\right)$ and the negative-ideal solution $\left(D_i^-\right)$ should be calculated as the calculation process mentioned in Sect. 3.2.3. Therefore, the positive-ideal (A^*) and negative-ideal (A^-) solution under each KPF should be determined using Eqs. (10.9) and (10.10). For example, under KPF "Bunker price", there are four alternative bunkering ports as solutions, the global ratings of them are (97.0,97.0,97.0), (98.0,98.0,98.0), (99.0,99.0,99.0), and (100.0,100.0,100.0).

$$\text{The positive-ideal}\left(A^*\right)\text{solution under KPF "Bunker price"} = (100.0, 100.0, 100.0)$$

$$\text{The negative-ideal}\left(A^-\right)\text{solution under KPF "Bunker price"} = (97.0, 97.0, 97.0)$$

In a similar way, all the positive-ideal (A^*) and negative-ideal (A^-) solution under each KPF can be obtained shown in Table 10.7.

The distances of each alternative from the positive-ideal solution $\left(D_i^*\right)$, the negative-ideal solution $\left(D_i^-\right)$ and the relative closeness of each alternative to the ideal solution (FC_i) can be calculated by Eqs. (10.13) and (10.14). All the relative closeness of each alternative to the ideal solution can be obtained shown in Table 10.8.

5 Discussion

The final ranking of alternatives is determined according to the FC_i values. It indicates that among the four alternative bunkering ports, the port of Busan is ranked as the most competitive bunkering port in the target China-Korea-Japan regular liner route, followed by the ports of Xingang, Niigata, and Dalian. In the actual operation, the liner shipping companies normally obtain bunkering services at the port of Busan. If the port of Busan is unavailable for vessel bunkering, bunker services are sometimes obtained from the ports of Xingang or Niigata. Such results illustrate that the developed benchmarking rule is appropriate and helpful for liner shipping companies to make optimal decisions on the choice of bunkering ports.

After the analysis, the strengths and weaknesses of the target bunkering ports and strategic recommendations can be identified. The results indicate that for most of the KPFs, the port of Busan is superior to other ports except for the KPF supply waiting time and the efficiency of bunker supply. The port of Busan should improve its efficiency of port operation under the limited port capacity to enhance its bunker supply efficiency and shorten its waiting time. The port of Xingang is taking advantage of its KPFs, ie.e, volume of containers and long port time, but it faces

Table 10.7 Fuzzy positive and negative ideal solution

KPFs	A^*	A^-
Bunker price	(100.0,100.0,100.0)	(97.0,97.0,97.0)
Geographical advantage	(52.8,63.2,77.6)	(16.0,24.9,36.0)
Port bunker fuel capacity	(42.0,53.9,64.0)	(14.0,19.0,29.1)
Supply waiting time	(48.0,65.1,80.0)	(25.4,39.2,52.4)
Bunker quality	(66.0,76.1,86.2)	(40.0,50.0,60.0)
Volume of containers	(80.0,80.0,80.0)	(10.4,10.4,10.4)
Safety of bunkering	(72.7,82.8,92.8)	(60.0,70.0,80.0)
Port bunker suppliers	(80.0,83.0,90.0)	(20.0,20.8,22.5)
Port bunkering supply regulations	(36.0,51.1,64.0)	(9.5,15.2,25.4)
Port tariffs	(31.7,42.2,52.4)	(20.0,24.7,36.3)
Information sharing among stakeholders	(54.0,63.0,72.0)	(22.7,32.0,41.2)
Port weather conditions	(36.0,55.3,72.0)	(31.4,49.4,65.4)
Efficiency of bunker supply	(65.4,77.0,92.8)	(36.0,46.5,60.0)
Environmental restrictions effects	(43.6,62.9,83.5)	(15.1,27.0,41.2)
Port time	(90.0,93.0,100.0)	(35.1,36.3,39.0)

Table 10.8 The final ranking of alternatives

	D*	D-	FCi	Ranking
Xingang	248.3783	234.6274	0.485765	2
Dalian	287.2261	195.6024	0.405118	4
Busan	83.36617	399.2777	0.827272	1
Niigata	168.5924	120.0256	0.415863	3

the problems of long supply waiting time and port bunker suppliers. The operators should improve the efficiency of the port operations, but simultaneously should secure some powerful bunker suppliers in the port; such partnerships will lead more liner shipping companies to obtain bunker services at this port. The port of Dalian has a bunker capacity advantage and a set number of bunker suppliers but lacks container volume and port time. The port of Niigata has high bunker quality and short supply waiting time but limited bunker capacity and container volume.

To determine the changes in the final alternative selection with variation taken from expert opinions, a sensitivity analysis was performed to determine the variation effects on the final selection of alternatives. The results of the sensitivity analysis in Table 10.9 show that, rather than the port of Xingang, the port of Niigata would rank as the second most competitive bunkering port if it were to increase it's the capacity of bunker fuel (KPF 3). At present, most of the bunkering capacity in the port of Niigata is used for domestic but not international vessels (Niigata Port Authority 2013). Thus, it limits the needs of liner shipping companies, and in the future, these companies will not consider the port of Niigata as an optional bunkering port for service. Therefore, expansion of its capacity of bunker fuel is crucial. Also, if the port of Dalian relaxes its strict regulations on port bunkering supply (KPF 9), it will overtake the port of Niigata in popularity. By 2012, only five

Table 10.9 Sensitivity analysis results

Alternative	*KPF 3: Port bunker fuel capacity*				
	D^*	D^-	FC_i	Ranking	Changed ranking
Xingang	263.41	212.82	0.45	2	3
Dalian	302.38	174.23	0.37	4	
Busan	83.37	377.67	0.82	1	
Niigata	135.82	120.03	0.47	3	2
	KPF 9: Port bunkering supply regulations				
	D^*	D^-	FC_i	Ranking	Changed ranking
Xingang	262.67	234.63	0.47	2	
Dalian	272.93	224.17	0.45	4	3
Busan	83.37	399.28	0.83	1	
Niigata	168.69	120.03	0.42	3	4

bunker suppliers in China have been approved by the Chinese government to supply bunker for international vessels. The strict control of foreign exchange and restricted bunker supply regulations, i.e., bunker fuel produced domestically cannot be supplied to international vessels, directly leads to the importation of bunker fuel for international vessels. The additional transport and transfer fees cause higher bunker prices in China, as compared to adjacent countries (Dong et al. 2011). Also the non-unified bunker supply management regulations may limit the efficiency of the bunkering process. In the future, port operators should loosen the restricted bunker supply regulations via negotiations with the government in order to provide more flexible and high quality bunker services to liner shipping companies.

6 Conclusions

With sustained high bunker prices, choosing optimal bunkering ports in order to save on total operating costs is an important issue for liner shipping companies. The bunkering port selection problem is solved by ship planning software that can only work optimally when ship arrivals can be forecasted rather accurately but that ignores liner shipping companies' preferences. By using a hybrid Fuzzy-Delphi-TOPSIS methodology, this chapter has developed a benchmarking framework for liner shipping companies operating along regular liner routes so as to evaluate the performance of bunkering ports. The results indicate that the KPFs of bunker price, bunker quality, safety of bunkering, and port tariffs are regarded as important factors in selecting bunkering ports. The port of Busan is ranked as the most competitive bunkering port, followed by the ports of Xingang, Niigata, and Dalian. In addition, the sensitivity analysis reveals that the port of Niigata can enhance its competitiveness by increasing its capacity of bunker fuel (KPF 3), and the port of Dalian should relax its strict regulations on port bunkering supply (KPF 9).

This chapter offers some important insight to liner shipping companies. The proposed framework enables them to: (1) better understand the complex relationships among the relevant KPFs; (2) more clearly understand the condition and changes of alternative bunkering ports; and (3) determine whether the bunkering services they receive are from the most efficient bunker ports and make prompt adjustments to meet their development strategies. Also, it enables bunkering port managers to: (1) comprehend the present strengths and weaknesses of their port; and (2) help them to establish future strategies to enhance the competitiveness of their port.

Admittedly, this chapter has its limitations, including: (1) during the process of solving the bunkering port selection problem, ship size factors are not considered; and (2) it is assumed that decision making mainly depends on users, i.e., liner shipping companies' decisions may be subjective and thus ignore the actual conditions of the ships and ports. We hereby propose some directions for future research, namely: (1) incorporate ship size into analysis; (2) develop a model that both considers the opinion of liner shipping companies and based on software for bunkering port selection decision making; and (3) compare the differences between regional routes and inter-continental routes. We strongly believe that this chapter represents the first step in exploring the KPFs for bunkering port selection.

References

Acosta M, Coronado D, Cerban MDM (2011) Bunkering competition and competitiveness at the ports of the Gibraltar Strait. J Transp Geogr 19:911–916

Adler M, Ziglio E (1996) Gazing into the oracle. Jessica Kingsley Publishers, London

Anfindsen OJ, Løvoll G, Mestl T (2012) Benchmarking of marine bunker fuel suppliers: the good, the bad, the ugly. BIJ 19(1):109–125

Bao Q, Ruan D, Shen Y, Hermans E, Janssena D (2012) Improved hierarchical fuzzy TOPSIS for road safety performance evaluation. Knowl-Based Syst 32:84–90

Besbes O, Savin S (2009) Going bunkers: the joint route selection and refueling problem. Manuf Serv Oper Manag 11(4):694–711

Bunker Ports News World- wide (2013) Available from. http://www.bunkerportsnews.com/. 23 Mar 2013

Bunker World (2013) Available from. http://www.bunkerworld.com/. 23 Mar 2013

Buyukozkan G, Cifci G (2012) A combined fuzzy AHP and fuzzy TOPSIS based strategic analysis of electronic service quality in healthcare industry. Expert Syst Appl 39:2341–2354

Chang Y, Chen C (2006) Knowledge-based simulation of bunkering services in the port of Kaohsiung. Civ Eng Environ Syst 23(1):21–34

Chang PC, Wang YW (2006) Fuzzy Delphi and back-propagation model for sales forecasting in PCB industry. Expert Syst Appl 30:715–726

Chong HY, Adnan H, Zin RM (2012) A feasible means of methodological advance from Delphi methods: a case study. Int JAcad Res 4(2):247–253

Chou CC (2010) An integrated quantitative and qualitative FMCDM model for location choices. Soft Comput 14(7):757–771

CK Line (2013) Available from. http://www.ckline.co.kr/. 23 Mar 2013

Corbett JJ, Wang H, Winebrake JJ (2009) The effectiveness and costs of speed reductions on emissions from international shipping. Transp Res Part D Transp Environ 14:593–598

Dinwoodie J, Tuck S, Knowles H, Benhin J, Sansom M (2012) Sustainable development of maritime operations in ports. Bus Strateg Environ 21(2):111–126

Dong ZH, Xu P, Feng B (2011) A research on China international voyage ship fuel supply development. China Ports 2011(09):53–55

Duru O, Bulut E, Yoshida S (2012) A fuzzy extended Delphi method for adjustment of statistical time series prediction: an empirical study on dry bulk freight market case. Expert Syst Appl 39:840–848

Du Y, Chen Q, Quan X, Long L, Fung RYK (2011) Berth allocation considering fuel consumption and vessel emissions. Transport Res E Log Transp Rev 47:1021–1037

Fu B (2009) The current state and future development prospects of the Chinese market for boat fuel. Int Pet Econ 17(12):23–26

Hsu TH, Yang TH (2000) Application of fuzzy analytic hierarchy process in the selection of advertising media. J Manag Syst 7(1):19–39

Huang SY, Hsu WJ, He Y (2011) Assessing capacity and improving utilization of anchorages. Transport Res E Log Transp Rev 47:216–227

Hu SJ (2005) Comprehensive transportation engineering. Beijing Jiaotong University Press, Beijing, pp 116–119

IMO (2009) International shipping and world trade facts and figures. Maritime Knowledge Center. URL Available from: http://www.imo.org. 24 Mar 2013

Jun KS, Chung ES, Kim YG, Kim Y (2012) A fuzzy multi-criteria approach to flood risk vulnerability in South Korea by considering climate change impacts. Expert Syst Appl 40 (4):1003–1013

Kim HJ, Chang YT, Kim KT, Kim HJ (2012) An epsilon-optimal algorithm considering greenhouse gas emissions for the management of a ship's bunker fuel. Transp Res Part D Transp Environ 17:97–103

Kim Y, Chung ES, Jun SM, Kim SU (2013) Prioritizing the best sites for treated wastewater instream use in an urban watershed using fuzzy TOPSIS. Resour Conserv Recycl 73:23–32

Kuo YF, Chen PC (2008) Constructing performance appraisal indicators for mobility of the service industries using fuzzy Delphi method. Expert Syst Appl 35:1930–1939

Mazraati M (2011) Challenges and prospects of international marine bunker fuels demand. OPEC Energ Rev 35(1):1–26

Ma Z, Shao C, Ma S, Ye Z (2011) Constructing road safety performance indicators using fuzzy Delphi method and Grey Delphi method. Expert Syst Appl 38:1509–1514

Ng AKY (2006) Assessing the attractiveness of ports in the north European container transshipment market: an agenda for future research in port competitiveness. Marit Econ Logist 8 (3):234–250

Niigata port authority (2013) Available from. http://www.pref.niigata.jp. 23 Mar 2013

Notteboom TE, Cariou P (2011) Bunker costs in container liner shipping: are slow steaming practices reflected in Maritime fuel surcharges? Paper presented at the European Conference on Shipping & Ports- (ECONSHIP) 2011, Chios, June 2011

Notteboom TE, Vernimmen B (2009) The effect of high fuel costs on liner service configuration in container shipping. J Transp Geogr 17:325–337

Prusty SK, Mohapatra PKJ, Mukherjee CK (2010) GOS tree (goal-objective- strategy tree) approach to strategic planning using a fuzzy-Delphi process: an application to the Indian shrimp industry. Technol Forecast Soc Chang 77:442–456

Psaraftis HN, Kontovas CA (2010) Balancing the economic and environmental performance of maritime transportation. Transp Res Part D Transp Environ 15:458–462

Qi X, Song DP (2012) Minimizing fuel emissions by optimizing vessel schedules in liner shipping with uncertain port times. Transport Res Part E Log Transp Rev 48:863–880

Schinas O, Stefanakos CN (2012) Cost assessment of environmental regulation and options for marine operators. Transp Res Part C Emerg Technol 25:81–99

Talley WK, Yip TL, Jin D (2012) Determinants of vessel-accident bunker spills. Transp Res Part D Transp Environ 17:605–609

Tsai HY, Chang CW, Lin HL (2010) Fuzzy hierarchy sensitive with Delphi method to evaluate hospital organization performance. Expert Syst Appl 37:5533–5541

Wang JW (2002) Prospect of maritime bunker supply of China. Maritime China 12:38–39

Wang ML, Lin YH (2008) To construct a monitoring mechanism of production loss by using fuzzy Delphi method and fuzzy regression technique: a case study of IC package testing company. Expert Syst Appl 35:1156–1165

Wang S, Meng Q (2012a) Sailing speed optimization for container ships in a liner shipping network. Transport Res Part E Log Transp Rev 48:701–714

Wang S, Meng Q (2012b) Robust schedule design for liner shipping services. Transport Res Part E Log Transp Rev 48:1093–1106

Wang S, Meng Q (2012c) Liner ship fleet deployment with container transshipment operations. Transport Res Part E Log Transp Rev 48:470–484

Wang S, Meng Q (2012d) Liner ship route schedule design with sea contingency time and port time uncertainty. Transp Res B Methodol 46:615–633

Wu YW (2011) The analysis of China's marine fuel oil market. Pet Petrochem Today 202:36–40

Yao Z, Ng SH, Lee LH (2012) A study on bunker fuel management for the shipping liner services. Comput Oper Res 39:1160–1172

You CD (2008) China ship fuel supply market analysis and prospect. Shipp Manag 30(03):19–22

Yuan S (2012) The managing status quo and countermeasures of fuel oil cost about domestic shipping companies. Logist Sci-Technol 2012(8):85–89

Chapter 11
Modern Heuristics of MCDM for the Operation Optimization in Container Terminals

Zhihong Jin, Na Li, Qi Xu, and Zhan Bian

Abstract This chapter systematically applies modern heuristics to solve multi-criteria decision making problems in the fields of container terminal, which consists of three geographically interrelated core areas: container terminal, anchorage ground on its sea side, and gateway on its land side. For the container terminal, the container loading sequence problem is considered and a hybrid dynamic programming approach is proposed. The considered problem aims at obtaining an optimized container loading sequence for a crane to retrieve all the containers from the yard to the ship. The proposed dynamic algorithms consist of two phases. A heuristic algorithm is developed to retrieve the containers subset which needs no relocation and may be loaded directly onto the ship at the first phase, and a dynamic programming with heuristic rules is applied to solve the loading sequence problem for the rest of the containers at the second phase. For the anchorage ground on the sea side of a container terminal, the tugboat scheduling problem is formulated as a multiprocessor tasks scheduling problem after analyzing the characteristics of tugboat operation. The model considers factors of multi-anchorage bases and three stages of operations (berthing/shifting-berth/unberthing). The objective is to minimize the total operation times for all tugboats and the waste of the tugboats horsepower in use at the same time. A hybrid simulated annealing algorithm is proposed to solve the addressed problem. For the gateway on the land side of a container terminal, resource deployment for truck appointment system on container terminals is solved as an optimization problem. A bi-objective model is set up to minimize resource input and balance workloads. Modern heuristics method based on non-dominated genetic algorithmII is proposed to solve difficulties of simultaneous optimization of resource input and appointment quotas. Three chromosomes representing quotas, yard cranes and gate lanes are set up, some of which are two

Z. Jin (✉) • N. Li • Q. Xu
College of Transportation Management, Dalian Maritime University, Dalian, China
e-mail: jinzhihong@dlmu.edu.cn

Z. Bian
College of Business Administration, Capital University of Economics and Business, Beijing, China

P.T.-W. Lee, Z. Yang (eds.), *Multi-Criteria Decision Making in Maritime Studies and Logistics*, International Series in Operations Research & Management Science 260,
DOI 10.1007/978-3-319-62338-2_11

dimensional. Numerical experiments are untaken to evaluate the effectiveness of the proposed algorithms and show the efficiency of the proposed algorithm. The three parts analyzed above cover all the core elements of modern heuristics of MCDM for the operation optimization in a container terminal from a container terminal to both its land side and its sea side.

Keywords Operation optimization • Container loading sequence problem • Dynamic programming • Tugboat scheduling problem • Simulated annealing • Resource deployment • Non-dominated genetic algorithm

1 Introduction

Container terminal is an important part in international logistics and plays a significant role in world trade. Recently, more and more people become to recognize the importance of global logistic business via container terminals. As the throughput of containers in container terminal increases and competition between ports becomes fierce, how to improve the efficiency in container terminal has become an important and immediate challenge for port managers. One of the most important performance measures in container terminals is to schedule all kinds of equipment at an optimum level and to reduce the turnaround time of vessels.

Practically, there are three geographically interrelated core areas: container terminal, anchorage ground on its sea side, and gateway on its land side. This chapter analyzes three optimization problems related to the three core areas mentioned above, i.e. (i) the container loading sequence problem in the terminal, (ii) the tugboat scheduling problem in the anchorage ground, (iii) the resource deployment in truck appointment system.

With the rapid development of container transport, most yards stack up containers to utilize more and more precious space. However, extra movements, that waste time and money, occur when a container is due to be retrieved from the yard but is under beneath other ones. Therefore, how to avoid or reduce the number of relocations to enhance the efficiency of handling activities has become a key issue for container terminals.

Besides, the tugboat scheduling optimization in container terminals is another problem that has to be solved so that the efficiency of container terminal can be improved. That is because the performance of the tugboat operation scheduling has a direct influence on time when a ship can start her handling operation and when a ship can leave the port.

Moreover, severe congestion of external trucks in and out has become a major problem for many terminals. Appointment is one of innovations of port practitioners and effective ways to solve the problem, and various forms of appointment systems have been adopted at many ports. But because of the lack of simultaneously scheduling of terminal operation with truck appointments, the effect of the appointed systems is not as good as anticipated. Basically, that is the problem of balancing between supply and demand. Terminal operators usually give out quotas according to their own capacity. The appointment quotas for each kind of containers or each yard block are decided by the input resources. Accordingly, the demand of external and internal trucks has to be satisfied by supply. The container terminal operators prefer the least input, while the truck companies prefer the most convenience. So, it is necessary and very important to balance the deployment of resources and the appointment quota.

2 Literature Review

About the container relocation and loading sequence problem, Kim (1997) developed a method for estimating the number of relocations for import containers. This method also applies to other papers, e.g. Kim and Kim (1999) addressed the relationship between storage height and relocations for import containers, and formulated mathematical models under several arrival strategies, then applied the Lagrangian relaxation and sub-gradient optimization to solve for the best storage height. The container storage location has a direct impact on the follow-up operations. Kim et al. (2000) proposed a dynamic programming model to minimize the number of relocations considering the factor of container's weight, and developed a decision tree to support real-time decisions. Yang and Kim (2006) suggested a genetic algorithm with simple heuristic rules to solve the dynamic location problem as well as the static one. To deal with the container retrieving problem during the loading process, Kim and Hong (2006) suggested a branch-and-bound algorithm for determining the locations of relocated blocks. Lee and Lee (2010) developed a three-phase heuristic algorithm to minimize the number of relocations. Caserta et al. (2011) applied dynamic programming to get a proper solution. Xu et al. (2008) mainly considered the determination of relocated container locations to reduce the relocation rate in container yards. Yi et al. (2010) formulated a gambling model to address the same issue. Wang et al. (2005) put forward a mathematic matrix model to deal with the container loading sequence problem. Zhu et al. (2010) proposed improvement strategies based on actual rules. Jin et al. (2011a, b) applied a heuristic algorithm to solve the problem. Existing researches in this field are trying hard to improve the applicability of optimization methods. Nevertheless, the quality of obtained solutions declines significantly owing to the expansion of the problems scale. What is more, the solutions are unable to display the process of retrieving operations which makes the algorithms lacking of practicability.

As to the tugboat scheduling problem, Liu (2009) established a mathematical model on the tugboat scheduling problem combined with the MTSP theory, and adopted the hybrid evolutionary strategy to solve the model. Liu (2011) established an tugboat scheduling model considering the minimum operation distance of the tugboats, and compared the performance of hybrid evolutionary strategy with the particle swarm optimization algorithm for solving the addressed problem. Wang and Meng (2007) used a hybrid method combined Ant Colony Optimization and

Genetic Algorithm to resolve the tugboat allocation problem. Wang et al. (2012) formulated a mix-integer model for the tugboat assignment problem combined with existing scheduling rules, and analyzed the effects of the number and service capacity of tugboats on the turnaround time of ships. Liu and Wang (2005) considered the tugboat operation scheduling problem as a parallel machines scheduling problem with special process constraint, and employed a hybrid algorithm based the evolutionary strategy to solve the problem. Dong et al. (2012) adopted the improved particle swarm optimization combined with dynamic genetic operators to solve the formulated tugboat dispatch model. Liu and Wang (2006) used the particle swarm optimization algorithm combined with the local search approach to solve the tugboat scheduling model they proposed. Xu et al. (2012, 2014a, b) presented a tugboat scheduling model with multi-anchorage bases, different operation modes, derived a lower bound of the flow time of the operation system, and adopted the hybrid simulated annealing to the problem. Liu et al. (2016) adopted the rolling scheduling strategy based on time window to the dynamic tugboat scheduling problem for the one-way channel port.

As for the scheme making of appointment system, Huynh and Walton (2008) proposed a methodology to assist terminal operators to determine the optimal number of trucks to accept. Huynh et al. (2004) studied the relationship between the number of yard cranes and truck turn time. Huynh and Walton (2011) compared the individual appointment system versus block appointment system and analyze their effects on resource utilization and truck turn time in grounded operations. Namboothiri and Erera (2008) studied the management of a fleet of trucks providing container pickup and delivery service at a container terminal with appointment system. Phan and Kim (2015, 2016) also proposed a decentralized decision-making model to support the negotiation process between truck companies and the terminal operator. It gives an acceptable solution to every player involved as well as the near optimal solution to the entire system from the global point of view.

Not only the use of an appointment system can facilitate the movement of trucks in and out of the terminal, it can also help the terminal to manage its labour and yard resources (Huynh and Walton 2011), as well as to reduce emissions. If the truck arrival information could be acquired even partly, a more efficient yard crane dispatching (Guo et al. 2011) as well as labour allocation for yard and gate could be realised. In 2003, Kim et al. (2003) studied the yard crane scheduling based on the sequence of external trucks. Zhao and Goodchild (2010) proposed a simple rule for using truck information to reduce container rehandling work and suggests a method for evaluating yard crane productivity and truck transaction time. In 2013, Zhao and Goodchild (2013) published another paper specifying the benefit of truck appointment to improve the yard crane efficiency. Van Asperen et al. (2011) used a discrete-event simulation model to evaluate the impact of a truck announcement system on the performance of online container stacking rules. Morais and Lord (2006) developed a quantitative method to assess the effect of appointment system to reduce emissions at Canadian ports. For American ports Los Angeles and Long Beach, Giuliano and O'Brien (2007) presented an evaluation of truck appointment to reduce emissions. Chen et al. (2013) proposed a methodology to optimize truck arrival patterns to reduce emissions from idling truck engines at marine container terminals. More practical and considering the disruption of late or early arrivals of trucks, Li et al. (2015) set up a simulation model to prove the effectiveness of truck appointment to reducing truck emissions. Besides, many researchers studied the arrival pattern of external truck at container terminals. Guan and Liu (2009) applied a multi-server queuing model to analyze gate congestion and to quantify the truck waiting cost. An optimization model was developed to minimize the total gate system cost with data from field observations.

It can be seen from the literature that more and more researchers have become interested in optimization problems in the field of container terminal. The application of modern heuristics to the optimization problems mentioned above is effective and efficient. The following sectors of this chapter will introduce the application of modern heuristics to the container loading sequence problem, the tugboat scheduling, and the resource deployment in truck appointment problem, respectively.

3 The Container Loading Sequence Problem

3.1 Problem Definition and Basic Assumptions

Given an initial layout of a bay in the yard (known as the yard plan) and a final layout of a bay on the containership (known as the stowage plan), the container loading sequence problem yields a container retrieval sequence that retrieves all the containers from the yard bay, one at a time in a specified order, such that the number of container relocations is minimized.

Figure 11.1a shows the layout of a yard and the initial storage state of yard bay 5. Each number represents a slot, e.g. number 43 represents the container which is stacked at the stack 4, tier 3, denoted by container 43. Figure 11.1b illustrates the stowage plan of the containers in yard bay 5 of Fig. 11.1a. The containers are loaded onto bay 07 of the containership. Numbers 02, 04, 06 of the 1st column represent the 1st, 2nd, 3rd tier of the hold respectively. Numbers 82, 84 of the same column denote the 1st and 2nd tier above the deck respectively. Numbers 02, 04, 06 of the 1st row indicate the 1st, 2nd, 3rd column of the portside respectively. Numbers 01, 03, 05 of the same row describe the 1st, 2nd, 3rd column of the starboard respectively. Number 43 means that container 43 of yard bay 5 is loaded to the 1st tier in the hold, the 1st column along the portside. If there is a × in the slot, it means the slot is under no consideration during the loading process.

Figure 11.2 illustrates the loading sequence of the containers in yard bay 5. Numbers in the slots represent the loading sequence, e.g., number 1 means that container 43 is the first one to be retrieved and loaded onto the ship during the process.

Two kinds of yard cranes are usually applied to container terminal operations, namely Rubber Tyred Gantry Crane (RTGC) and Rail Mounted Gantry Crane

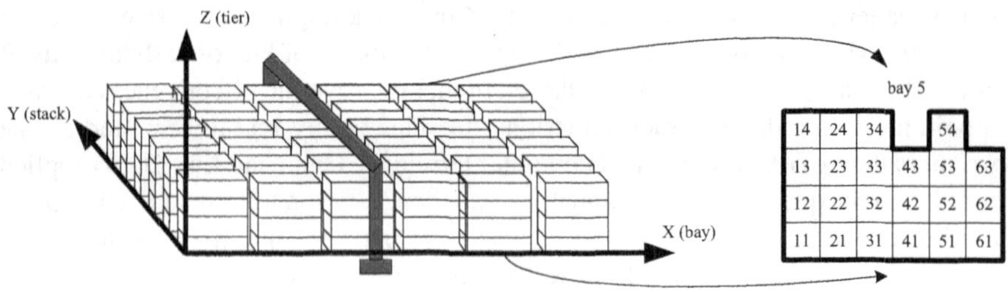

(a) The yard plan of bay 5

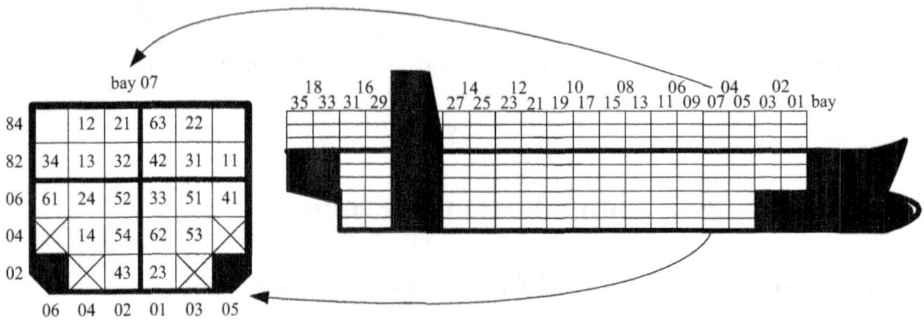

(b) The stowage plan of yard bay 5

Fig. 11.1 Illustrations of the yard plan and stowage plan of a bay (**a**) The yard plan of bay 5. (**b**) The stowage plan of yard bay 5

Fig. 11.2 The loading sequence of a yard bay

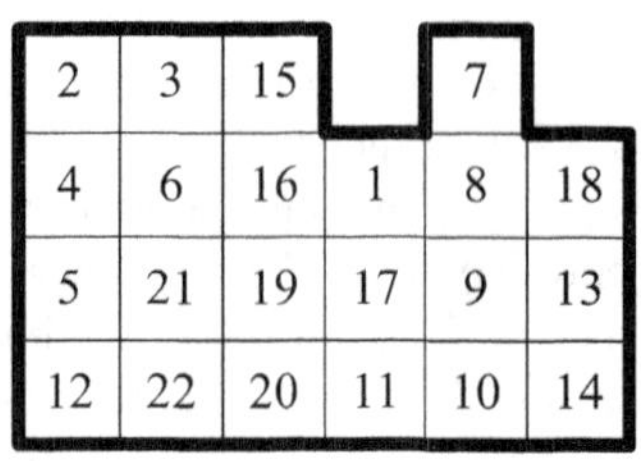

(RMGC). Their performances are summarized in Table 11.1. Besides, we focus on loading operations for only one bay in this research.

Furthermore, during the retrieving process, the yard crane can only retrieve containers from the top of a stack, i.e. late-in-first-out (LIFO) rule. For the loading operation, the containers have to be firstly loaded onto the hold then onto the deck by quay cranes. Based on the industry practices and on the analysis of the actual conditions of the container terminal operations, basic assumptions used in this research are made as follows:

Table 11.1 The performance of yard cranes

Device	Working span (Stack)	Stacking height (Tier)	Working capacity (TEU/Bay)
RTGC	6	3 ~ 4	Maximum 24
RMGC	10	6	Maximum 60

Source: Dalian Container Terminal CO., LTD

1. A yard bay consists of several stacks and each stack permits several containers to stack up within the maximal tier (i.e., 6 tiers).
2. Only containers of same dimensions and of the same vessel are stacked within a yard bay. To simplify notations and explanations, it is assumed that all the containers are of the same length (i.e., 20-feet long).
3. All the containers in the bay have accomplished the Customs examination, i.e., there is no relocation that results from Customs un-examination.
4. Without consideration of new containers arriving, the initial state of a yard bay is given and the stowage plan which is approved by the shipping line is known.
5. Only at the moment when a container, which is not on the top tier, is to be retrieved, relocations of above containers within the bay occur.
6. Each relocation operation moves the top container to the empty slot of another stack, and the empty slot does not hang in the air.
7. The initial yard bay has enough space to locate all containers relocated for retrieving a container.
8. Under the consideration of convenience and safety, relocation takes place in the same bay.

3.2 *Notations and Model Development*

It is obvious that the states of the yard storage and the ship stowage change once a container is retrieved and loaded onto the ship, therefore the loading process can be transformed into a dynamic one. Thus, we develop a two-phase algorithm which consists of a traverse heuristic and a hybrid dynamic programming to solve the problem in Sect. 4. In this section, the parameters and variables involved in the model are proposed in order to coincide with the algorithm, shown as Table 11.2.

The optimization goal of the loading sequence problem is to minimize the total number of relocations during the retrieving operation. Let M denotes the total number of relocations. The objective function can be formulated as follows:

$$M = \min \sum_{k=1}^{N-l+2} \sum_{s=1}^{p_k} a_{ks} m_{ks} \tag{11.1}$$

During each stage of the loading process, we should make sure that containers are not hung in the air when loaded on board. Eq. (11.2) ensures containers handled

Table 11.2 Definitions of parameters and variables

Name	Property	Definition
a_{ks}	Decision variable	Whether a relocation proposal is accepted as a result of state transition from stage $k-1$ to stage k under the condition of state s at stage k. $k \in \{1, 2, \ldots, N-l+2\}$, $s \in \{1, 2, \ldots, p_k\}$, $a_{ks} = \begin{cases} 0, \text{the relocation proposal is not accepted} \\ 1, \text{the relocation proposal is accepted} \end{cases}$
C_{lij}	Integer variable	Loading sequence of the container in column i, tier j on the ship after $l-1$ containers are loaded onto the ship. $l \in \{1, 2, \ldots, N\}$, $i \in \{1, 2, \ldots, I'\}$, $j \in \{1, 2, \ldots, J'\}$, $C_{lij} = \begin{cases} 0, \text{column i, tier j has not been occupied yet} \\ n, n \in \{1, 2, \ldots, l-1\} \text{ load a container to column i, tier j at the n}^{\text{th}} \text{ time} \\ \text{-1, no container loaded to column i, tier j} \end{cases}$
C_{ksij}	Integer variable	The loading sequence of the container in column i, tier j on the ship under the condition of state s at stage k. $k \in \{1, 2, \ldots, N-l+2\}$, $s \in \{1, 2, \ldots, p_k\}$, $i \in \{1, 2, \ldots, I'\}$, $j \in \{1, 2, \ldots, J'\}$, $C_{ksij} = \begin{cases} 0, \text{column i, tier j has not been occupied yet} \\ n, n \in \{1, 2, \ldots, k+l-2\} \text{load a container to column i, tier j at the n}^{\text{th}} \text{ time} \\ \text{-1, no container loaded to column i, tier j} \end{cases}$
D_{lij}	Integer variable	The serial number of the container in stack i, tier j in the yard bay after $l-1$ containers are loaded onto the ship. $l \in \{1, 2, \ldots, N\}$, $i \in \{1, 2, \ldots, I\}$, $j \in \{1, 2, \ldots, J\}$
D_{ksij}	Integer variable	The serial number of the container in stack i, tier j in the yard under the condition of state s at stage k. $k \in \{1, 2, \ldots, N-l+2\}$, $s \in \{1, 2, \ldots, p_k\}$, $i \in \{1, 2, \ldots, I\}$, $j \in \{1, 2, \ldots, J\}$
I	Constant	The number of stacks in a yard bay
I'	Constant	The number of columns of the stowage plan
J	Constant	The maximal tier of stacks in a yard bay
J'	Constant	The number of tiers of the stowage plan
k	Integer variable	The number of stages of the dynamic programming. $k \in \{1, 2, \ldots, N-l+2\}$
l	Integer variable	The number of stages of the traverse heuristic. $l \in \{1, 2, \ldots, N\}$
m_{ks}	Integer variable	The number of relocations as a result of state transition from stage $k-1$ to stage k under the condition of state s at stage k. $k \in \{1, 2, \ldots, N-l+2\}$, $s \in \{1, 2, \ldots, p_k\}$
N	Constant	The total number of containers
p_k	Integer variable	The total number of states at stage k, $k \in \{1, 2, \ldots, N-l+2\}$

by the traverse heuristic are not hung in the air and Eq. (11.3) ensures containers satisfy the constraint during each sub-stage of the hybrid dynamic programming.

$$C_{lij} \leq C_{li(j-1)}\left(\text{where } l \in \{1, 2, \ldots, N\}, i \in \left\{1, 2, \ldots, I'\right\}, j \in \left\{2, \ldots, J'\right\}\right) \tag{11.2}$$

$$C_{ksij} \leq C_{ksi(j-1)}\left(\text{where } k \in \{1, 2, \ldots, N-l+2\},\ s \in \{1, 2, \ldots, p_k\},\ i \in \left\{1, 2, \ldots, I'\right\},\ j \in \left\{2, \ldots, J'\right\}\right) \tag{11.3}$$

When a yard crane conducts a retrieving operation, only the container on the top tier can be retrieved. Equations (11.4) and (11.5) ensure that during each sub-stage of the proposed algorithm, containers to be retrieved satisfy the constraint.

$$\prod_{j=1}^{k} D_{lij} \neq 0(\text{where } D_{lik} \neq 0,\ l \in \{1, 2, \ldots, N\},\ i \in \{1, 2, \ldots, I\}) \tag{11.4}$$

$$\prod_{j=1}^{k} D_{ksij} \neq 0(\text{where } D_{ksik} \neq 0,\ k \in \{1, 2, \ldots, N-l+2\},\ s \in \{1, 2, \ldots, p_k\},\ i \in \{1, 2, \ldots, I\}) \tag{11.5}$$

At each sub-stage of the hybrid dynamic programming, only one relocation proposal is accepted. The constraint can be described as Eq. (11.6).

$$\sum_{s=1}^{p_k} a_{ks} = 1 \ \ (\text{where } \forall k,\ k \in \{1, 2, \ldots, N-l+2\}) \tag{11.6}$$

And finally, decision variables should be assigned and represented by Eq. (11.7).

$$a_{ks} \in \{0, 1\}(\text{where } k \in \{1, 2, \ldots, N-l+2\},\ s \in \{1, 2, \ldots, p_k\}) \tag{11.7}$$

3.3 A Hybrid Optimization Algorithm Based on Dynamic Programming

In view of dynamic characteristics of container loading sequence problem, we suggest dynamic programming combined with heuristic rules to solve the problem. The optimality principle was first proposed by Bellman. Later the theory of dynamic programming was developed (Bellman 1952 1953, 1955), and then applied to several research fields (Bellman 1965; Feldmann 1967; Li and Glazebrook 2010; Sanaye and Mahmoudimehr 2012). There are a few researches on container terminals using dynamic programming (Jin and Gao 2006; Lam et al. 2007; Alessandri et al. 2009; Jin et al. 2011a, b; Meng and Wang 2011) of which various heuristic rules are generally applied to reduce the combinatorial complexity.

The hybrid dynamic programming consists of two phases, namely, the traverse phase and dynamic programming phase.

At the traverse phase, a traverse algorithm based on heuristic rules is developed to retrieve container subsets which need no relocation directly onto the ship. Let D_l, C_l, and C denote the set of D_{lij}, the set of C_{lij} and the set of C_{ij} respectively. The concrete solution procedure of this method is explained as follows:

Step 1: Set the stage number l as 1, the stack number i as 1, i.e., $l \leftarrow 1, i \leftarrow 1$. Turn to Step 2.

Step 2: From left to right, compare the top non-zero figure in stack i of D_l with the figure of C which corresponds with the bottom zero figure of C_l. Then turn to ***Step 3***.

Step 3: If the figures are equal, then retrieve the container corresponding with the figure and load it onto the ship, $l \leftarrow l+1$, both C_l and D_l change into new states, then turn to ***Step 2***; otherwise, turn to ***Step 4***.

Step 4: $i \leftarrow i+1$, and turn to ***Step 5***.

Step 5: If $i \in \{1, 2, \ldots, I\}$, then turn to ***Step 2***; otherwise, turn to ***Step 6***.

Step 6: If $l = N$, then the phase ends, output the result; otherwise, turn to ***Step 2***.

The dynamic programming phase begins when the first relocation occurs and ends when all the containers are loaded onto the ship. Let B denotes the set of containers in the yard under the present state. *TOP*(B) is the list of containers in B which can be directly loaded onto the ship. *Ship*(B, *TOP*(B)) represents loading the containers of *LC* onto the ship. *destno*(m) is the target stack with the lowest relocation cost for container m. Here, the relocation cost can be defined as follows:

$$Relocation\ cost = 1\, n_1 + 10\, |n_1 - n_2| + 100\, h \tag{11.8}$$

Therein, n_1 is the serial number of the target stack, n_2 is the serial number of the retrieve stack and h is the height of the target stack, $n_1, n_2 \in \{1, 2, \ldots, I\}$. 1, 10, 100 represent the weights of distances between the target stack and the truck lane, between two stacks, and between two tiers respectively.

Move(B, m, n) indicates moving container m to stack n. $C(B)$ is the smallest number of relocations when the loading operation ends.

When *Ship*(B, *TOP*(B)) is done, we can obtain a new yard storage state presented by B', and obviously, $|B'| < |B|$; and *Move*(B, m, n) turns $TOP(B) = \varnothing$ to$TOP(B) \neq \varnothing$. Therefore, *Ship*(B, *TOP*(B)) and *Move*(B, m, n) can both change the state of B, and B is dynamically changing. The dynamic equation can be formulated as follows:

$$C(B)=\begin{cases} 0 & \text{where } B=\varnothing \\ C(Ship(B,TOP(B))) & \text{where } TOP(B)\neq\varnothing \\ \min(1+C(Move(B,m_1,destno(m_1))),1+C(Move(B,m_2,destno(m_2)))), & \\ \qquad \text{where } TOP(B)=\varnothing, m_1,m_2: \text{containers to be relocated} \end{cases} \tag{11.9}$$

Regard each container loaded onto the ship as a stage. When a relocation occurs, several new states will be generated. Thus, the number of states during the computing process will explosively increase. Considering the complexity, we suggest the following heuristic rules, shown as Fig. 11.3, and propose an example to explain the rules, shown as Fig. 11.4.

Heuristic rules:

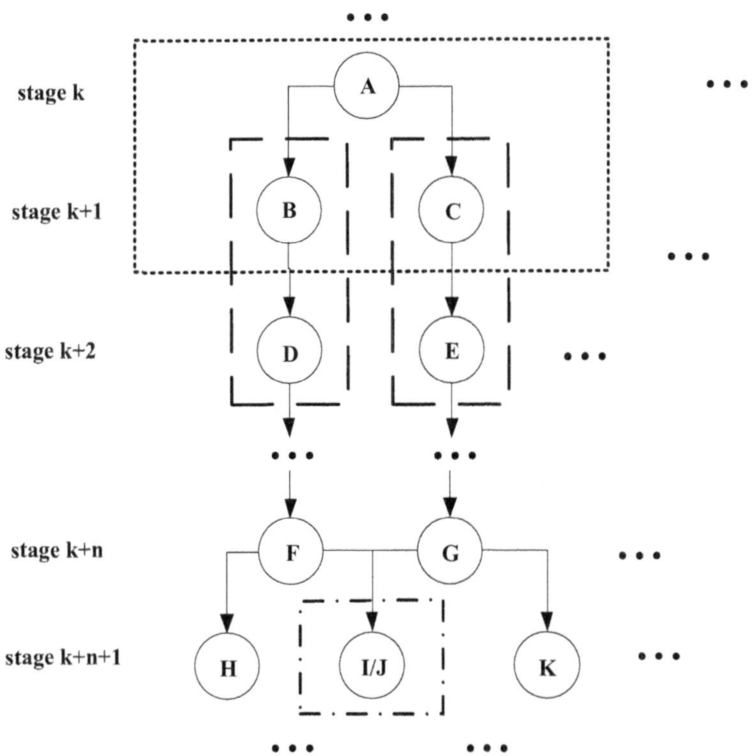

Fig. 11.3 The illustration of heuristic rules

1. Rule 1: if any relocation is needed, pick out the first and the second lowest cost choices, see the following description of the target stack choosing process (i.e., from stage k to stage $k+1$) in Fig. 11.4.
2. Rule 2: if there are containers that can be loaded without relocation, only one new state should be generated from stage $k+1$ to stage $k+2$. It is discussed in detail by the process from stage $k+1$ to stage $k+2$of the following example, shown as Fig. 11.4.
3. Rule3: if there are identical states in stage $k+1+n$, only one can be retained. As shown in Fig. 11.3, the state F of stage $k+n$ generates two new states H and I at stage $k+1+n$, and similarly, states J and K are generated by another state G, if I and J are exactly the same, then only one state can be kept at stage $k+1+n$, so we delete either of them.

Figure 11.4 illustrates the state transition process of a certain stage of the example showed in Fig. 11.1. From the final stowage plan in Fig. 11.1b and the stowage plan of stage k, we can find 5 containers can be retrieved. These containers can be represented as the set of $\{61, 32, 62, 31, 41\}$, and the corresponding retrieve stacks of the yard plan can be represented as the set of $\{6, 3, 6, 3, 4\}$. It is easy to calculate that the number of containers located in the upper tier is 2, 2, 1, 3 and 1 respectively. Therefore, stack 6 and stack 4 are chosen as the retrieve stack, i.e., container 62 and container 41 are selected to be retrieved at stage $k+1$. When retrieving container 62, container 63 should be moved to another stack first. According to Eq. (11.8) and Rule 1, we can find that stack 5 is of the lowest relocation cost, thus it is chosen to be the target stack. Similarly, stack 5 is chosen as the target stack for container 42. Then, two new states B and C are generated at state $k+1$.

As for state B, according to the method used in the last paragraph, it is obvious that container 11 can be loaded without relocation, and based on Rule 2, only one new state should be generated for stage $k+2$, shown as state D. Similarly, state E is the one and only state generated by state C for stage $k+2$.

In order to generalize the proposed algorithm precisely, we describe it with the framework, shown as Fig. 11.5. It clearly shows that the algorithm begins with the Initialization, and after the traverse phase and then the hybrid dynamic programming phase, ends with the output of loading proposals and the number of relocations.

3.4 Numerical Experiments

In this section, computational examples are conducted to demonstrate the performance of the algorithm developed in this research. The proposed algorithm has been implemented by Microsoft Visual C++ and run on a personal computer which has a Core I5 CPU running at 2.50 GHz and with 4.0 GB memory. In all the cases, the containers are generated, and randomly placed in the yard, subject to

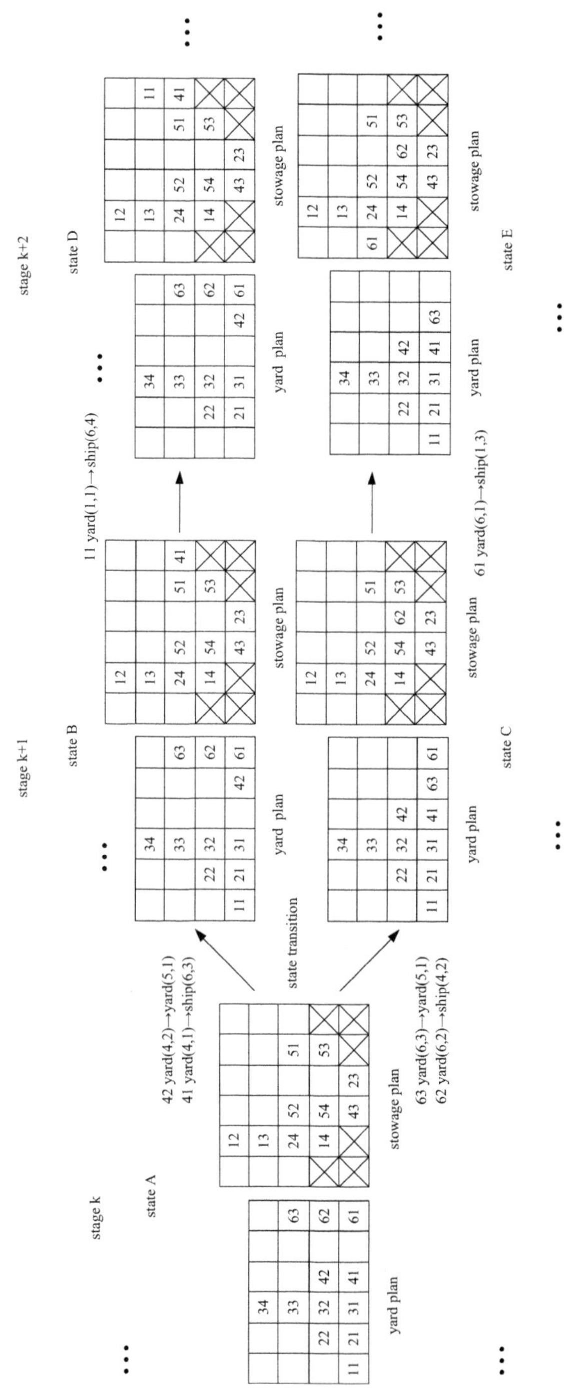

Fig. 11.4 The state transition of dynamic programming

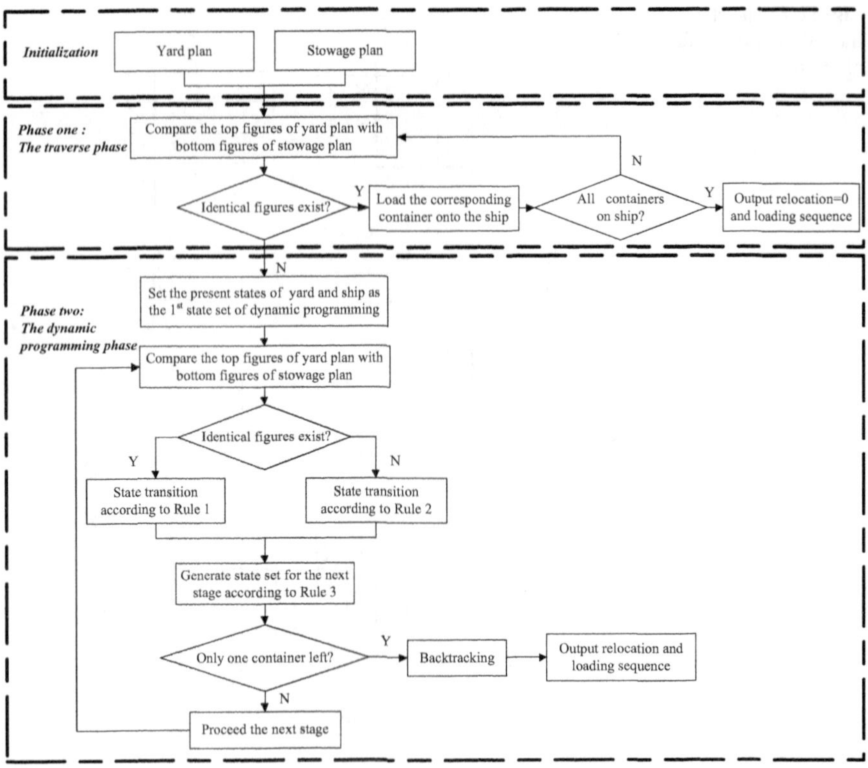

Fig. 11.5 The state transition of dynamic programming

pre-determined number of rows, stacks, and maximum stack height, and stowage plan is generated in the same way.

Moreover, we compare the results of this research with others obtained under the following circumstances:

1. Actual scheduling rules: choose the stack with fewest blocked containers as the retrieve stack, and choose the target stack randomly.
2. Improved heuristic rules: applied to an existing study (Zhu et al. 2010).

As mentioned before, container terminals always use two kinds of devices, RTGC and RMGC, to execute loading/unloading container operations. As is shown in Table 11.1, RTGC and RMGC are of different working capacity, so it is necessary to conduct numerical experiments respectively. Thus, the following session is divided into two parts, one is the numerical experiment part of the RTGC, and the other is of the RMGC.

Example 1 presents a small RTGC retrieving instance with only 17 containers spread over 6 stacks with maximum height of 5, as shown in Fig. 11.6a. The corresponding stowage plan can be seen in Fig. 11.6b.

Fig. 11.6 The yard plan and stowage plan of a bay (**a**) The yard plan of a bay. (**b**) Corresponding stowage plan of the bay

14					
13	23	33		53	
12	22	32	42	52	62
11	21	31	41	51	61

(**a**) The yard plan of a bay

08	13	12		33	23	62
06	11	22	53	14	52	61
04	X	21	32	42	51	X
02	X	X	31	41	X	X
	06	04	02	01	03	05

(**b**) Corresponding stowage plan of the bay

The instance was solved in 0.01 s with 6 relocations using actual scheduling rules and obtained the loading sequence, as shown in Fig. 11.7a. However, it took 0.02 s to execute improved heuristic rules and the number of relocations was 6. Figure 11.7b illustrates the loading sequence of the algorithm.

The hybrid dynamic programming developed in this research consumed 0.05 s to complete, a litter longer than the other two different rules discussed above. With the same amount of relocations, known as 6, we finally came up with as many as 15 different loading proposals, which include the solutions under the two different circumstances. Furthermore, the relocation process was visualized so that it could directly conduct the operation site for the workers. Thus, workers can select the relatively easier proposals according to the situation. Table 11.3 shows the details of each procedure. Obviously, relocations occurred when containers 42, 32, 52, 62, 22, 13 were loaded onto the ship.

In order to test the efficiency of the proposed algorithm for solving RTGC operations comprehensively, we conducted 20 random experiments, setting the number of containers N as 20. Let A, B, C denote actual scheduling rules, improved heuristic rules and hybrid optimization algorithm proposed in this research respectively. We present the results in Table 11.4.

As it can be seen that, Table 11.4 illustrates the amount of relocations, number of proposals and CPU times respectively obtained by three different algorithms. We obtain 99 relocations altogether by A, 72 relocations by B, and 53 relocations by C. Therefore, the number of relocations by C decreases 46.5%, 26.4% respectively

Fig. 11.7 The loading sequences based on two different rules (**a**) The loading sequence based on actual scheduling rules. (**b**) The loading sequence based on improved heuristic rules

3					
17	12	4		9	
15	14	8	2	11	6
16	13	7	1	10	5

(a) The loading sequence based on actual scheduling rules

3					
17	10	4		7	
15	14	6	2	9	12
16	13	5	1	8	11

(b) The loading sequence based on improved heuristic rules

compared with A and B. It is particularly noteworthy that we obtain total 44 proposals as alternative choices by C. Owing to the lower complexities of A and B, the search spaces of the algorithms are smaller than that of C. As the result of that, it took 0.38 s, 0.66 s and 1.01 s respectively for three methods dealing with 20 instances, with 0.019 s, 0.033 s and 0.05 s on average respectively. In general, the hybrid optimization algorithm proposed in this research can solve RTGC loading sequence problem effectively.

In the following section, we executed 20 random experiments of RMGC loading sequence problem, setting the number of containers N as 50. The results are enumerated in Table 11.5.

It is generally known that the number of relocations increases with more containers in a bay. As Table 11.5 shows, figures in relocations columns are much larger than that in Table 11.4. The optimizations on the number of relocations and proposals are still conspicuous. We obtain 993 relocations altogether by A, 932 relocations by B, and 701 relocations by C. So the number of relocations by C decreases 29.4%, 24.7% respectively compared with A and B. We obtain 1006 proposals by C. Besides, it cost 0.039 s, 0.08 s and 0.2 s on average respectively to complete the 50 instances.

From different scaled experiments discussed above, we can come to a conclusion that the algorithm developed in this research can solve container loading sequence problems efficiently.

Table 11.3 Details of loading sequence based on hybrid dynamic programming

Sequence	Container	State change (Yard → Yard)	(Yard → Ship)	Illustration of state change	Relocation
1	42	(4,2) → (6,3)		Container 42, from yard stack 4, tier 2 to yard stack 6, tier 3	√
2	41		(4,1) → (4,1)	Container 41, from yard stack 4, tier 1 to ship column 4, tier 1	
3	42		(6,3) → (4,2)	Container 42, from yard stack 6, tier 3 to ship column 4, tier 2	
4	14		(1,4) → (4,3)	Container 14, from yard stack 1, tier 4 to ship column 4, tier 3	
5	33		(3,3) → (4,4)	Container 33, from yard stack 3, tier 3 to ship column 4, tier 4	
6	32	(3,2) → (4,1)		Container 32, from yard stack 3, tier 2 to yard stack 4, tier 1	√
7	31		(3,1) → (3,1)	Container 31, from yard stack 3, tier 1 to ship column 3, tier 1	
8	32		(4,1) → (3,2)	Container 32, from yard stack 4, tier 1 to ship column 3, tier 2	
9	53		(5,3) → (3,3)	Container 53, from yard stack 5, tier 3 to ship column 3, tier 3	
10	52	(5,2) → (4,1)		Container 52, from yard stack 5, tier 2 to yard stack 4, tier 1	√
11	51		(5,1) → (5,1)	Container 51, from yard stack 5, tier 1 to ship column 5, tier 1	
12	52		(4,1) → (5,2)	Container 52, from yard stack 4, tier 1 to ship column 5, tier 2	

(continued)

Table 11.3 (continued)

Sequence	Container	State change (Yard → Yard)	(Yard → Ship)	Illustration of state change	Relocation
13	23		(2,3) → (5,3)	Container 23, from yard stack 2, tier 3 to ship column 5, tier 3	
14	62	(6,2) → (5,1)		Container 62, from yard stack 6, tier 2 to yard stack 5, tier 1	√
15	61		(6,1) → (6,1)	Container 61, from yard stack 6, tier 1 to ship column 6, tier 1	
16	62		(5,1) → (6,2)	Container 62, from yard stack 5, tier 1 to ship column 6, tier 2	
17	22	(2,2) → (3,1)		Container 22, from yard stack 2, tier 2 to yard stack 3, tier 1	√
18	21		(2,1) → (2,1)	Container 21, from yard stack 2, tier 1 to ship column 2, tier 1	
19	22		(3,1) → (2,2)	Container 22, from yard stack 3, tier 1 to ship column 2, tier 2	
20	13	(1,3) → (2,1)		Container 13, from yard stack 1, tier 3 to yard stack 2, tier 1	√
21	12		(1,2) → (2,3)	Container 12, from yard stack 1, tier 2 to ship column 2, tier 3	
22	11		(1,1) → (1,1)	Container 11, from yard stack 1, tier 1 to ship column 1, tier 1	
23	13		(2,1) → (1,2)	Container 13, from yard stack 2, tier 1 to ship column 1, tier 2	
Total relocations					6

Table 11.4 Comparisons of performances for RTGC instances

	No. of relocations			Optimization rate (%)		Total no. of obtained proposals			CPU time (s)		
Test no.	A	B	C	(A – C)/A	(B – C)/B	A	B	C	A	B	C
1	8	6	4	50	33	1	1	4	0.04	0.05	0.05
2	5	5	3	40	40	1	1	1	0.03	0.03	0.04
3	3	1	1	67	0	1	1	1	0.01	0.03	0.04
4	3	3	2	33	33	1	1	2	0.01	0.03	0.05
5	6	3	3	50	0	1	1	2	0.02	0.02	0.04
6	3	3	3	0	0	1	1	1	0.01	0.03	0.05
7	8	6	5	38	17	1	1	2	0.01	0.03	0.05
8	2	2	2	0	0	1	1	2	0.02	0.04	0.04
9	7	7	5	29	29	1	1	1	0.03	0.04	0.10
10	2	0	0	100	–	1	1	1	0.01	0.03	0.04
11	6	4	4	33	0	1	1	6	0.02	0.03	0.08
12	3	3	2	33	33	1	1	1	0.01	0.03	0.04
13	8	7	5	38	29	1	1	10	0.03	0.04	0.10
14	2	1	1	50	0	1	1	2	0.02	0.03	0.04
15	5	3	3	40	0	1	1	2	0.02	0.04	0.04
16	9	6	1	89	83	1	1	1	0.02	0.02	0.04
17	3	3	2	33	33	1	1	1	0.02	0.03	0.05
18	4	4	2	50	50	1	1	1	0.01	0.03	0.04
19	4	1	1	75	0	1	1	1	0.02	0.04	0.04
20	8	4	4	50	0	1	1	2	0.02	0.04	0.04

4 The Tugboat Scheduling Problem

4.1 Problem Definition and Basic Assumptions

A typical tugboat operation process is illustrated in Fig. 11.8. As Fig. 11.8 shows, the duration from the time when a tugboat starts tugging a ship to the finishing time of the berthing operation is treated as stage 1, the duration when a tugboat starts tugging the exact ship leaving the first berth to the finishing time when that ship enter into the second target berth is treated as stage 2, and the duration from the starting time of the unberthing operation to the time when the ship leaves the port is looked upon as stage 3.

The tugboat scheduling optimization problem is the problem that determines how to allocate the most suitable tugboats to fulfill the tugging task of each ship, and schedules all the operation tasks of every tugboats, so that some specific targets can be met. As the tugboats resource is limited to serve all the ships calling at a port, it is necessary to schedule all the tugboats at an optimum level if the port operation efficiency need to be improved.

In practice, tugboats scheduling managers allocate suitable tugboats to ships according to their length. Each ship can have one or more tugboats serving for it

Table 11.5 Comparisons of performances for RMGC instances

Test no.	No. of relocations			Optimization rate (%)		Total no. of obtained proposals			CPU time (s)		
	A	B	C	(A – C)/A	(B – C)/B	A	B	C	A	B	C
1	21	19	13	38	32	1	1	3	0.05	0.10	0.12
2	18	19	14	22	26	1	1	4	0.04	0.07	0.10
3	25	20	11	56	45	1	1	49	0.04	0.09	0.24
4	19	17	15	21	12	1	1	229	0.03	0.09	0.80
5	24	20	16	33	20	1	1	1	0.05	0.09	0.14
6	30	26	21	30	19	1	1	34	0.06	0.12	0.20
7	20	21	17	15	19	1	1	3	0.04	0.07	0.10
8	20	17	13	35	24	1	1	6	0.05	0.08	0.10
9	16	13	13	19	0	1	1	3	0.04	0.05	0.08
10	17	18	15	12	17	1	1	4	0.04	0.08	0.10
11	23	25	18	22	28	1	1	7	0.05	0.13	0.10
12	16	13	15	6	−15	1	1	49	0.03	0.05	0.44
13	22	25	14	36	44	1	1	2	0.05	0.14	0.10
14	23	17	10	57	41	1	1	3	0.06	0.05	0.14
15	20	24	12	40	50	1	1	5	0.03	0.10	0.10
16	30	28	15	50	46	1	1	6	0.07	0.14	0.10
17	20	16	14	30	13	1	1	28	0.04	0.05	0.27
18	16	13	16	0	−23	1	1	10	0.04	0.05	0.14
19	34	27	19	44	30	1	1	2	0.09	0.12	0.30
20	21	19	17	19	11	1	1	4	0.04	0.08	0.10
21	23	24	15	35	38	1	1	2	0.05	0.10	0.10
22	15	14	13	13	7	1	1	64	0.02	0.08	0.40
23	17	20	17	0	15	1	1	3	0.03	0.07	0.08
24	17	17	15	12	12	1	1	3	0.02	0.05	0.08
25	25	22	14	44	36	1	1	5	0.06	0.09	0.09
26	20	18	13	35	28	1	1	1	0.04	0.07	0.10
27	20	20	15	25	25	1	1	8	0.04	0.09	0.10
28	17	15	13	24	13	1	1	18	0.03	0.06	0.14
29	23	17	12	48	29	1	1	1	0.05	0.08	0.12
30	10	11	10	0	9	1	1	8	0.02	0.05	0.08
31	18	20	16	11	20	1	1	19	0.03	0.07	0.16
32	12	14	10	17	29	1	1	36	0.02	0.07	0.30
33	20	16	13	35	19	1	1	12	0.03	0.07	0.12
34	10	10	9	10	10	1	1	22	0.01	0.05	0.30
35	21	18	15	29	17	1	1	7	0.03	0.07	0.08
36	16	15	13	19	13	1	1	207	0.04	0.07	0.62
37	17	20	15	12	25	1	1	31	0.03	0.10	0.50
38	11	9	9	18	0	1	1	2	0.01	0.07	0.10
39	15	19	13	13	32	1	1	1	0.01	0.07	0.07
40	22	19	13	41	32	1	1	6	0.03	0.09	0.14
41	20	19	14	30	26	1	1	16	0.03	0.07	0.12

(continued)

Table 11.5 (continued)

Test no.	No. of relocations			Optimization rate (%)		Total no. of obtained proposals			CPU time (s)		
	A	B	C	(A – C)/A	(B – C)/B	A	B	C	A	B	C
42	20	19	16	20	16	1	1	4	0.04	0.07	0.08
43	27	23	18	33	22	1	1	7	0.07	0.13	0.10
44	19	21	15	21	29	1	1	15	0.04	0.08	0.14
45	16	17	13	19	24	1	1	28	0.03	0.06	0.50
46	26	21	10	62	52	1	1	1	0.05	0.08	0.14
47	16	18	15	6	17	1	1	16	0.02	0.05	0.14
48	16	17	12	25	29	1	1	6	0.02	0.06	0.10
49	22	19	13	41	32	1	1	3	0.05	0.08	0.12
50	27	23	14	48	39	1	1	2	0.06	0.15	0.20

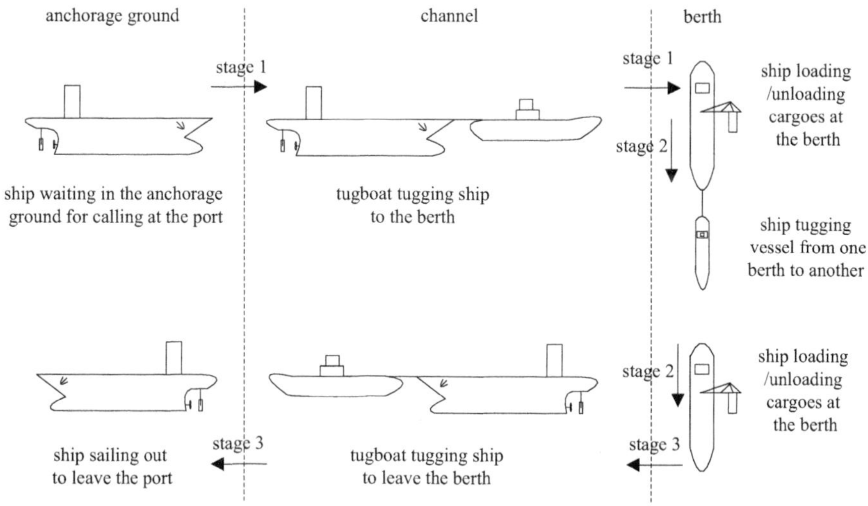

Fig. 11.8 Illustration of typical tugboat operation process

simultaneously by the scheduling rules. The main idea of the scheduling rules is as follows: big ships should be served by big tugboats (as to the unit of 'horsepower'), and small ships should be served by small tugboats; if more than one tugboat with the same horsepower are available, the allocation among the available tugboats is made by some heuristics rules. For example, there are six types of tugboats in a port according to the horsepower unit, such as 1200PS, 2600PS, 3200PS, 3400PS, 4000PS and 5000PS. The scheduling rules for allocating tugboats to ships are as follows:

1. *S1*(less than 100 meter): 1200PS (or bigger) *1
2. *S2*(100–200 meter): 2600PS(or bigger)*2
3. *S3*(200–250 meter): 3200PS (or bigger)*2

4. *S4*(250–300 meter): 3400PS (or bigger)*2
5. *S5*(greater than 300 meter): 4000PS (or bigger)*2

And the heuristics rules concluded from real-world practice include:

1. TSD rule: choosing the tugboat with the shortest distance from the scheduled ship to serve for it;
2. FAT rule: choosing the tugboat which is the first available one for the scheduled ship;
3. UWAT rule: from the perspective of balancing all tugboats' working amount, choosing the tugboat with the minimum working amount up till now to serve for the scheduled ship.

According to the hybrid flow shop theory, the tugboat scheduling can be considered as a Multiprocessor Tasks Scheduling Problem (MTSP) with 3 stages. In the scheduling system, tugboats are taken as movable 'machines', and ships have to experience the berthing, shifting-berth(if there exists this operation) and unberthing operations operated by tugboats sequentially.

On the other hand, compared with a typical MTSP, the tugboat scheduling problem has its own characteristics. Firstly, the exact same tugboat can provide all the three types of service (berthing, shifting berth, and unberthing), which means that the machine set for all the three stages are the same. This is different from a typical MTSP in which the available machine set in each stage are not the same. Besides, not all ships have to experience the shifting-berth operation, which makes the problem different from a typical MTSP with the characteristics that all jobs have to experience all the stages.

The following assumptions are introduced for the formulation of the problem:

1. The planning horizon is one day.
2. Three operation stages (i.e. berthing, shifting-berth and unberthing) are taken into consideration, but not all ships have to experience the shifting-berth operation. For ship which doesn't have to experience the second operation, assume there is a virtual shifting-berth operation, and the operation time for that is zero.
3. The ready times for all the tugboats are 0, and all the tugboats are at the anchorage bases at time 0; all the ships to be served have arrived at the anchorage ground at time 0.
4. There are three types of locations in a port: berths for ships to load/unloading cargoes; meeting locations where ships meet tugboats at the entrance of port; and the anchorage bases.
5. All the ships enjoy the same precedence.
6. The scheduling rules for allocating tugboats to ships include what are mentioned before.
7. The sailing speeds of all tugboats whenever sailing are the same.
8. The tugboats may return to the anchorage base during the planning horizon according to scheduling plans.

4.2 Notations and Model Development

1. Notations

(a) *Parameters*

j, l: stage index, j, $l \in J = \{1,2,3\}$, in which 1–3 represents berthing, shifting-berth and unberthing operation

i,k: ship index

cy_i: the descriptive binary parameter that illustrates whether ship i will experience the shifting-berth operation (if $cy_i = 1$, it means that ships i will experience the shifting-berth operation, otherwise will not experience the operation).

m: tugboat index

hp_m: the horsepower of tugboat m

M: the set of all the tugboats

ta_m: style of tugboat m (which may be 1–6, representing 1200PS, 2600PS, 3200PS,3400PS, 4000PS and 5000PS respectively)

N: the set of all ships, $N = \{1, 2, \ldots n\}$

S_i: style of ship i

set_i: set of tugboat style which can provide the related service for ship i

O_{ij}: operation of ship i at stage j

lp_{ij}: the lowest horsepower needed to fulfill the operation of O_{ij}

CM_b: set of tugboats in the anchorage base b ($b \in B$, B is the set of all the anchorage bases), thus we can get $\bigcup_{b \in B} CM_b = M$

M_{ijb}: set of tugboats in base b that can serve for operation O_{ij} based on the scheduling rules; thus the set of tugboats in all the bases that can serve or perform operation O_{ij} can be expressed as $M_{ij} = \bigcup_{b \in B} M_{ijb} = \{m | ta_m = set_i, \forall m \in CM_b\}$

E_{jm}: the set of ships that might be served by tugboat m at stage j

LOS_{ij}: location where operation O_{ij} starts (if $j = 1$, LOS_{ij} is the meeting place where ship i meet tugboat at the entrance of the port; else if $j = 2$, LOS_{ij} is the first berth where ship i loads/unloads its cargo; else if $j = 3$, LOS_{ij} is the second berth where ship i loads/unloads its cargo, while $LOS_{i3} = LOS_{i2}$ if $cy_i = 0$)

LOF_{ij}: location where operation O_{ij} finishes (if $j = 1$, LOF_{ij} is the first berth where ship i loads/unloads its cargo; else if $j = 2$, $LOF_{ij}$is the second berth where ship i loads/unloads its cargo, while $LOF_{i2} = LOF_{i1}$; else if $j = 3$, LOF_{ij} is the meeting place where ship i meet tugboat at the entrance of the port)

$ST(a, b)$: duration for sailing between location a and b

p_{ij}: processing time of operation O_{ij}

tb_i: sailing time of ship i from the waiting place to the berthing place, and$tb_i = ST(LOS_{i1}, LOF_{i2})$

te_i: berthing time of ship $i$at berth

toa_i: duration of ship i for loading and unloading cargoes at the first target berth

tob_i: duration of ship i for loading and unloading cargoes at the second target berth (if there exists a shifting-berth operation)

tu_i: unberthing time of ship i at berth

tl_i: sailing time of ship i from the unberthing place to the place where ship i leaves the port

s_{ijkl}^m: setup time between task O_{ij} and O_{kl} by tugboat m

bp_m: the anchorage base where tugboat m belongs

H: a sufficiently large constant

(b) *Decision Variables*

$$x_{ijm} = \begin{cases} 1, & \text{if } O_{ij} \text{ is assigned to tugboat } m \\ 0, & \text{otherwise} \end{cases}$$

$$y_{ijkl}^m = \begin{cases} 1, & \text{if } O_{ij} \text{ and } O_{kl} \text{ are assigned to the same tugboat } m \\ 0, & \text{otherwise} \end{cases}$$

$$u_{ijkl}^m = \begin{cases} 1, & \text{if } O_{ij} \text{ precedes } O_{kl} (\text{not necessarily immediately}) \text{ on tugboat } m \\ 0, & \text{otherwise} \end{cases}$$

$$z_{ijkl}^m = \begin{cases} 1, & \text{if } O_{ij} \text{ immediately precedes } O_{kl} \text{ on tugboat } m \\ 0, & \text{otherwise} \end{cases}$$

$$w_{ijm} = \begin{cases} 1, & \text{if tugboat } m \text{ goes back to the anchorage base after completing operation } O_{ij} \\ 0, & \text{otherwise} \end{cases}$$

(c) Derived variables.

TS_{ij}: the starting time of O_{ij}

TF_{ij}: the finishing time of O_{ij}

BT_m: the setting out time of tugboat m from its anchorage base in the planning horizon

FT_m: the returning time of tugboat m after finishing its last task in the planning horizon

sh_{mh}: the duration of the h th scheduling round for tugboat m in the planning horizon

g_m: number of the scheduling round for tugboat m in the planning horizon

2. Model

(a) *Objectives*

The proposed model aims at minimizing: (i) the total operation times of tugboats; (ii) the waste of the tugboats horsepower in use.

For the first objective, the total operation time of tugboats is equal to the total duration for all the scheduling rounds of all tugboats. In practice, a scheduling round is used to define the duration from the time when a tugboat leaves for its target place from the anchorage base to the time when it returns to the base after finishing a certain amount of tasks (maybe one task, maybe more than one). Define the set of tasks right before which tugboat m returns to the base as OS_m, and all the tasks in OS_m are ordered by the operation sequence. It is concluded that the duration of each scheduling round of tugboat m as follows.

$$\begin{aligned} sh_{m1} &= TF_{OS_m\{1\}} + ST\left(LOF_{OS_m\{1\}}, bp_m\right) - BT_m \\ sh_{m2} &= TF_{OS_m\{2\}} + ST\left(LOF_{OS_m\{2\}}, bp_m\right) - \left(TS_{ij} - ST\left(bp_m, LOS_{ij}\right)\right) \\ &\quad \left\{(i,j) | z_{ijkl}^m = 1, (k,l) = OS_m\{1\}\right\} \\ &\cdots\cdots\cdots\cdots \\ sh_{mh} &= TF_{OS_m\{h\}} + ST\left(LOF_{OS_m\{h\}}, bp_m\right) - \left(TS_{ij} - ST\left(bp_m, LOS_{ij}\right)\right) \\ &\quad \left\{(i,j) | z_{ijkl}^m = 1, (k,l) = OS_m\{h-1\}\right\} \\ &\cdots\cdots\cdots\cdots \\ sh_{mg_m} &= TF_{OS_m\{g_m\}} + ST\left(LOF_{OS_m\{g_m\}}, bp_m\right) - \left(TS_{ij} - ST\left(bp_m, LOS_{ij}\right)\right) \\ &\quad \left\{(i,j) | z_{ijkl}^m = 1, (k,l) = OS_m\{g_m - 1\}\right\} \end{aligned} \tag{11.10}$$

Thus the first objective function can be expressed as Eq. (11.11).

$$f_1 = \sum_{m \in M} \sum_{h \in g_m} sh_{mh} \tag{11.11}$$

For the second objective, the waste of the tugboats horsepower should be the sum of the lowest horsepower of tugboats needed for operation from the total horsepower of tugboats in use. Thus the second objective function can be expressed as Eq. (11.12).

$$f_2 = \sum_{i \in N} \sum_{j \in J} \sum_{m \in M} x_{ijm} \cdot hp_m - \sum_{i \in N} \sum_{j \in J} lp_{ij} \tag{11.12}$$

If weights are applied to both of objective functions, the weighted objective function of the model can be concluded as Eq. (11.13).

$$\text{Minimize } f = \alpha f_1 + (1 - \alpha) f_2 \tag{11.13}$$

(b) *Constraints*

The constraints in the proposed model include (11.14, 11.15, 11.16, 11.17, 11.18, 11.19, 11.20, 11.21, 11.22, 11.23, 11.24, 11.25, 11.26, 11.27, 11.28, 11.29, 11.30, 11.31 and 11.32).

$$TS_{ij} \geq 0, \quad \forall i \in N, \forall j \in J \tag{11.14}$$

$$TS_{i1} + p_{i1} + toa_i \cdot cy_i \leq TS_{i2}, \quad \forall i \in N \tag{11.15}$$

$$TS_{i2} + p_{i2} + tob_i \cdot cy_i + toa_i \cdot (1 - cy_i) \leq TS_{i3}, \quad \forall i \in N \tag{11.16}$$

$$\sum_{m \in M} x_{ijm} = \begin{cases} 1, & S_i = S1 \\ 2, & \text{otherwise} \end{cases} \quad \forall i \in N, \forall j \in J \tag{11.17}$$

$$set_i = \begin{cases} \{1200\text{PS}, 2600\text{PS}, 3200\text{PS}, 3400\text{PS}, 4000\text{PS}, 5000\text{PS}\}, & S_i = S1 \\ \{2600\text{PS}, 3200\text{PS}, 3400\text{PS}, 4000\text{PS}, 5000\text{PS}\}, & S_i = S2 \\ \{3200\text{PS}, 3400\text{PS}, 4000\text{PS}, 5000\text{PS}\}, & S_i = S3 \\ \{3400\text{PS}, 4000\text{PS}, 5000\text{PS}\}, & S_i = S4 \\ \{4000\text{PS}, 5000\text{PS}\}, & S_i = S5 \end{cases} \tag{11.18}$$

$$lp_{ij} = \begin{cases} 0, & j = 2, and\ cy_i = 0 \\ set_i\{1\}, & S_i = S1 \\ 2 \cdot set_i\{1\}, & \text{otherwise} \end{cases} \quad \forall i \in N, \forall j \in J \tag{11.19}$$

$$y_{ijkl}^m \leq 0.5\left(x_{ijm} + x_{klm}\right) \leq y_{ijkl}^m + 0.5, \quad \forall i,k \in E_{jm}, \forall m \in \bigcup_{b \in B} M_{ijb}, \forall j,l \in J \tag{11.20}$$

$$y_{ijkl}^m = y_{klij}^m, \quad \forall i,k \in E_{jm}, \forall m \in \bigcup_{b \in B} M_{ijb}, \forall j,l \in J \tag{11.21}$$

$$u_{ijkl}^m + u_{klij}^m = y_{ijkl}^m, \quad \forall i,k \in E_{jm}, \forall m \in \bigcup_{b \in B} M_{ijb}, \forall j,l \in J \tag{11.22}$$

$$u_{ijkl}^m - z_{klij}^m \geq 0, \quad \forall i,k \in E_{jm}, \forall m \in \bigcup_{b \in B} M_{ijb}, \forall j,l \in J \tag{11.23}$$

$$\sum_{k \in E_{lm}} z_{ijkl}^m \leq 1, \quad \forall i \in E_{jm}, \forall m \in \bigcup_{b \in B} M_{ijb}, \forall j,l \in J \tag{11.24}$$

$$\sum_{k \in E_{lm}} z_{klij}^m \leq 1, \quad \forall i \in E_{jm}, \forall m \in \bigcup_{b \in B} M_{ijb}, \forall j,l \in J \tag{11.25}$$

$$TS_{ij} + p_{ij} + s_{ijkl}^m \leq TS_{kl} + H\left(1 - z_{ijkl}^m\right), \quad \forall i,k \in N, \forall m \in \bigcup_{b \in B} M_{ijb}, \forall j \in J \tag{11.26}$$

$$TS_{kl} + p_{kl} + s_{klij}^m \leq TS_{ij} + H\left(1 - z_{klij}^m\right), \quad \forall i,k \in N, \forall m \in \bigcup_{b \in B} M_{ijb}, \forall j \in J \tag{11.27}$$

$$p_{ij} = \begin{cases} tb_i + te_i, & j = 1 \\ (tu_i + ST(LOS_{i2}, LOF_{i2}) + te_i) \cdot cy_i, & j = 2 \\ tu_i + tl_i, & j = 3 \end{cases} \quad \forall i \in N \tag{11.28}$$

$$s_{ijkl}^m = ST\left(LOF_{ij}, LOS_{kl}\right) \cdot z_{ijkl}^m \quad \forall i,k \in N, \forall j,l \in J, \forall m \in \bigcup_{b \in B} M_{ijb} \tag{11.29}$$

$$w_{ijm} \cdot H \geq z_{ijkl}^m \cdot \left[\left(TS_{kl} - TF_{ij}\right) - \left(ST\left(LOF_{ij}, bp_m\right) + ST\left(bp_m, LOS_{kl}\right)\right)\right]$$
$$\forall i,k \in E_{jm}, \forall m \in \bigcup_{b \in B} M_{ijb}, \forall j,l \in J \tag{11.30}$$

$$w_{ijm} < \left(z^m_{ijkl} \cdot \frac{TS_{kl} - TF_{ij}}{2 \times \left(ST\left(LOF_{ij}, bp_m\right) + ST\left(bp_m, LOS_{kl}\right)\right)} + 0.5 \right)$$
$$\forall i, k \in E_{jm}, \forall m \in \bigcup_{b \in B} M_{ijb}, \forall j, l \in J \qquad (11.31)$$

$$x_{ijm}, y^m_{ijkl}, u^m_{ijkl}, z^m_{ijkl}, w_{ijm} = 0 \; or \; 1, \quad \forall i, k \in N, \forall m \in \bigcup_{b \in B} M_{ijb}, \forall j, l \in J \qquad (11.32)$$

Constraint (11.14) guarantees that each operation begins after time zero. Constraints (11.15) and (11.16) ensure that, for every ship, the shifting-berth operation begins only after the berthing and handling operations are completed, and the unberthing operation begins only after the shifting-berth and handling operations are completed. Constraint (11.17) means that if the style of the ship is *S1*, only one tugboat is needed; otherwise, two tugboats are needed. Constraint (11.18) defines the available set of tugboat style which can serve for ship i according to the scheduling rules. Constraint (11.19) defines the lowest horsepower of tugboats needed to fulfill the operation of stage j for ship i. Constraints (11.20) and (11.21) define $y_{ijkl} = y_{klij} = 1$ when $x_{ijm} = x_{klm} = 1$. Constraint (11.22) guarantees that every tugboat can only serve for one operation at any time. Constraint (11.23) is set to make sure that $u^m_{ijkl} = 1$ when $z^m_{ijkl} = 1$. Constraint (11.24) and (11.25) guarantee that there are at most one predecessor and successor for operation O_{ij} on tugboat m. Constraint (11.26) and (11.27) simultaneously determine that the starting time of any operation has to be after the time when its immediately preceded task finishes. Constraint (11.28) defines the processing time for each task. Constraint (11.29) defines the setup time for each operation O_{ij}. Constraint (11.30) and (11.31) simultaneously determine when tugboat m should return to the anchorage base: if the sum of the sailing time from the finishing place of O_{ij} to the base and the sailing time from the base to the starting place of O_{ij}'s successor task on tugboat m (i.e. O_{kl}) is less than the time cost if m directly sails to O_{kl}'s starting place and waits there until the task begins, then tugboat m should return to the base; otherwise, m should sail directly to O_{kl}'s starting place. Constraint (11.32) specifies the binary property of the decision variables.

4.3 A Hybrid Simulated Annealing Algorithm

In this section, a proposed HSA is introduced to solve the formulated tugboat scheduling problem.

1. Individual coding

The real integer method is adopted to code for an individual. As every ship may experience 3 stages of operation at most, we set the number of columns as three

ship index	2	2	3	2	1	3	4	3	1	4	1	4
tugboat 1	1	0	3	1	2	1	2	1	0	2	2	1
tugboat 2	0	0	2	0	0	3	3	2	0	3	0	3
whether tugboat 1 returns to the base after the task	1	0	0	0	1	0	0	1	0	1	1	1
whether tugboat 2 returns to the base after the task	0	0	1	0	0	1	1	0	0	1	0	1

Fig. 11.9 Illustration of coding for an individual

times of the number of ships. Assume that there are four ships to be served (ship 1,2 don't have to shift a berth, while ship 3,4 will experience a shifting-berth operation), and three available tugboats, then the coding expression of the individuals should be a $5 \times (3 \times 6)$ matrix, which can be illustrated as Fig. 11.9.

The first row of the coding representation means the service order for ships, and the next two rows are the indexes of tugboats serving for ships in the first row. Note that each index appears three times in the first row: if it is the first time an index appears, it means that the ship is berthing; the second time it appears, it may be a virtual or real shifting-berth operation; otherwise, the unberthing operation. The fourth and fifth rows are descriptive parts which tell us whether tugboat 1 and 2 return to the base after finishing the task.

As ship 1 and 2 don't have to shift a berth, the virtual shifting-berth operations are proposed to keep the total operations as three times of the number of ships. That can be illustrated as the shadow parts with diagonal lines in Fig. 11.2. Besides, if the ship style is 1, then an index of tugboat is generated from the available tugboat set to fill in the corresponding second row, and the third row is zero (as shown in the shadow parts with grids); otherwise, two indexes of tugboats are generated to fill in the two rows. Thirdly, as all tugboats have to return to the base after finishing their last tasks, the corresponding symbols in the fourth or fifth rows should be 1 (as the shadow parts with dots).

According to that individual coding, the service order for ships in Fig. 11.5 is: ship 2 (berthing)- ship 2 (virtual shifting-berth)- ship 3 (berthing)- ship 2 (unberthing) – ship 1 (berthing) – ship 3 (shifting-berth) – ship 4 (berthing) – ship 3 (unberthing) – ship 1 (virtual shifting-berth) – ship 4 (shifting-berth) – ship 1 (unberthing) – ship 4 (unberthing). The tugboat providing berthing service for ship 2 is tugboat 1, and after finishing the berthing service for ship 2, tugboat 1 returns to the anchorage base, and so on.

2. Initial individuals generation

The procedure for generating the initial schedule can be described as Fig. 11.10.

As it can be seen from Fig. 11.10, the procedure for the initial individuals generation mainly include three parts: randomly generating the service order for ships; allocate the tugboat serving for ships; deciding whether tugboats should return to the base after completing the operation.

3. Neighborhood search scheme

The procedure for the neighborhood search scheme is as Fig. 11.11.

Given a solution p, a neighbor of p can be obtained by using the three-point interchanging scheme proposed in this section. The main idea is as follows: randomly generate three positions in the original solution, so that the original solution is divided into five parts; let a, b, c, d, s be the four partial solutions of p; a temporary solution is obtained by interchanging a and b, c and d; based on the three rows of the temporary solution, calculate part s' according to the rules expressed by (11.30) and (11.31).

However, during the neighborhood search process, the temporary solution may be an infeasible solution. For example, the virtual shifting-berth operation (the shadow parts in Fig. 11.12) is after the unberthing operation, which is infeasible.

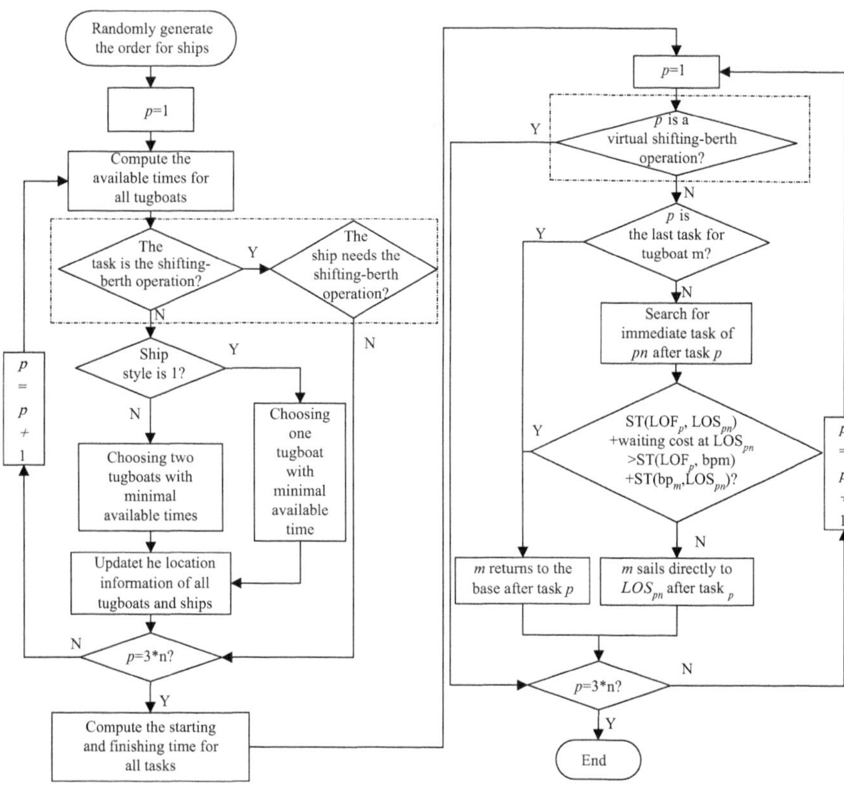

Fig. 11.10 The generation procedure of the initial individuals

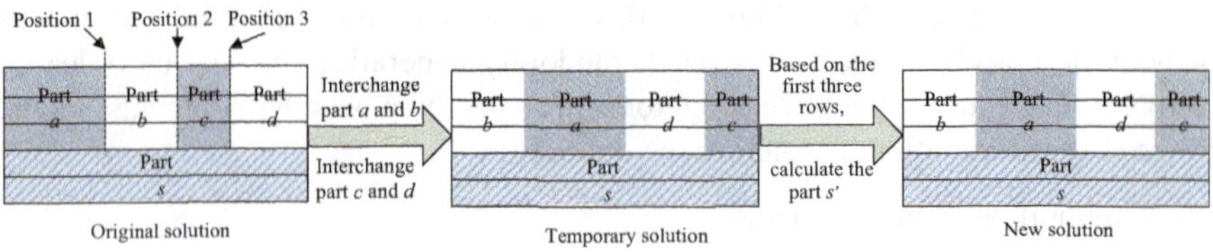

Fig. 11.11 The neighborhood search scheme

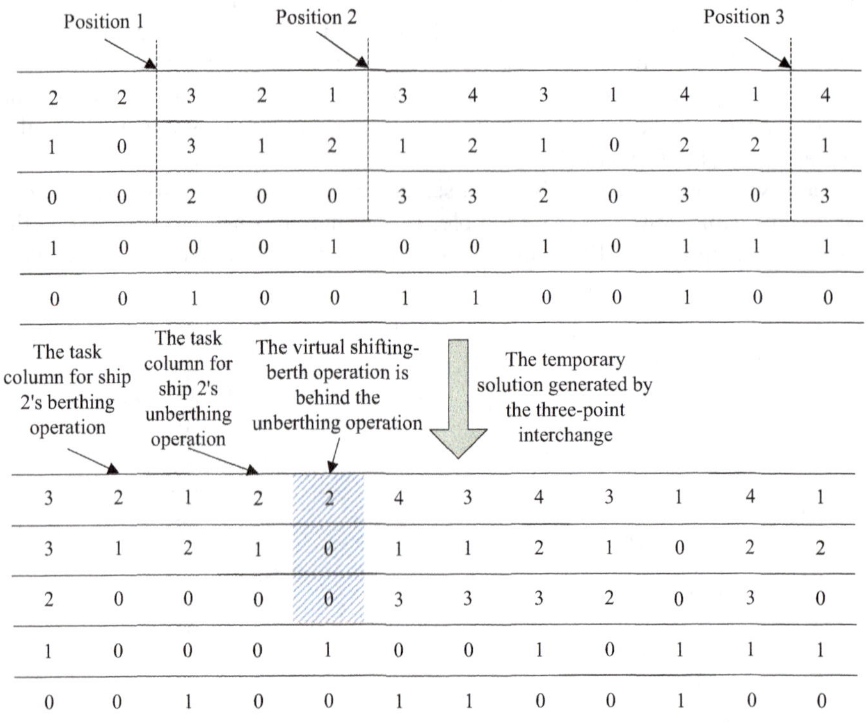

Fig. 11.12 The infeasible solution generated by the neighborhood search scheme

Thus it is necessary to modify the temporary solution. Steps for modifying the temporary solution are as follows:

Step1 Initialize $p = 1$.
Step 2 Judge if the second and third row of the pth column are both zero:

1. if both the values are zero, which means that the task in the pth column is a virtual shifting-berth operation
   (a) search for two columns: one for the berthing operation for ship served in the$p$th column; one for the unberthing operation for the same ship. Define the places of the two columns as $p1$ and$p2$.

(b) if p is less than $p1$, interchange values of the first three rows in the two columns, then go to ***Step 3***.
(c) if p is larger than $p2$, interchange values of the first three rows in the two columns, then go to ***Step 3***.

2. if the two values are not both zero, then go to ***Step 3***.

Step 3 Judge if p is equal to 3^*n:

1. if p is equal to 3^*n, then the modification is completed;
2. else, set $p = p + 1$, and go to ***Step 2*** .

After modified according to the steps introduced above, the temporary solution can be changed to a new solution by deciding whether tugboats should return to the base according to (11.21) and (11.22).

4. Parameters setting

The related parameters of the Simulated Annealing algorithm are set as follows: the initial temperature $t = 100^{\circ}C$, the cooling operation rule $t = 0.95t$, the length of inner loop at each temperature $L = 100$, stop criteria $t < 1 \times 10^{-4^{\circ}}C$, constant $k_b = 30 - 80$ according to different number of ships.

4.4 Numerical Experiments

To implement a comparison of the findings from the proposed algorithm, some experimental data were randomly generated, and details of which are as follows:

1. location data: the sailing times between each location(P1-P8, M1-M2, B1-B2) are as Table 11.6. Therein, P1-P8 are locations of 8 berths; M1 is the location where ships whose target berths are P1-P4 meet tugboats at the entrance of port; M2 is the location where ships whose target berths are P5-P5 meet tugboats at the entrance of port; B1 and B2 are two anchorage bases of tugboats whose service area are P1-P4 and P5-P8 respectively.
2. ship data: styles of ships are generated to $S1, S2, S3, S4, S5$ which takes up about $10\%, 20\%, 40\%, 20\%, 10\%$ of the total ships, berthing/unberthing times, loading & unloading times of ships are normally distributed in N(35, 25), N(300, 1600), and the berthing locations of ships are uniformly distributed to P1-P8. The proportion of ships that need the shifting-berth operation is 5%.
3. tugboat data: quantities of the six kinds of tugboats in the two anchorage bases are all one.

Based on the proposed algorithm, the tugboat scheduling problem can be solved when the number of ships is 10–30, with different weights of the objective functions. The results can be shown as Table 11.7.

Table 11.6 Sailing times between each location

	P1	P2	P3	P4	P5	P6	P7	P8	M1	M2	B1	B2
P1	0	18	14	20	32	34	35	30	19	29	15	32
P2	18	0	23	15	31	33	27	35	21	33	12	35
P3	14	23	0	12	39	34	30	32	15	38	16	31
P4	20	15	12	0	35	38	31	39	12	31	18	34
P5	32	31	39	35	0	18	12	19	31	12	29	11
P6	34	33	34	38	18	0	13	15	34	11	36	15
P7	35	27	30	31	12	13	0	12	29	18	25	19
P8	30	35	32	39	19	15	12	0	33	15	39	12
M1	19	21	15	12	31	34	29	33	0	30	15	25
M2	29	33	38	31	12	11	18	15	30	0	28	16
B1	15	12	16	18	29	36	25	39	15	28	0	26
B2	32	35	31	34	11	15	19	12	25	16	26	0

Table 11.7 Results of the problem using the proposed algorithm

Number of ships	Value of α	$f = \alpha f_1 + (1 - \alpha) f_2$	Number of ships	Value of α	$f = \alpha f_1 + (1 - \alpha) f_2$
10	0	8400	15	0	15,200
	0.2	7745		0.2	13,312
	0.4	7930		0.4	10,523
	0.6	7254		0.6	8644
	0.8	4661		0.8	6429
	1	2866		1	4213
20	0	20,600	25	0	20,800
	0.2	16,487		0.2	18,252
	0.4	13,864		0.4	14,437
	0.6	8542		0.6	10,985
	0.8	6430		0.8	7435
	1	5491		1	6862
30	0	21,300			
	0.2	18,903			
	0.4	15,228			
	0.6	11,745			
	0.8	10,079			
	1	8367			

If only the total tugboat operation time is considered as the objective, then the results solved by the algorithm can be compared with results from existing scheduling rules, as shown in Table 11.8. As Table 11.8. illustrates, the proposed algorithm can get a better solution compared with existing scheduling rules.

As we can see from Fig. 11.13 that the total operation times of tugboats increase dramatically with the increase of the proportion of the shifting-berth operation. That is because a single shifting-berth operation contains an unberthing operation, a

Table 11.8 Results from HSA*vs.* existing scheduling rules

Number of ships	HSA	TSD	GAP_1^a/%	FAT	GAP_2^b/%	UWAT	GAP_3^c/%
10	2866	3402	18.70	3267	13.99	3551	23.90
15	4213	5027	19.32	4806	14.08	5306	25.94
20	5491	6567	19.60	6301	14.75	6925	26.12
25	6862	8324	21.31	8105	18.11	8804	28.30
30	8367	10,252	22.53	9842	17.63	10,942	30.78
Average Gap	–	–	20.29	–	15.71	–	27.01

$GAP_1^a = (TSD - HSA)/HSA \times 100\%$
$GAP_2^b = (FAT - HSA)/HSA \times 100\%$
$GAP_3^c = (UWAT - HSA)/HSA \times 100\%$

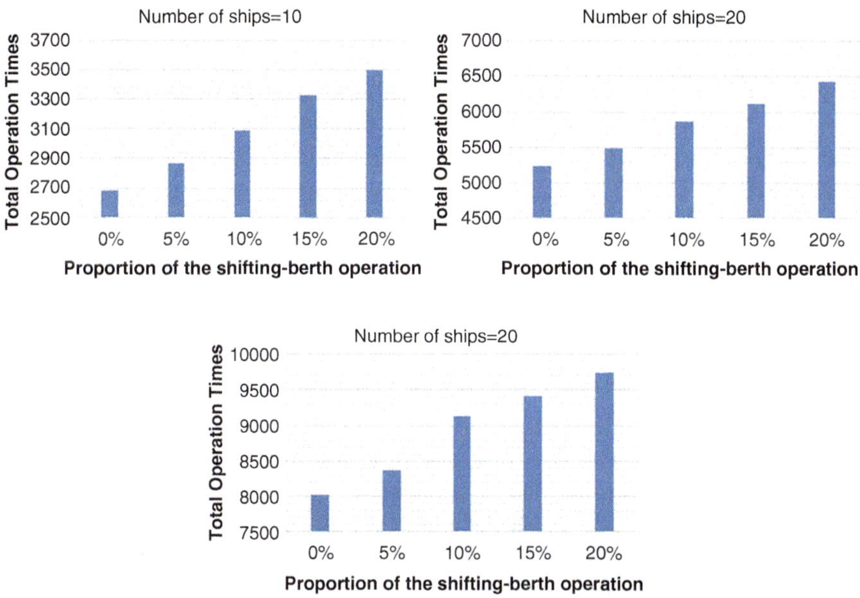

Fig. 11.13 Results with different proportion of the shifting berth operation

shift between the berths, and a berthing operation, thus needs more tugboats resource than normal berthing and unberthing operations. So it is necessary to reduce the number of shifting-berth operation in practice, so that the full utilization of limited tugboats resources.

Besides, we assume different deployment scheme of the available tugboats in the port (i.e. Scheme 1: the number of all types are 1; Scheme 2: the number of type 6 are 2, others are 1; Scheme 3: the number of type 5 and 6 are 2, others are 1; Scheme 4: the number of type 4, 5 and 6 are 2, others are 1). The results solved by the HSA are summarized in Table 11.9.

As Table 11.9 shows, the total operation times of all tugboats reveal a mild trend of decrease as the number of tugboats deployed increase. That is to say, the total

Table 11.9 Results with different tugboat deployment scheme

	Results with different tugboat deployment scheme						
Number of ships	Scheme 1[1]	Scheme 2[2]	GAP1[a] (%)	Scheme 3[3]	GAP2[b] (%)	Scheme 4[4]	GAP3[c] (%)
10	2866	2848	−0.63	2840	−0.91	2835	−1.08
15	4213	4196	−0.40	4172	−0.97	4167	−1.09
20	5491	5472	−0.35	5455	−0.66	5421	−1.27
25	6862	6835	−0.39	6791	−1.03	6774	−1.28
30	8367	8341	−0.31	8305	−0.74	8294	−0.87
Average	/	/	−0.59	/	−0.55	/	0.70

[a](value of 2 − value of 1)/value of 1 × 100%
[b](value of 3 − value of 1)/value of 1 × 100%
[c](value of 4 − value of 1)/value of 1 × 100%

operation times of tugboats can only be slightly reduced by simply increasing the number of tugboats deployed, and the cost of increasing tugboats may well be larger than the time cost saved by that. Under that circumstance, adding extra tugboats is not advised.

5 The Resource Deployment Problem

5.1 Problem Definition and Basic Assumptions

As we can see in Fig. 11.14, in a typical Asian container terminal, trucks have to drive to the yard block to be loaded or unloaded.

For outbound trucks, the 1st step of a truck is waiting for gate-in. At the gate, full containers would be checked up of integrity and Customs. Then the truck enters and stops at buffer area. As the driver got the information of block number, it goes to the yard block to unload a container. According to type of the container, it can go to general full container area, empty container area, hazardous or refrigerated area. After unloading service, it finishes tasks and gates out. For full container area, the loading machines are mostly yard cranes. While in empty container area, the equipment are usually forklifts, which are much more economic.

Inbound procedure is quite similar. A truck with an empty chassis arrives at the gates and waits in the lanes for pickups. It usually takes few minutes because the truck doesn't carry a container. And then it goes to buffer area or pick-up zones directly. After loaded with a container, it goes to gate-out picking-up lanes. There empty containers are fumigated and full ones are checked up. In recent years, more and more trucks delivers a container and then picks up another imported one in one shuttle. It would save one gate-in and one gate-out time periods.

The berth template and yard template is tactic level planning and are known ahead. While the quota of appointment system and the allocation of yard equipment

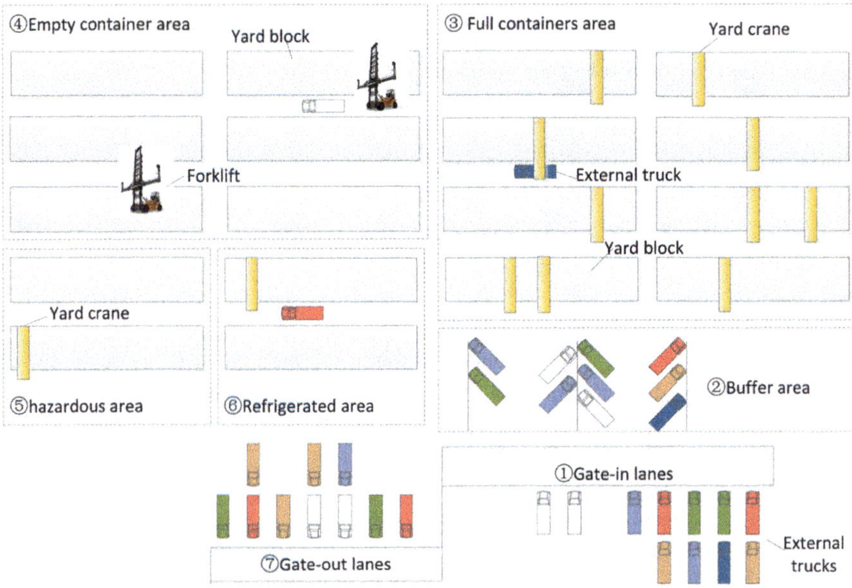

Fig. 11.14 Process of external trucks in a container terminal

and gate lanes are operational ones, which is on one day decision and to be solved in this section.

In this problem, the multiple criteria contains benefits of terminal operation as well as drayage companies. However, the demand of truck arrivals from drayage companies are solved by the deployment of yard cranes, which is treated as constraints. The total number of allocated equipment and the workload balance are objectives from terminal's perspective. Formulation and heuristic algorithm are used to solve the multi-criteria decision making problem.

According to the practical situation in most Chinese terminals, the following assumptions are made:

1. All the quotas would be used up by external trucks. But they may not come even with a reservation.
2. The outbound containers are full containers. And the inbound ones contain both full container and empty container.
3. There are separate blocks for outbound and inbound containers to be stored in yard. And gate-in or gate-out lanes are separate for each type of containers. So they have almost independent service system from each other.
4. For outbound containers, CY closing time of the vessel line is known. And the block location for storage is randomly chosen from the line's outbound blocks.
5. For inbound containers, they are stored in a fixed location before the reservation.
6. The loading and unloading workloads from vessels in the planning horizon are known.

7. For each truck, the loading and unloading service is assumed to be one time, neither 40 ft, 45 ft nor two 20 ft containers.
8. The planning horizon is 24 h in one day.

5.2 Notations and Model Development

1. Notations

(a) *Sets and indices*

$B = \{b|b = 1, 2, 3, \cdots, |B|\}$ Set of yard blocks, specifically B^O is the block set of outbound containers, while B_F^I is the block set of inbound full ones, and B_E^I is the block set of inbound empty ones;
$V = \{v|v = 1, 2, 3, \cdots, |V|\}$ Set of vessels in the planning horizon;
$W = \{w|w = 0, 1, 2, 3, \cdots, |W|\}$ Set of time windows;
$S = \{s|s = 1, 2, \cdots, |S|\}$ Set of work shifts in the planning horizon

(b) *Parameters*

T_Y^I Total number of yard cranes for inbound blocks;
T_Y^O Total number of yard cranes for outbound blocks;
T_F^I Total number of fork lifts;
T_G^I Total number of Gate-out lanes for inbound blocks;
T_G^O Total number of Gate-in lanes for outbound blocks;
a Average processing rate of a yard crane (number of trucks in each time window)
c Average processing rate of a gate lane (number of trucks in each time window)
O_{bv}^w Vessel side operation amount of block b for vessel v in time window w
F_{bv}^w Amount of appointed trucks arriving in time w for vessel $v$in block b
S_{bv} Amount of outbound containers that have already been stored in block$b$ for vessel v
p_v Remaining time windows to receive outbound containers for vessel v
N_v^O Total amount of outbound containers which would be loaded onto vessel v
N_v^I Total amount of inbound containers stored on terminals from vessel v
β Arrival rate of trucks with appointments
$\omega_j, j = 1, 2, 3, 4, 5$parameter of objectives

(c) *Decision variables.*

Q_{bv}^w Appointment quota for block $b$in time period w for vessel v
Y_b^s Number of yard cranes deployed in block b in work shift$s$
L_b^sNumber of forklifts deployed in block$b$ in work shift$s$
G_w^I Gate-in lanes for outbound check-up in period w

G_w^o Gate-out lanes for inbound check-up in period w

2. Model

(a) *Integer planning model for inbound container pick-up*

$$\text{Min } Z_1 = \omega_1 \sum_s \sum_b \left(Y_b^s + L_b^s\right) + \omega_2 \sum_w G_w^I, \ \forall b \in B_F^I \tag{11.33}$$

$$\text{Min } Z_2 = \frac{\omega_3 \sum_s \left(\sum_b Y_b^s - |B| \cdot \bar{Y}_b^s\right)^2 + \omega_5 \left(\sum_b L_b^s - |B| \cdot \overline{L_b^s}\right)^2}{|S|} + \omega_4 \frac{\sum_w \left(G_w^I - \overline{G_w^I}\right)^2}{|W|}, \ \forall b \in B_F^I \tag{11.34}$$

Subject to

$$\beta \cdot \sum_w \sum_v \left(F_{bv}^w + Q_{bv}^w\right) + \sum_w \sum_v O_{bv}^w \le a \cdot (|W|/|S|) \cdot \sum_s Y_b^s, \ \forall b \in B_F^I \tag{11.35}$$

$$\sum_b Y_b^s \le T_Y^I, \ \forall s \in S \tag{11.36}$$

$$\beta \cdot \sum_w \sum_v \left(F_{bv}^w + Q_{bv}^w\right) + \sum_w \sum_v O_{bv}^w \le a \cdot (|W|/|S|) \cdot \sum_s L_b^s, \ \forall b \in B_E^I \tag{11.37}$$

$$\sum_b L_b^s \le T_L^I, \ \forall s \in S \tag{11.38}$$

$$\beta \cdot \sum_w \sum_b \sum_v \left(F_{bv}^w + Q_{bv}^w\right) \le c \cdot \sum_w G_w^I \tag{11.39}$$

$$G_w^I \le T_G^I, \ \forall w \in W \tag{11.40}$$

$$\beta \cdot \sum_w^{|W|} \sum_b Q_{bv}^w \le N_v^I \text{-} \sum_w^{|W|} \sum_b F_{bv}^w, \ \forall b \in B_E^I \cup B_F^I, \ \forall v \in V \tag{11.41}$$

$$0 \le Y_b^s, L_b^s \le 4, \ \text{and are integers}, \ \forall b \in B_E^I, \ \forall s \in S \tag{11.42}$$

$$Q_{bv}^w, \ G_w^I \ge 0, \ \text{and are integers}, \ \forall b \in B_E^I, \ \forall v \in V, \ \forall w \in W \tag{11.43}$$

The first objective is to minimize the amount of equipment deployed in the planning horizon. Although the pool of equipment and gate-out lanes are known, the input of equipment as well as labors in each day could be different according to the demand. The second is to balance the equipment input in different work shifts and gate lanes in different time windows. In this way, the amount of the equipment pool would be smaller. And labors could get job chance mostly the same in different work shifts and equipment is fully used.

Constraints (11.35, 11.36, 11.37, 11.38, 11.39 and 11.40) are capacity constraints of equipment and gate. The yard cranes or forklifts deployed should satisfy the demand of workload of the block, while does not exceed the maximum number.

Formula (11.41) is the constraint of the total pick-up demand.

It is noted that trucks might not come even with an appointment. So for constraints (11.35), (11.37), (11.39) and (11.41), the appointed arrivals are multiplied byβ, which is an empirical parameter acquired from historical data.

At last, constraints (11.42) and (11.43) are variables definition.

(b) *Integer planning model for outbound container delivery.*

The differences between inbound and outbound containers appointment include:

(i) The outbound containers have to cope with the CY closing time. But the inbound ones are not in such a tense situation.
(ii) The inbound containers contain two types, which require different equipment. But for outbound ones, they are almost full ones. The yard service requires only yard cranes.
(iii) The outbound delivery requires some check-up while entering the gate. So the gate-in lanes are the scarce resource, which should be optimized. The inbound pick-up requires checking at the gate-out lanes instead.

Based on those real circumstances, the formulation for outbound container delivery is as follows.

$$\text{Min } Z_1 = \omega_1 \sum_s \sum_b Y_b^s + \omega_2 \sum_w G_w^O,\ \forall b \in B^O \tag{11.44}$$

$$\text{Min } Z_2 = \frac{\omega_3 \sum_s \left(\sum_b Y_b^s - |B| \cdot \bar{Y}_b^s \right)^2}{|S|} + \omega_4 \frac{\sum_w \left(G_w^O - \overline{G_w^O} \right)^2}{|W|},\ \forall b \in B^O \tag{11.45}$$

Subject to

$$\beta \cdot \sum_w \sum_v \left(F_{bv}^w + Q_{bv}^w \right) + \sum_w \sum_v O_{bv}^w \le a \cdot (|W|/|S|) \cdot \sum_s Y_b^s,\ \forall b \in B^O \tag{11.46}$$

$$\sum_b Y_b^s \le T_Y^O,\ \forall s \in S,\ \forall s \in S \tag{11.47}$$

$$\beta \cdot \sum_w \sum_b \sum_v \left(F_{bv}^w + Q_{bv}^w \right) \le c \cdot \sum_w G_w^O \tag{11.48}$$

$$G_w^O \le T_G^O,\ \forall w \in W \tag{11.49}$$

$$\beta \cdot \sum_w^{p_v} \sum_b Q_{bv}^w = \left[N_v \text{-} \left(\sum_w^{|W|} \sum_b F_{bv}^w + \sum_b S_{bv} \right) \right],\ p_v < |W|\ \forall b \in B^O,\ \forall v \in V \tag{11.50}$$

$$\beta \cdot \sum_{w}^{|W|} \sum_{b} Q_{bv}^{w} = \frac{|W|}{p_v}\left[N_v - \left(\sum_{w}^{|W|} \sum_{b} F_{bv}^{w} + \sum_{b} S_{bv}\right)\right], \quad p_v \geq |W|, \ \forall b \in B^O, \ \forall v \in V \tag{11.51}$$

$$0 \leq Y_b^s \leq 4, \ \text{and are integers}, \ \forall b \in B^O, \ \forall s \in S \tag{11.52}$$

$$Q_{bv}^{w}, \ G_{w}^{O} \geq 0, \ \text{and are integers}, \ \forall b \in B^O, \ \forall v \in V, \ \forall w \in W \tag{11.53}$$

Specifically, for outbound containers, to make sure that all of them could arrive before CY closing time, constraints (11.50) or (11.51) must be satisfied. The outbound delivery appointment quotas are set according to whether CY closing time is within the planning horizon.

5.3 A Non-dominated Genetic Algorithm

From the formulation of inbound and outbound appointment, it can be found out that all in all they use separate utility and equipment. So the optimization could be taken separately. Take the outbound equipment deployment as an example as follows.

1. Chromosome representations and initial solutions

In the chromosome of quotas, as in Fig. 11.15, each gene represents a quotaQ_v^w for each vessel in every time window. The volume is required randomly from the total amount. For that the blocks to stack containers for a vessel is known, Q_{bv}^w is randomly acquired from the amount of each time window.

The initial solutions of equipment and gate-in lanes are generated randomly under the constraints (11.46, 11.47, 11.48 and 11.49). And the representations of chromosomes are shown in Figs. 11.16 and 11.17. Because the planning horizon is 24 h and there are 3 work shifts in one day, there are three columns in the equipment allocation chromosome. Genes of each row show the deployed number of yard cranes or forklifts of one block in three work shifts. For gate-in lanes chromosome, each gene represents the number of lanes in each time window.

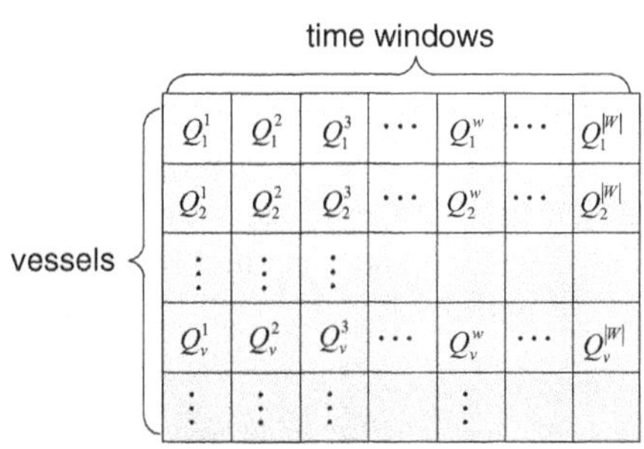

Fig. 11.15 The chromosome of quotas

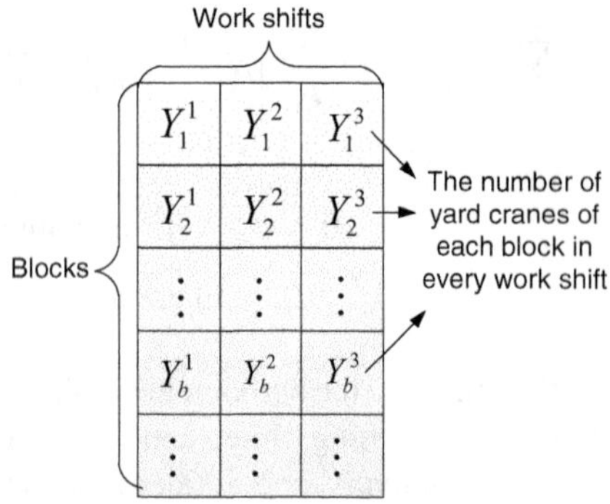

Fig. 11.16 The chromosome of yard crane deployment

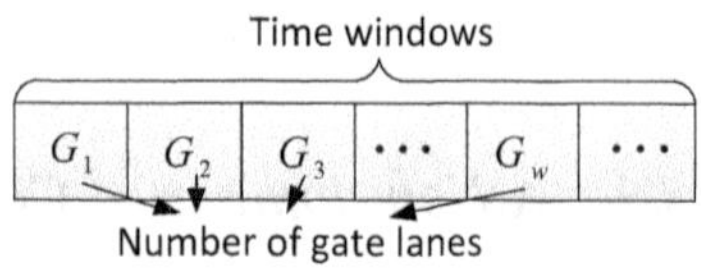

Fig. 11.17 The chromosome of gate lanes

The initial solutions are obtained as follows:

Step 1: parameters initialization.
Step 2: randomly spread the appointment quotas of each vessel to time windows.
Step 3: randomly allocate equipment to each block of every work shift.
Step 4: evaluate whether the allocation satisfy the capacity constraints, if not, adjust the number or delete the unfeasible solution.
Step 5: randomly assign the gate lanes of each time window.
Step 6: whether the assignment satisfy the capacity constraints, if not, increase the lanes or treat it as an unfeasible solution.
Step 7: if satisfy the population size? If not, go back to ***step 1***; if yes, stop.

The flowchart of initial solution can be seen in Fig. 11.18.

2. Genetic operations

There are three chromosomes in the algorithm. However, the genetic operation for equipment and gate-in lanes would generate large unfeasible solutions. And the mutual genetic operations would weaken the evolution effect of quota chromosomes and make severe fluctuation of the objectives in the evolution. So the quota chromosomes are chosen for the genetic operation.

The total amount of appointment quota for each vessel is fixed ahead. The crossover of every two chromosomes would cause the total amount changed, which need either reassignment or deletion and cause time waste. As a result, only mutations are operated for the quota chromosome, as in Figs. 11.19, 11.20, 11.21 and 11.22.

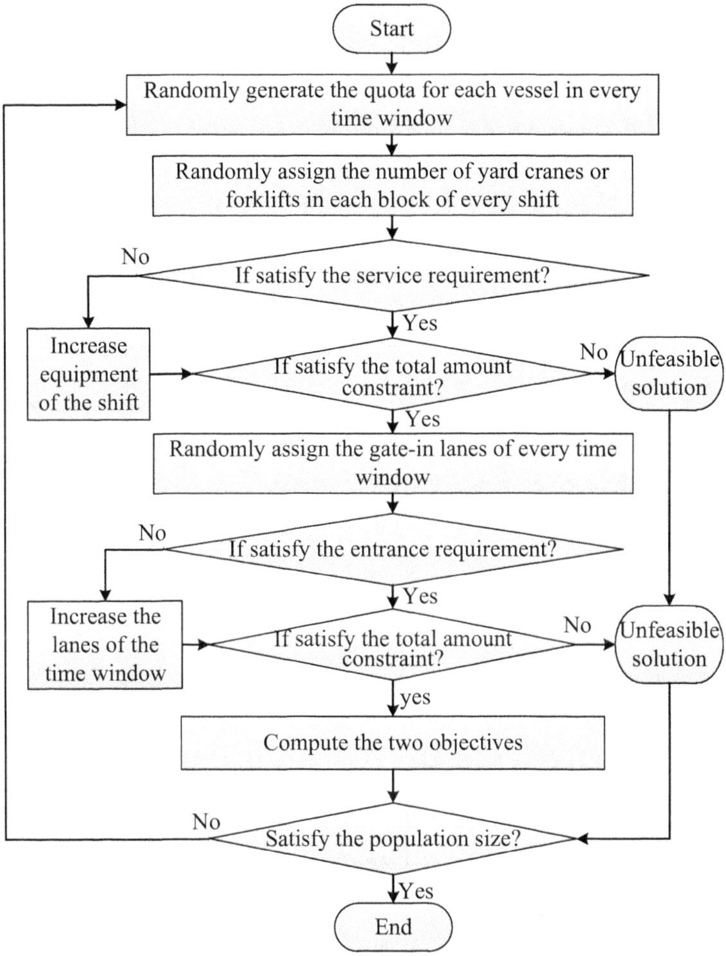

Fig. 11.18 The flowchart of initial solutions

Four different mutations, which include swap, insertion, combine and spread mutations, are chosen to generate offspring (Chen et al. 2013). However, the differences from Chen et al. are as follows.

First, the structure of the chromosome is two dimensional. So, the mutation would be operated in each vessel, which is in every row orderly. Second, to enlarge the effect of mutation, swap and insert operations are dealt in different work shifts. The reason is that the assignment of yard cranes are calculated again by constraint (11.46) after chromosome mutation, where the total work load of a shift are put together as the service demand. The mutation procedure of swap and insert are illustrated as following:

Step 1: randomly choose two different work shifts.

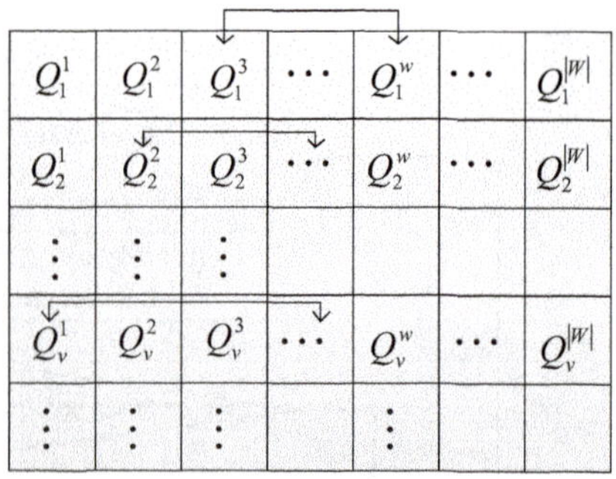

Fig. 11.19 Swap mutation operation

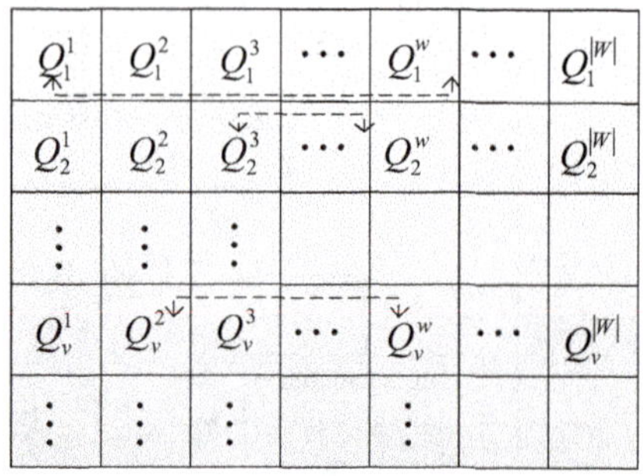

Fig. 11.20 Inserting mutation operation

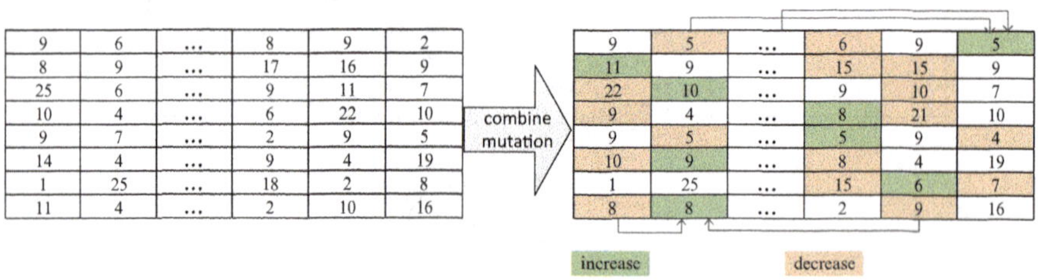

Fig. 11.21 Combine mutation operation

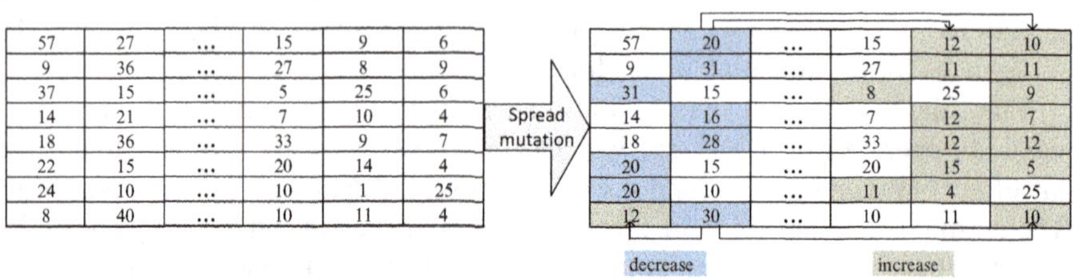

Fig. 11.22 Spread mutation operation

Step 2: randomly choose one gene in either work shift.
Step 3: swap the two genes. (One gene is inserted to the other gene's location, while genes between them are moved forward or backward.)
Third, the amounts of quota to be combined or spread are set as a variable ratio, as. $\Delta Q = Q_v^w \times r^G, r = random(0, 1)$, GG is the iteration times.

3. Deep mutation

To increase the number of Pareto Front (PF) solutions as well as to address the discontinuity in the Pareto Front, deep mutations including the greedy and mean mutations are applied to all the Pareto Front solutions (Chen et al. 2013). These two mutations are operated in a slight way in order to generate another PF near the former one.

The greedy mutation is to randomly choose two genes of one vessel. And then let part of two genes' difference spread to the minimum gene of the row.

The mean mutation is to choose randomly two genes of one row, and then compute the mean value and replace the former two quotas with the mean value. If it is not an integer, randomly allocate a pair of consecutive odd and even number to the two quotas.

The two mutations are based on the idea of homogenization, which is also a way to keep diversity.

4. Revised NSGAII procedure Fig. 11.23

The procedure of revised NSGAII flowchart can be seen in Fig. 11.23. The improvement from the famous NSGAII (Deb et al. 2002) is the deep mutation to increase the quality and quantity of Pareto Front. And the mutations are revised according to the characteristics of the problem and the model.

The convergence criterion is the number of Pareto Front exceeds 90% or the iterations exceed the maximum 1000. The minimum iteration is 100.

5.4 Numerical Experiments

Because the outbound containers and inbound ones have almost separate systems, they could be dealt with separately. Experiments of outbound delivering appointments are taken as example to illustrate the results.

The numerical experiments are generated based on the practical operation of a container terminal in Shenzhen, China. Four types of vessels are chosen as the experiments setting, which are distributed uniformly in [100,200],[200,500], [500,800], [800, 1200] respectively. The CY closing times for those vessels are 12 h, 14 h, 16 h, etc. which is the double of vessel serial number. The containers that have already arrived at the terminal are 20%. Those that have already appointed to arrive in the planning horizon are about 20%. For each type of vessels, 3, 3, 4 and 5 blocks are assigned to receive outbound containers. The block serial numbers to

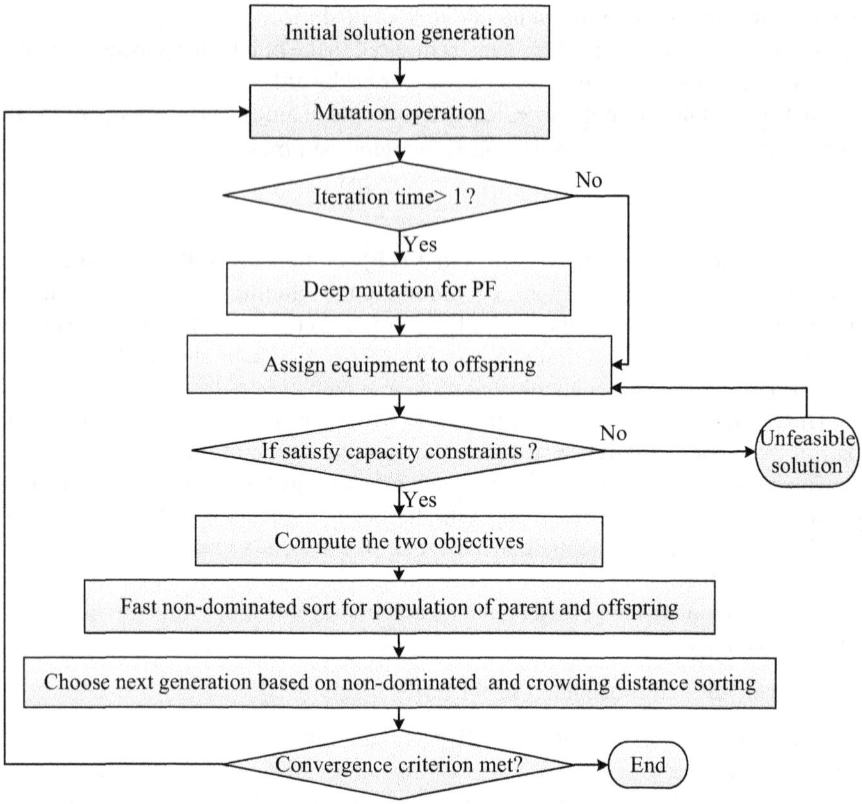

Fig. 11.23 Revised NSGAII flowchart

locate these outbound containers are randomly chosen several consecutive ones. The number of vessels to receive containers in each type of cases is shown in table10. Numbers are generated randomly according to the real situation, which reflect four sets of terminal situations.

And vessels that would be loaded in the planning horizon are also shown in Table 11.10. For example, if vessel 1 and 2 will be under service in the planning, they will occupy a certain amount of workload of the yard cranes. Other vessels have not arrived and would receive outbound containers. Loading amount of each time window in each block is randomly generated.

The planning horizon is one day, which is divided by 24 time windows. The terminal is assumed to have 20 blocks for outbound containers. The total number of yard cranes that could be allocated is 40. The total number of gate-in lanes is 10. The average operation ratio of yard crane is 20 containers in an hour. And the gate-in lanes' operation rate is 40 trucks one hour.

Based on these regulations, 10 instances are generated for each type. So totally 40 instances are generated.

Table. 11.10 Experimental setting of vessels

		Vessels to receive containers				Vessels to be loaded			
Amount of containers		U[100, 200]	U[200, 500]	U[500, 800]	U[800, 1200]	U[100, 200]	U[200, 500]	U[500, 800]	U[800, 1200]
No. of vessels	Set 1	0	6	4	3	3	3	2	1
	Set 2	2	5	5	2	2	2	1	1
	Set 3	2	2	2	1	2	1	1	1
	Set 4	5	1	1	1	1	1	1	0

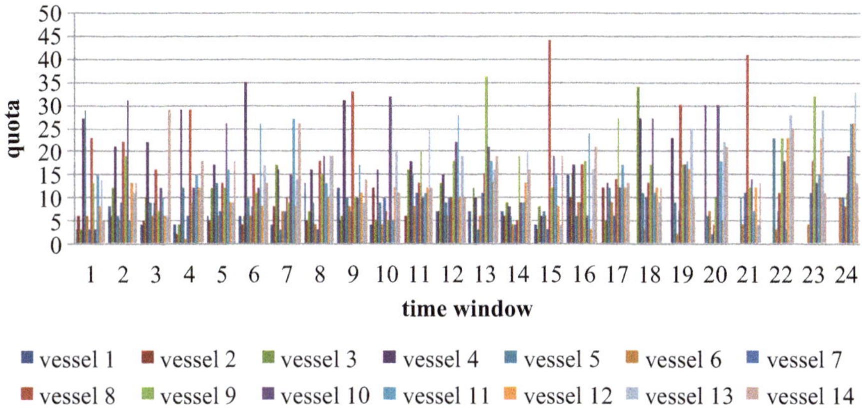

Fig. 11.24 The result of appointment quota for each vessel in every time window of an instance in set 2

For the algorithm, the maximum iteration is 1000. The population size is 100. And rate for mutation is 100, which means each parent will have at least four offspring. If the parent chromosome is Pareto Front, it will have another two children chromosome. So for the following iteration, the population for selection is about 500 or more individuals.

To illustrate the result of each instance, an instance in set 2 is taken for example. This is a result with Z_1 is 407 and Z_2 is 10, which is relatively low in total number and equilibrium of resource input. is 4, is 1, is 50 and is 10. Figure 11.24 shows the appointment quota for each vessel in every time window. Also, the work load of yard cranes is split by truck side and vessel side. Those with high volume of truck arrivals in some time windows mean that, the block has spare service capability for the vessel at those times. The equilibrium of objective 2 insures that trucks outside the terminal are mostly the same situation, which would not cause huge congestion for the city. And the equilibrium of the yard crane means the input of crane drivers are the least for a low cost. See Figs. 11.25 and 11.26.

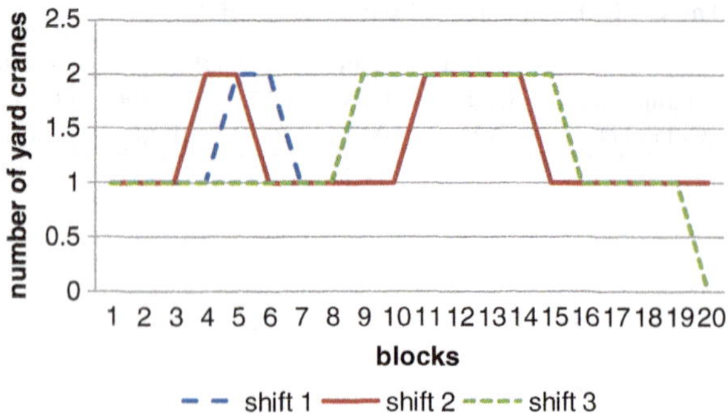

Fig. 11.25 Allocation of yard cranes in 20 blocks for 3 work shifts of an instance in set 2

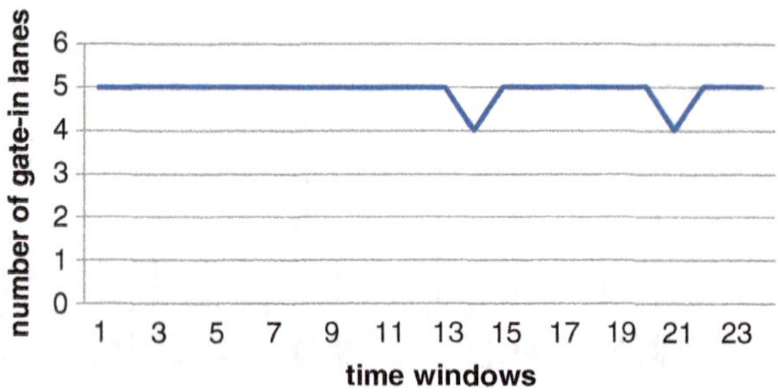

Fig. 11.26 Allocation of gate lanes for 24 time windows of an instance in set 2

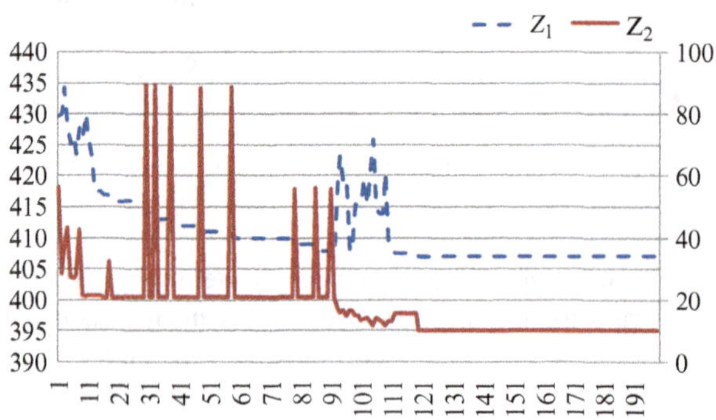

Fig. 11.27 Convergence of average of Pareto Front of an instance in set 2

The convergence of the algorithm is shown in Fig. 11.27, which illustrates the average of Pareto Front.

To verify the efficiency of the algorithm, Pareto Front Genetic Algorithm (Chen et al. 2013) is also run for numerical instances. In Pareto Front Genetic Algorithm (PFGA), the new population is formed by Pareto Front only. And genetic operations are applied to Pareto Front. The greedy and mean mutations are performed to those less crowed based on Euclidean distance. Stopping criterion for PFGA is no improvement of PF after 100 consecutive iterations, or iteration exceeds the maximum 1000.

The solution quality comparison are shown in Table 11.11, which shows the average value of Pareto front solutions in each instance by revised NSGA II and

Table 11.11 Comparison with PFGA

Instances		Revised NSGA		PFGA		Improve	
		Z_1	Z_2	Z_1	Z_2	Z_1 (%)	Z_2 (%)
Set 1	1	463.00	11.10	465.00	12.90	0.43	13.95
	2	460.00	10.00	476.00	11.60	3.36	13.79
	3	463.00	10.00	479.00	12.07	3.34	17.13
	4	465.00	10.00	466.00	10.00	0.21	0.00
	5	483.00	11.60	501.00	12.07	3.59	3.87
	6	483.00	11.10	495.00	105.34	2.42	89.46
	7	455.00	10.00	458.00	10.00	0.66	0.00
	8	485.00	10.57	489.00	52.01	0.92	79.68
	9	450.00	10.00	471.00	11.07	4.46	9.64
	10	459.00	10.57	466.00	11.10	1.50	4.80
Set 2	1	458.00	12.90	465.00	16.07	1.51	19.71
	2	413.00	10.00	413.00	10.00	0.00	0.00
	3	462.00	14.17	467.00	14.77	1.07	4.06
	4	430.00	13.27	439.00	14.40	2.05	7.87
	5	413.00	10.00	419.00	10.23	1.43	2.28
	6	426.00	11.60	436.00	12.50	2.29	7.20
	7	449.00	12.90	461.00	14.17	2.60	8.94
	8	407.00	10.00	415.00	11.07	1.93	9.64
	9	410.00	10.00	416.00	10.00	1.44	0.00
	10	481.00	14.40	486.00	41.79	1.03	65.55
Set 3	1	318.00	3.60	332.00	16.34	4.22	77.97
	2	339.00	3.90	347.00	16.88	2.31	76.89
	3	333.00	3.60	344.00	51.38	3.20	92.99
	4	332.00	4.43	343.00	50.84	3.21	91.28
	5	318.00	4.17	324.00	15.71	1.85	73.48
	6	334.00	4.93	336.00	41.67	0.60	88.16
	7	335.00	3.60	346.00	18.04	3.18	80.05
	8	341.00	16.38	343.00	18.18	0.58	9.90
	9	326.00	3.90	330.00	37.73	1.21	89.66
	10	335.00	3.60	340.00	16.38	1.47	78.02
Set 4	1	294.00	6.43	300.00	6.67	2.00	3.50
	2	293.00	5.57	294.00	6.27	0.34	11.17
	3	294.00	5.57	294.00	5.83	0.00	4.57
	4	294.00	17.54	301.00	40.07	2.33	56.21
	5	300.00	17.38	301.00	17.78	0.33	2.25
	6	295.00	17.38	299.00	40.00	1.34	56.56
	7	300.00	6.27	306.00	17.78	1.96	64.75
	8	289.00	16.68	290.00	16.68	0.34	0.00
	9	296.00	6.27	304.00	7.50	2.63	16.44
	10	290.00	16.68	293.00	39.40	1.02	57.67
Average		381.78	9.80	388.76	22.11	1.76	34.73

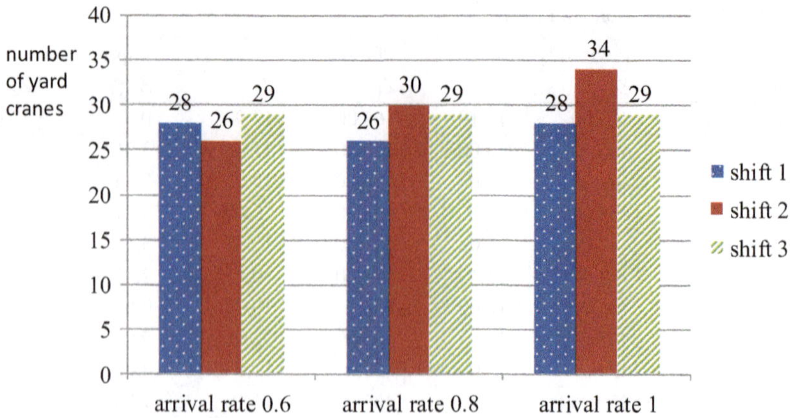

Fig. 11.28 Parameter analysis of truck arrival rate for number of yard cranes

Fig. 11.29 Parameter analysis of truck arrival rate for gate-in lanes

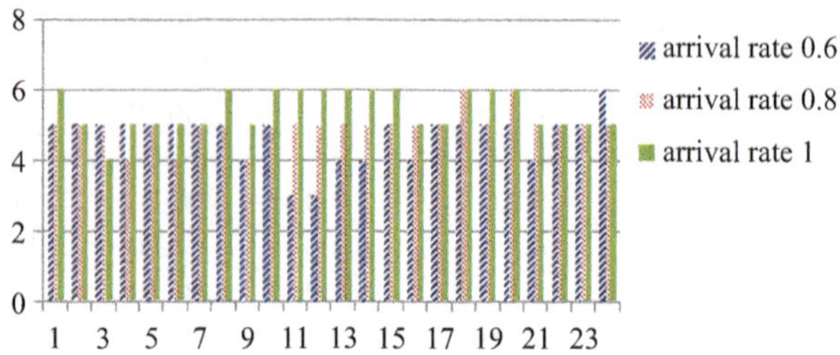

PFGA. As we can see from the results, revised NSGA II outperforms PFGA by Chen et al. (2013). For objective Z_1, results of NSGA II are 1.76% better those of PFGA. For objective Z_2, results of NSGA II are 34.73% better.

Both algorithms have a reasonable computation time which is within several minutes. However, in most cases, PFGA has a much less time to convergence.

In practice, external trucks sometimes could not exist at the appointed time for various reasons, such as congestion on road, not ready of goods, or other human factors. And sometimes, the arrival quota may be not totally used up. So the model has a parameter of arrival rate, which is an empirical statistic data. Different arrival rate has been set to an instance in set 1. We can see from the results in Fig. 11.28, the deployment of yard cranes increases as the arrival rate change from 0.6 to 0.8 and 1.

The similar situation happens to the deployment of gate-in lanes. The total number of gate-in lanes change from 112, to 119 and 130. The allocation of gate-in lanes in each time window is shown in Fig. 11.29.

In objectives Z_1 and Z_2, ratios of ω_1/ω_2 and ω_3/ω_4 are set to 2, 3, 4, 5 respectively. The ratios could reflect the importance of yard input compared with gate input. Usually, the yard cranes are more expensive and the salary of crane drivers is higher than that of inspecting worker at gate. So ω_1 and ω_3, which represent the

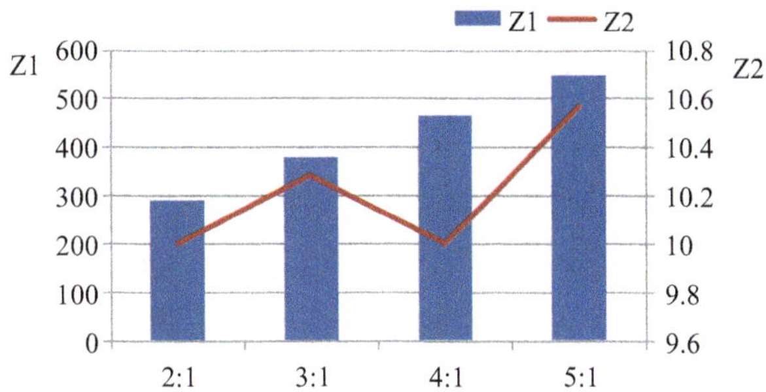

Fig. 11.30 Parameter analysis ω_1/ω_2 and ω_3/ω_4 for objectives

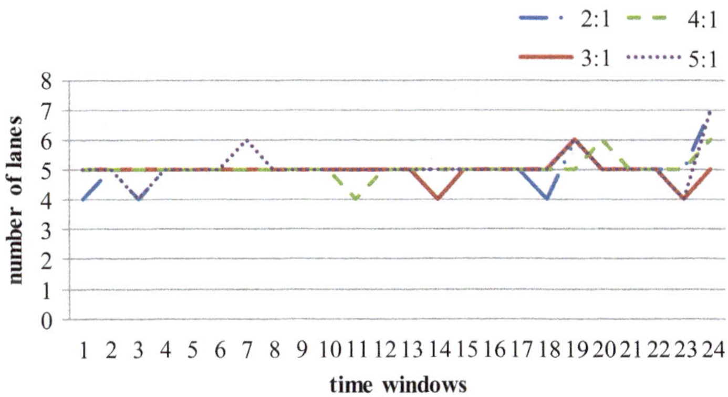

Fig. 11.31 Parameter analysis ω_1/ω_2 and ω_3/ω_4 for gate-in lanes

input of yard resource are given 2, 20; 3, 30; 4, 40; 5, 50 respectively. And ω_2, ω_4 is set to 1 and 10. An instance from set 1 is performed and the results are shown in Fig. 11.30.

As the ratios increase, the value of Z_1 rises up, while Z_2 fluctuates. The deployment of yard cranes is the same in four cases. The allocation of lanes has a bit difference, though the total number of gate-in lanes is almost the same in Fig. 11.31.

6 Conclusions

This chapter discussed applications of modern heuristics to container operation optimization problems, i.e. the container loading sequence problem, the tugboat scheduling problem, and the resource deployment problem in truck appointment system.

For the container loading sequence problem, a two-phase hybrid dynamic algorithm which aims to generate an optimal movement sequence for the crane to retrieve all the containers from a given yard to the ship was developed. Numerical results showed that the two-phase hybrid dynamic algorithm was able to solve loading sequence instances of both RTGC and RMGC, which was within the range of real cases, and thus of practical use to the industry. The number of relocations is much smaller than that of actual scheduling rules and improved heuristic rules. Therefore, it is proved that the algorithm in this research can tackle the practical scheduling problem efficiently.

For the tugboat scheduling problem, a model considering multi-anchorage bases, three stages of operations (berthing/shifting-berth/unberthing) were established based on the MTSP theory. A hybrid Simulated Annealing algorithm was proposed to solve the addressed problem. By the numerical experiments with the shifting-berth operation, it is proved that the total operation times of tugboats is most sensitive to the proportion of the shifting-berth operation, and influenced slightly by the tugboat deployment scheme.

For the resource deployment problem, a revised NSGAII is proposed to solve the problem. The chromosome representation and genetic operation are designed according to the characteristics of the problem. The numerical results show the feasibility and efficiency of the model and algorithm. To evaluate the effectiveness and efficiency of the algorithm, comparison with PFGA is performed. The results show that the revised NSGA II has a relative advantageous than PFGA in solution quality. And at last the parameter analysis of arrival rate and objective weights reflect the variation trend of objectives and solutions.

References

Alessandri A, Cervellera C, Cuneo M, Gaggero M, Soncin G (2009) Management of logistics operations in intermodal terminals by using dynamic modelling and nonlinear programming. Marit Econo Logist 11(1):58–76

Bellman R (1952) On the theory of dynamic programming. Proc Natl Acad Sci U S A 38 (8):716–719

Bellman R (1953) Bottleneck problems and dynamic programming. Proc Natl Acad Sci U S A 39 (9):947–951

Bellman R (1955) Functional equations in the theory of dynamic programming V. positivity and quasi-linearity. Proc Natl Acad Sci U S A 41(10):743–746

Bellman R (1965) ON the application of dynamic programming to the determination of optimal play in chess and checkers. Proc Natl Acad Sci U S A 53(2):244–247

Caserta M, Voß S, Sniedovich M (2011) Applying the corridor method to a blocks relocation problem. Eur J Oper Res 33(4):915–929

Chen G, Govindan K, Golias MM (2013) Reducing truck emissions at container terminals in a low carbon economy: proposal of a queueing-based bi-objective model for optimizing truck arrival pattern. Transp Res E Log 55:3–22

Deb K, Pratap A, Agarwal S, Meyarivan T (2002) A fast and elitist multiobjective genetic algorithm: NSGA-II. IEEE Transactions on Evolutionary Computation 6:182–197

Dong LC, Xu ZQ, Mi WJ (2012) The dynamic tugboat schedule based on particle swarm algorithm combined with genetic operators. Math Pract Theory 42(6):122–133

Feldmann EG (1967) Dynamic program for drug quality. J Am Pharm Assoc 7(6):301–302

Giuliano G, O'Brien T (2007) Reducing port-related truck emissions: the terminal gate appointment system at the Ports of Los Angeles and Long Beach. Transp Res D Transp Environ 12 (7):460–473
Guan C, Liu R (2009) Container terminal gate appointment system optimization. Marit Econo Log 11(4):378–398
Guo X, Huang SY, Hsu WJ, Low MYH (2011) Dynamic yard crane dispatching in container terminals with predicted vehicle arrival information. Adv Eng Inform 25(3):472–484
Huynh N, Walton CM (2008) Robust scheduling of truck arrivals at marine container terminals. J Transp Eng 134(8):347–353
Huynh N, Walton CM (2011) Improving efficiency of Drayage operations at seaport container terminals through the use of an appointment system. In: Böse JW (ed) Handbook of terminal planning, vol 49. Springer, New York, pp 323–344
Huynh N, Walton C, Davis J (2004) Finding the number of yard cranes needed to achieve desired truck turn time at marine container terminals. Transp Res Rec (J Transp Res Board) 1873:99–108
Jin C, Gao P (2006) Container berth expansion planning with dynamic programming and fuzzy set theory. 2006 I.E. international conference on service operations and logistics, and informatics (SOLI 2006), Proceedings, pp 260–265
Jin ZH, Lan H, Bian Z, Ji MJ (2011a) Optimization on containership loading scheduling based on actual constraints. J Dalian Marit Univ 37(1):71–74
Jin ZH, Mao J, Li N (2011b) Scheduling of relocating containers within a bay in container yard based on hybrid dynamic programming. J Transp Syst Eng Inf Technol 11(6):131–136
Kim KH (1997) Evaluation of the number of re-handles in container yards. Comput Ind Eng 32 (4):701–711
Kim KH, Hong GP (2006) A heuristic rule for relocating blocks. Comput Oper Res 33(4):940–954
Kim KH, Kim HB (1999) Segregating space allocation models for container inventories in port container terminals. Int J Prod Econ 59(1–3):415–423
Kim KH, Park YM, Ryu KR (2000) Deriving decision rules to locate export containers in container yard. Eur J Oper Res 124(1):89–101
Kim KH, Lee KM, Hwang H (2003) Sequencing delivery and receiving operations for yard cranes in port container terminals. Int J Prod Econ 84(3):283–292
Lam SW, Lee LH, Tang LC (2007) An approximate dynamic programming approach for the empty container allocation problem. Transp Res C Emerg Technol 15(4):265–277
Lee YS, Lee YJ (2010) A heuristic for retrieving containers from a yard. Comput Oper Res 37 (6):1139–1147
Li D, Glazebrook KD (2010) An approximate dynamic programming approach to the development of heuristics for the scheduling of impatient jobs in a clearing system. Nav Res Logist 57 (3):225–236
Li N, Chen G, Govindan K, Jin Z (2015) Disruption management for truck appointment system at a container terminal: a green initiative. Transportation Research Part D
Liu ZX (2009) Hybrid evolutionary strategy optimization for port tugboat operation scheduling. The third international symposium on intelligent information technology application, pp 511–515
Liu ZX (2011) Port tugboat operation scheduling optimization considering the minimum operation distance. J Southwest Jiaotong Univ 46(5):875–881
Liu ZX, Wang SM (2005) Research on bi-objectives parallel machines scheduling problem with special process constraint. Comput Integr Manuf Syst 11(11):1616–1620
Liu ZX, Wang SM (2006) Research on parallel machines scheduling problem based on particle optimization algorithm. Comput Integr Manuf Syst 12(2):183–187+296
Liu ZX, Li J, Shao ZY, He JJ (2016) Design of hybrid evolutionary strategy algorithm of dynamic tugboat scheduling problem. Comput Eng Des 37(2):519–524+529
Meng Q, Wang TS (2011) A scenario-based dynamic programming model for multi-period liner ship fleet planning. Transp Res E-Log 47(4):401–413

Morais P, Lord E (2006) Terminal appointment system study
Namboothiri R, Erera AL (2008) Planning local container drayage operations given a port access appointment system. Transp Res E-Log 44(2):185–202
Phan M-H, Kim KH (2015) Negotiating truck arrival times among trucking companies and a container terminal. Transp Res E-Log 75:132–144
Phan M-H, Kim KH (2016) Collaborative truck scheduling and appointments for trucking companies and container terminals. Transp Res B Methodol 86:37–50
Sanaye S, Mahmoudimehr J (2012) Minimization of fuel consumption in cyclic and non-cyclic natural gas transmission networks: assessment of genetic algorithm optimization method as an alternative to non-sequential dynamic programming. J Taiwan Inst Chem Eng 43(6):904–917
Van Asperen E, Borgman B, Dekker R (2011) Evaluating impact of truck announcements on container stacking efficiency. Flex Serv Manuf J 25(4):543–556
Wang S, Meng B (2007) Resource allocation and scheduling problem based on genetic algorithm and ant colony optimization. Lect Notes Artif Intell 4426:879–886
Wang X, Chen HY, Wang C, Liu D, Lv C (2005) The algorithm of getting the reasonable harbor's shipping order. Math Econ 22(3):284–290
Wang S, Kaku I, Chen GY, Zhu M (2012) Research on the modeling of tugboat assignment problem in container terminal. Adv Mat Res 433-440:1957–1961
Xu Y, Chen QH, Long L, Yang LZ, Liu LY (2008) Heuristics for container relocation problem. J Syst Simul 20(14):3666–3669
Xu Q, Mao J, Jin ZH (2012) Simulated annealing–based ant colony algorithm for tugboat scheduling optimization. Math Probl Eng 246978
Xu Q, Bian Z, Chen Y, Jin ZH (2014a) Scheduling optimization of port tugboat operation considering multi-anchorage. J Shanghai Jiaotong Univ 48(1):132–139
Xu Q, Shao QQ, Jin ZH (2014b) Optimization on tugboat operation scheduling based upon the hybrid flow shop arrangement. Syst Eng Theory Pract 34(2):485–493
Yang JH, Kim KH (2006) A grouped storage method for minimizing relocations in block stacking systems. J Intell Manuf 17(4):453–463
Yi ZJ, Li BS, Li XQ (2010) Game heuristic optimization algorithm for reshuffle in container yards. J Shanghai Marit Univ 31(3):47–51
Zhao W, Goodchild AV (2010) The impact of truck arrival information on container terminal rehandling. Transport Res E-Log 46(3):327–343
Zhao W, Goodchild AV (2013) Using the truck appointment system to improve yard efficiencyin container terminals. Marit Econ Log 15(1):101–119
Zhu MH, Fan XM, Cheng HC, He Q (2010) Heuristics for export container loading sequence problem. China Mech Eng 21(9):1066–1070

Chapter 12
Modelling Interdependency Among Attributes in MCDM: Its Application in Port Performance Measurement

Min-Ho Ha and Zaili Yang

Abstract The measurement of port and terminal performance may require an essential understanding of the cause-effect relationship among the influencing factors and criteria. Port performance indicators (PPIs) can interact with and feedback themselves (inner dependency) and/or each other (outer dependency). However previous studies have done little on the analysis of interdependency among the PPIs. This chapter aims to propose a new conceptual PPIs' interdependency model using a hybrid approach of a fuzzy logic based evidential reasoning (FER), a decision making trial and evaluation laboratory (DEMATEL) and an analytic hierarchy process (AHP). The combined approach of DEMATEL and AHP is applied to calculate the weights of dependent PPIs which are used as a part of the FER model to measure and analyse the performance of six container terminals in Korea from different port stakeholders' perspectives. The empirical results indicate that the hybrid approach offers a diagnostic instrument to container terminals in identifying the particular areas for improvement to enhance their competitiveness.

Keywords Port performance • Interdependency • Fuzzy logic • Evidential reasoning • DEMATEL • AHP

1 Introduction

Seaports present very complex and interdependent systems with a number of firms providing products and services. Even within a single port the associated activities can be very broad. Hence, it is not a straightforward task to develop powerful assessment instruments capable of dealing with the complexity (Yeo et al. 2014). In such systems, decision-makers typically need to assess the level of uncertainty in a

M.-H. Ha • Z. Yang (✉)
Liverpool Logistics, Offshore and Marine Research Institute (LOOM), Liverpool John Moores University, Liverpool, UK
e-mail: Z.Yang@ljmu.ac.uk

P.T.-W. Lee, Z. Yang (eds.), *Multi-Criteria Decision Making in Maritime Studies and Logistics*, International Series in Operations Research & Management Science 260,
DOI 10.1007/978-3-319-62338-2_12

port or terminal environment. Furthermore, they also require an essential understanding of the cause-effect relationships among the influencing factors and criteria (Lee et al. 2013). A number of port performance indicators (PPIs) may interact with and feedback themselves (inner dependency) or each other (outer dependency).

The multiple criteria decision making (MCDM) approach has been intensively conducted by researchers from both decision sciences on modelling (Yang and Xu 2002; Yang et al. 2009; Yeo et al. 2014) and port studies from empirical perspectives (Lirn et al. 2004; Yeo et al. 2014). Further, in the past three decades studies on seaport performance measurement have adopted more advanced methodologies, however scholars and practitioners have done little on the analysis of interdependencies among PPIs. Hence, it needs new solutions and supporting tools. This chapter aims at modelling interdependency among the PPIs in MCDM using a hybrid approach of a fuzzy logic based evidential reasoning (FER) method (Yang and Xu 2002), a decision making trial and evaluation laboratory (DEMATEL) tool (Gabus and Fontela 1973), and an analytic hierarchy process (AHP) (Saaty 1980). The combined method of DEMATEL and AHP is applied to evaluate relative weights of PPIs. Furthermore, the FER is applied for dealing with uncertainties presented in the evaluations of the selected PPIs. This model is applied to major container terminals in South Korea to demonstrate and validate the proposed framework. The performance measurement practices were conducted by taking perspectives from different port stakeholders. The hybrid approach attempting to use quantitative modelling for dealing with the uncertainties and interdependency problems can fulfil the aforementioned research gap.

In the next section, PPIs are identified from multi-stakeholder perspectives. In Sect. 3, the applied methodology for modelling interdependency of PPIs for port performance measurement is presented. The calculation examples of identification of PPIs' interdependency and evaluation of weights using DEMATEL and AHP are demonstrated in Sect. 4. In Sect. 5, a case study of Korean container terminal performance measurement is conducted. Finally, the chapter concludes with a discussion of implications in Sect. 6.

2 Port Performance Indicators from Multi-stakeholder Perspectives

This section is to incorporate multiple objectives of key stakeholders in a port performance measurement (PPM) model. A stakeholder-driven approach in PPM is useful to cover the different objectives and desired results of stakeholders (Dooms and Verbeke 2007). This can be achieved by integrating a multi-stakeholder dimension in a PPM framework which takes into account the corresponding port performance indicators (PPIs). PPIs which are most crucially needed to be used for measuring port performance were identified. Moreover, PPIs evaluations need to be conducted with inputs from associated stakeholders. This may assist decision makers not only in diagnosing both the efficiency and effectiveness aspects of performance but also in identifying the strengths and weaknesses of ports. Previous studies suggest that port performance should come across the range of port activities to cope with new evolutionary changes (Marlow and Paixão Casaca 2003; Bichou 2006; Brooks 2006; Woo et al. 2011). In addition, the PPIs should allow the ports to measure and communicate their impacts on society, economy and environment (ESPO 2010) and to be consistent with their goals (Kaplan and Norton 2004). On top of that we identified crucial interests in major container ports investigating their missions, visions, goals, and objectives and discussed them with port stakeholders. Therefore, the selection of PPIs has been done through a careful literature review (of more than 120 journal papers from Web of Knowledge using key words such as "port choice", "port selection", "port competitiveness", "port management and strategy") and industrial practices in a pre-selection phase and then confirmed by a panel of ten experts (i.e. 2 academia and 8 port stakeholders) to assess the suitability of the identified indicators and to test the feasibility of the selected indicators. They include (1) 6 industrial experts who have been working in the shipping and port industries for more than 15 years with PhD (1 expert from a shipping line), MSc (3 experts from terminal operators, a shipping line and a forwarder) and BA (1 from a terminal operator and a forwarder, respectively) degrees participated in the judgements. (2) 2 professors who have more than 15 years teaching and research experience participated in the survey. (3) 2 experts from government/port authorities (1 department manager and 1 managing director) who have been working for port logistics departments participated in the survey. Based on this, 6 dimensions, 16 principal-PPIs and 60 PPIs are defined. The dimensions relate to (1) the extent to which the container port/terminal operates effectively and efficiently in its basic role regarding cargo/vessel handling (core activities, CA); (2) the extent to which the container port/terminal has reliable resources (e.g. HR and technology) in order to support core activities (supporting activities, SA); (3) the extent to which the container port/terminal indicates its financial condition (financial strength, FS); (4) the extent to which the port users are satisfied with port/terminal services delivered and service price (users satisfaction, US); (5) the extent to which the port/terminal achieves its supply chain integration (terminal supply chain integration, TSCI); (6) the extent to which the port/terminal contributes to socio-economic sustainable growth (sustainable growth, SG) (Table 12.1).

3 Modelling Interdependency of PPIs for Port Performance Measurement

PPIs are often deemed as interdependent factors in decision problems on port performance, port selection and port competitiveness. However, many decision problems need to be explained using a network (interdependency) because PPIs interplay each other within a cluster and between clusters at the same level or

Table 12.1 Port performance indicators (PPIs)

Dimensions	Principal-PPIs	PPIs	Sources
Core activities (CA)	Output (OPC)	Throughput growth, Vessel call size growth	UNCTAD (1976), De monie (1987), Roll and Hayuth (1993), Tongzon (1995), Cullinane et al. (2006), Brooks (2007), Woo et al. (2011)
	Productivity (PDC)	Ship load rate, Berth utilization, Berth occupancy, Crane productivity, Yard utilization, Labour productivity	
	Lead time (LTC)	Vessel turnaround, Truck turnaround, Container dwell time	
Supporting activities (SA)	Human capital (HCS)	Knowledge and skills, Capabilities, Training and education, Commitment and loyalty	Barney (1991), Marlow and Paixão Casaca (2003), Kaplan and Norton (2004), Albadvi et al. (2007), Woo et al. (2013)
	Organisation capital (OCS)	Culture, Leadership, Alignment, Teamwork	
	Information capital (ICS)	IT systems, Database, Networks	
Financial strength (FS)	Profitability (PFF)	Revenue growth, EBIT(operating profit) margin, Net profit margin	Su et al. (2003), Bichou and Gray (2004), Brooks (2006)
	Liquidity & Solvency (LSF)	Current ratio, Debt to total asset, Debt to equity	
Users' satisfaction (US)	Service fulfilment (SFU)	Overall service reliability, Responsiveness to special requests, Accuracy of documents & information, Incidence of cargo damage, Incidence of service delay	Marlow and Paixão Casaca (2003), Woo et al. (2011), Brooks and Schellinck 2013
	Service costs (SCU)	Overall service cost, Cargo handling charges, Cost of terminal ancillary services	
Terminal supply chain integration (TSCI)	Intermodal transport systems (ITST)	Sea-side connectivity, Land-side connectivity, Reliability for multimodal operations, Efficiency of multimodal operations	Song and Panayides (2008), Panayides and Song (2009), ESPO (2010), Woo et al. (2013)
	Value-added services (VAST)	Facilities to add value to cargoes, service adaptation to customers, Capacity to handle different types of cargo, Tailored services to customers	
	Information/communication integration (ICIT)	Integrated EDI for communication, Integrated IT to share data, Collaborate with Channel members for channel optimisation, Latest port IT systems	

(continued)

Table 12.1 (continued)

Dimensions	Principal-PPIs	PPIs	Sources
Sustainable growth (SG)	Safety and security (SSS)	Identifying restricted areas and access control, Formal safety and security training practices, Adequate monitoring and threat awareness, Safety and security officers and facilities	De Langen (2002), IMO (2002), Peris-Mora et al. (2005), Darbra et al. (2009), ESPO (2010), Woo et al. 2011
	Environment (EVS)	Carbon footprint, Water consumption, Energy consumption, Waste recycling, Environment management programmes	
	Social engagement (SES)	Employment opportunity, Regional GDP, Reporting corporate and social responsibility	

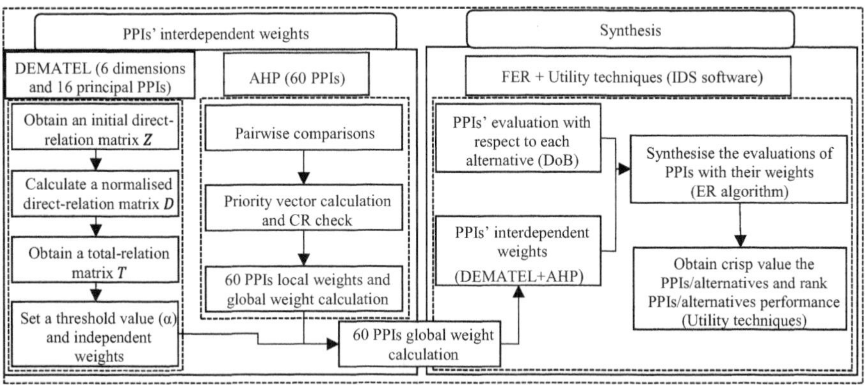

Fig. 12.1 A hybrid methodology for port performance measurement

different levels (Saaty 2004). Unlike a hierarchy a network structure allows for feedback between clusters, which makes it possible to identify and analyse interdependency both within a cluster and between clusters (Saaty 2004). The former is called an inner dependence and the latter is called an outer dependence.

In this section we present a hybrid approach to the measurement of port performance. FER is applied for dealing with uncertainties presented in the evaluations of the selected PPIs. The combined approach of DEMATEL and AHP is applied to evaluate the relative interdependent importance of the selected PPIs. The proposed framework using a hybrid approach of FER, DEMATEL and AHP follows the steps presented in Fig. 12.1. The evaluations of quantitative and qualitative PPIs at the bottom level and their associated weights need to be conducted in the first phase. The performance of PPIs with their weights can be synthesised using the ER and a utility technique. The synthesis processes can be conducted from PPIs at the bottom level to the goal.

3.1 DEMATEL

DEMATEL is useful to analyse interdependency of PPIs through converting qualitative designs into quantitative analysis (Liou et al. 2007; Buyukozkan and Cifci 2012). The method can be applied as follows (Liou et al. 2007).

Step 1. Pairwise comparisons: obtain an initial direct-relation matrix (Z).

The initial direct-relation matrix Z is an average $n \times n$ matrix constructed by pairwise comparisons in terms of directions and strength of influences between PPIs. As shown in Eq. (12.1), the initial direct-relation matrix $(Z) = [z_{ij}]_{n \times n}$, where z_{ij} is denoted as an average direct-relation value of x_{ij} and all principal diagonal $z_{ij}\,(i = j)$ are equal to zero, $X^k = \left[x_{ij}^k\right]$ is a judgement on causal relationship between x_{ij} by the kth expert.

$$Z = \left[z_{ij}\right]_{n\times n} = \frac{1}{m}\sum\nolimits_{k=1}^{m} x_{ij}^k,\ \ i,j = 1\ldots n \tag{12.1}$$

Step 2. Calculate a normalised direct-relation matrix (D).

The normalised direct-relation matrix $D = [d_{ij}]_{n \times n}$, where the value of each PPI in matrix D is $0 \le d_{ij} \le 1$, can be obtained through Eq. (12.2). In order to obtain a coefficient s, the maximum value of the sums of each row and column is used.

$$D = s \cdot Z \quad or \quad \left[d_{ij}\right]_{n\times n} = s \cdot \left[z_{ij}\right]_{n\times n},\ \ s > 0$$

$$s = \min\left[\frac{1}{max_{1\le i\le n}\sum_{j=1}^{n} \mid z_{ij} \mid}, \frac{1}{max_{1\le j\le n}\sum_{i=1}^{n} \mid z_{ij} \mid}\right]\ \ i,j = 1, 2, \ldots, n \tag{12.2}$$

Step 3. Obtain a total-relation matrix (T) and its sum of rows and columns.

The total-relation matrix T is obtained by operation of the normalised direct-relation matrix D using Eq. (12.3), in which I is denoted as the identity matrix. In Eq. (12.4), R_i and C_j denote the sums of rows and columns in the matrix T in which t_{ij} indicates the interdependent value of each pair of the investigated PPIs. Furthermore, the horizontal axis value pr_i^+ indicates how crucial the ith PPI is, whilst the vertical axis value pr_i^- classifies the PPIs into the cause and effect group. If the value of pr_i^- is positive, the PPI is classified into the cause group. Alternatively, when the value of pr_i^- is negative, the PPI is grouped into the effect group.

$$T = \lim_{m=\infty}\left(D^1 + D^2 + \ldots + D^m\right) = \sum\nolimits_{m=1}^{\infty} D^i = D(I - D)^{-1} \tag{12.3}$$

$$R_i = \sum\nolimits_{j=1}^{n} t_{ij} \quad C_j = \sum\nolimits_{i=1}^{n} t_{ij} \qquad (i,j=1,2,\ldots\ldots,n).$$
$$pr_i^+ = R_i + C_i \quad pr_i^- = R_i - C_i \tag{12.4}$$

Step 4. Obtain a threshold value (α).

The aim of setting a threshold value (α) is to filter and eliminate the PPIs that have trivial influence on others in the matrix T. The threshold value is computed by the average value of t_{ij}, where N indicates the total number of elements ($i \times j$). Only the PPIs whose influence values of t_{ij} are higher than the threshold value can be considered as interdependency among the PPIs.

$$\alpha = \frac{\sum_{i=1}^{n} \sum_{j=1}^{n} t_{ij}}{N} \tag{12.5}$$

3.2 *AHP*

The AHP is useful for dealing with MCDM problems and aids decision maker to capture both subjective and objective aspects of a decision (Saaty 2004). Through the pairwise comparisons based on the Saaty's nine-point scale ranging from 1 (equal) to 9 (extreme), the weights of PPIs can be obtained (Saaty 1980). Let e_{ij}^{l} be the relative importance judgement on the pair of PPIs P_i and $P_j(i,j=1,2,\ldots.. n)$ by the lth expert. Then, the aggregated weight comparison between P_i and P_j by m experts ($l \in m$) can be obtained by $e_{ij} = \frac{1}{m}\left(e_{ij}^{1} + \ldots + e_{ij}^{l} + \ldots + e_{ij}^{m}\right)$. Next, the synthesised i^{th} criterion weight comparison between P_i and P_j by m experts can be calculated using $w_i = \frac{1}{n}\sum\nolimits_{j=1}^{n}\left(\frac{e_{ij}}{\sum\nolimits_{i=1}^{n} e_{ij}}\right)$, where n means the total number of indicators and $\sum\nolimits_{i=1}^{n} w_i = 1$. In the AHP method, the consistency of the pairwise judgements (judgement reliability) can be obtained by calculating a Consistency Ratio (CR) using $CR = \frac{CI}{RI}$, where $CI = \frac{\lambda_{max}-n}{n-1}$, $\lambda_{max} = \frac{\sum\nolimits_{j=1}^{n} \frac{\sum\nolimits_{i=1}^{n} w_i e_{ji}}{w_j}}{n}$. CI is consistency index, λ_{max} is the principal eigenvalue of the comparison matrix, RI is average random index and n is the number of PPIs. The value of CR is greater than 0.1 indicates an inconsistency in the pairwise judgements and the experts needs to revise their pairwise judgements. Therefore, the judgements should inform an acceptable level with the CR of 0.10 or less.

3.3 Evidential Reasoning

FER makes it possible to model uncertainties of various natures in a flexible manner, requiring analysts to derive rational decisions from uncertain and incomplete data related to different quantitative and qualitative determinants (Yang and Xu 2002). It has therefore been applied in areas where uncertainty in data is high such as maritime security analysis (Yang et al. 2016). The ER approach is used to aggregate all output of degrees of belief (DoBs; β_j^k) from each rule (R_k) and to generate a conclusion (Yang and Xu 2002). The first step of the ER algorithm is to transform the DoBs (β_j^k) into two parts of basic probability mass (i.e. individual assigned DoBs and individual remaining (unassigned) DoBs) to aggregate all the output from R_k to generate combined DoB (β_j) in each possible D_j of D using following equations (Yang and Xu 2002; Yeo et al. 2014).

$$m_j^k = w_k \beta_1^n \tag{12.6}$$

$$m_D^k = \bar{m}_D^k + \tilde{m}_D^k \tag{12.7}$$

$$\bar{m}_D^k = 1 - w_k \tag{12.8}$$

$$\tilde{m}_D^k = w_k \left(1 - \sum_{j=1}^{n} \beta_j^k \right) \tag{12.9}$$

where $m_j^k (j = 1, \ldots, N; k = 1, \ldots, L)$ denotes individual degrees to which the rules (R_k) support the aggregated result D that is assessed to the assessment terms with DoBs; $w_k \left(\sum_{k=1}^{l} w_k = 1 \right)$ indicates relative importance of PPI in R_k; m_D^k represents the individual remaining belief degrees that are not yet assigned for m_j^k that is spilt into $\bar{m}_D^k$ (i.e. the remaining belief values unassigned to any individual evaluation grade caused by relative importance) and $\tilde{m}_D^k$ (i.e. the remaining belief values unassigned to any individual evaluation grade caused by incomplete assessment).

Next, suppose $m_j^{c(k)}$ represents the combined belief degree in D_j by aggregating in R_k, $\tilde{m}_D^{c(k)}$ represents the combined remaining belief degree to any D_j caused by the possible incompleteness in R_k and $\bar{m}_D^{c(k)}$ represents the combined relative importance of PPI in R_k (Eqs. 12.10, 12.11, 12.12 and 12.13). Finally after all assessments are aggregated, the overall combined DoB is generated using normalization process (Eqs. 12.14 and 12.15).

$$\{D_j\} : m_j^{c(k+1)} = K_{c(k+1)} \left(m_j^{c(k)} m_j^{k+1} + m_j^{c(k)} m_D^{k+1} + m_D^{c(k)} m_j^{k+1} \right) \tag{12.10}$$

$$\{D\} : \tilde{m}_D^{c(k+1)} = K_{c(k+1)} \left(\tilde{m}_D^{c(k)} \tilde{m}_D^{k+1} + \tilde{m}_D^{c(k)} \bar{m}_D^{k+1} + \bar{m}_D^{c(k)} \tilde{m}_D^{k+1} \right) \tag{12.11}$$

$$\bar{m}_D^{c(k+1)} = K_{c(k+1)} \left(\bar{m}_D^{c(k)} \bar{m}_D^{k+1} \right) \tag{12.12}$$

$$K_{c(k+1)} = \left[1 - \sum_{j=1}^{N} \sum_{\substack{t=1 \\ j \neq t}}^{N} m_j^{c(k)} m_t^{k+1}\right]^{-1}, k = 1, 2, \ldots, L-1 \quad (12.13)$$

$$\{D_j\} : \beta_j = \frac{m_j^{c(L)}}{1 - \bar{m}_D^{c(L)}} \quad (12.14)$$

$$\{D_j\} : \beta_D = \frac{\tilde{m}_D^{c(L)}}{1 - \bar{m}_D^{c(L)}} \quad (12.15)$$

where β_j represents the normalized DoB assigned to D_j in the final synthesized conclusion D and β_D indicates the normalized remaining DoB unassigned to any D_j.

It is not straightforward to use the overall result obtained using ER to rank each candidate port/terminal. Thus, utility techniques are used in order to obtain a single crisp value for the top-level PPI (goal) of each alternative (port/terminal) from the aggregated values (Yeo et al. 2014). D_j needs to be given utility values U_j for a crisp ranking index result, R_C and β_D requires to be assigned back to β_1 and β_N for the possible most preferred R_B and the possible worst preferred R_W. Consequently, the larger R_C, the more preferred the associated port/terminal is.

$$R_B = \sum_{j=2}^{N} \beta_j U_j + (\beta_1 + \beta_D) U_1, R_W = \sum_{j=1}^{N-1} \beta_j U_j + (\beta_N + \beta_D) U_N \quad (12.16)$$

$$R_C = \frac{R_B + R_W}{2}, \text{when} \sum_{j=1}^{N} \beta_j < 1 \text{ or } R_C = \sum_{j=1}^{N} \beta_j U_j, \text{when} \sum_{j=1}^{N} \beta_j = 1 \quad (12.17)$$

4 Identification of PPIs' Interdependency and Evaluation of Weights Using DEMATEL and AHP

4.1 DEMATEL

DEMATEL is applied to evaluate interdependent weights of six dimensions and 16 principal-PPIs.

Step 1. Pairwise comparisons: obtain an initial direct-relation matrix (Z).

First, the interdependency among six dimensions were determined by eight experts[1] through pairwise comparisons. The pair-wise comparison scale for this

[1]Eight experts (2 terminal operators, 1 liner company, 1 forwarder, 2 academics and 2 government representatives) among the 10 panel experts.

Table 12.2 The initial influence matrix

	CA	SA	FS	US	TSCI	SG	Sum
CA	0	1.80	2.20	3.00	1.80	1.60	**10.40**
SA	1.80	0	1.80	2.10	1.80	1.70	9.20
FS	1.60	1.80	0	1.20	1.30	2.50	8.40
US	2.20	1.90	1.80	0	1.70	1.30	8.90
TSCI	1.70	1.70	1.70	1.80	0	1.20	8.10
SG	2.20	2.00	1.20	0.90	1.10	0	7.40
Sum	9.50	9.20	8.70	9.00	7.70	8.30	

Table 12.3 The normalised direct-relation matrix

	CA	SA	FS	US	TSCI	SG
CA	0	0.17	0.21	0.29	0.17	0.15
SA	0.17	0	0.17	0.20	0.17	0.16
FS	0.15	0.17	0	0.12	0.13	0.24
US	0.21	0.18	0.17	0	0.16	0.13
TSCI	0.16	0.16	0.16	0.17	0	0.12
SG	0.21	0.19	0.12	0.09	0.11	0

Table 12.4 The total influence matrix

	CA	SA	FS	US	TSCI	SG	R_i	pr_i^+	Normalised pr_i^+	pr_i^-
CA	**0.97**	**1.09**	**1.08**	**1.16**	**0.96**	**1.00**	6.25	12.02	18.61%	0.48
SA	**1.02**	0.84	**0.96**	**1.01**	**0.88**	**0.92**	5.62	11.23	17.38%	0.01
FS	**0.93**	**0.92**	0.74	0.87	0.78	**0.91**	5.14	10.52	16.28%	(0.23)
US	**1.03**	**0.98**	**0.94**	0.83	0.86	0.87	5.51	11.08	17.14%	(0.05)
TSCI	**0.92**	**0.90**	0.87	**0.90**	0.66	0.80	5.06	9.89	15.31%	0.22
SG	**0.91**	0.87	0.79	0.79	0.71	0.65	4.72	9.87	15.27%	(0.43)
C_j	5.77	5.61	5.38	5.57	4.84	5.15	**32.30**			

chapter ranges from 0 to 4 with ‘0 (no influence)’, ‘1 (low influence)’, ‘2 (medium influence), ‘3 (high influence)’ and ‘4 (very high influence)’, respectively. Surveys were conducted in the form of questions, for example, such as “to what extent (i.e. from the ‘no influence (0)’ to the ‘very high influence (4)’) dimension A (i.e. CA) affects dimension B (i.e. SA)?” as well as their bidirectional influences, for example, “to what extent SA affect CA?” The initial average direct-relation 6×6 *matrix* (Z) is obtained as shown in Table 12.2.

Step 2. Calculate a normalised direct-relation matrix (D).

The normalised direct-relation matrix D is calculated by Eq. 12.2. The maximum value of the sums of each row and column is identified as 10.4 (Table 12.2) which can be used to obtain the normalised direct-relation matrix D as shown in Table 12.3.

Step 3–4. Obtain a total-relation matrix (T) and a threshold value (α).

The total-relation matrix T and sum of influence given and received by each dimension are obtained by Eqs. 12.3 and 12.4. (Table 12.4). A threshold value of

0.90 (=32.30/36) is calculated using Eq. 12.5. Only for the dimensions whose influence values of t_{ij} are higher than the threshold value we can determine that there are interdependent relationships among the six dimensions in Table 12.4. Accordingly, we can minimise the number of pairwise comparisons for the 16-principal PPIs. Hence, a zero value is given in the initial 16 × 16 matrix because no pairwise comparisons are conducted (Table 12.5).

In terms of pr_i^+(factors importance) in Table 12.4, core activities are the most important dimension, followed by supporting activities and users' satisfaction. The normalised pr_i^+ values ($\sum_{i=1}^{n} w_i = 1$) can be used for the interdependent weights of six dimensions. On top of that, core activities, supporting activities and terminal supply chain integration are identified as cause dimensions (i.e. positive pr_i^- value) while financial strength, users' satisfaction and sustainable growth are classified in effect dimensions (i.e. negative pr_i^- value). This classification is in line with previous studies. For instance, the literature on port performance measurement has used a technical or physical container terminal specification such as berth length, terminal area, number of cranes in berth and yard, labour, transport modes' turnaround as input data to measure efficiency and productivity of the container port industry (Tongzon 1995; Cullinane et al. 2002). Tangible and intangible resources such as human resources, information/ communication technology and organisational values cannot be overlooked as cause factors to investigate a firm's performance (Bagozzi et al. 1991; Barney 1991; Albadvi et al. 2007). Furthermore, it is empirically recognised that a higher integration between the players in supply chains leads to a higher competitiveness (Song and Panayides 2008; Panayides and Song 2009; Woo et al. 2013). Financial performance is denoted as the monetary units of tangible and intangible values yielded by a company's core business operations and any earning from the company's investments using resources such as land, labour and capital. Customer satisfaction can be measured by the perceived service qualities delivered by service providers (Brooks and Schellinck 2013). The internal and external effectiveness outcomes are driven by a series of value creation activities. Hence, there is no doubt that the CA, SA and TSCI are belonging to cause factors while FS and US are effect factors.

In a similar way, interdependency among 16 principal-PPIs and their interdependent weights can be obtained (Table 12.6). In terms of normalised pr_i^+ value, productivity is the most important principal-PPI with a value of 9.7%, followed by lead-time (9.64%), output (9.42%), information capital (7.6%), human capital (7.13%), organisation capital (7.04%) and service fulfilment (6.84%). Furthermore, in terms of pr_i^- value, the principal-PPIs including output, profitability, liquidity and solvency, service fulfilment, service costs safety and security, environment and social engagement are classified in effect factors (i.e. negative pr_i^- value). The results are also fully or partially in line with previous studies, including input data (cause factors) of the technical or physical container terminal specification for terminal efficiency and productivity (Cullinane et al. 2002); cause factors of the tangible and intangible resources for firm's performance (Alavi et al. 2006); terminal supply chain integration for port competitiveness

Table 12.5 The initial influence matrix (16 principal-PPIs)

		CA			SA			FS		US		TSCI			SG			
		OPC	PDC	LTC	HCS	OCS	ICS	PFF	LSF	SFU	SCU	ITST	VAST	ICIT	SSS	EVS	SES	Sum
CA	OPC	0	2.63	2.50	2.38	2.50	2.63	2.75	2.25	2.75	3.13	2.38	2.00	2.38	2.13	1.75	1.88	36.00
	PDC	3.38	0	3.25	2.75	2.75	2.63	2.63	2.25	2.75	3.00	2.50	2.00	2.25	1.75	2.00	1.75	**37.63**
	LTC	3.00	3.13	0	2.38	2.38	2.75	2.50	2.00	3.38	2.88	2.63	1.88	2.75	2.25	2.00	1.50	37.38
SA	HCS	2.50	3.00	2.75	0	0	0	2.38	2.38	2.75	2.38	2.00	2.50	2.50	2.50	2.13	1.88	31.63
	OCS	2.63	2.75	2.50	0	0	0	2.63	2.13	3.00	2.25	2.13	2.25	2.50	2.13	2.00	2.00	30.88
	ICS	2.88	3.25	3.50	0	0	0	2.38	2.13	2.63	2.38	2.25	2.38	2.75	2.50	1.75	1.50	32.25
FS	PFF	1.88	1.88	1.88	2.13	2.13	2.13	0	0	0	0	0	0	0	2.38	2.13	3.13	19.63
	LSF	1.88	1.75	1.75	2.25	2.38	2.25	0	0	0	0	0	0	0	2.13	2.75	2.50	19.63
US	SFU	3.00	2.88	2.75	2.75	2.88	2.88	2.50	2.50	0	0	0	0	0	0	0	0	22.13
	SCU	2.88	2.88	2.63	1.63	1.63	1.63	3.13	3.00	0	0	0	0	0	0	0	0	19.38
TSCI	ITST	2.88	2.38	2.88	2.13	2.13	2.88	0	0	2.63	2.00	0	0	0	0	0	0	19.88
	VAST	2.38	2.25	2.00	2.00	1.88	2.25	0	0	2.75	2.75	0	0	0	0	0	0	18.25
	ICIT	2.63	2.75	3.00	2.75	2.38	3.38	0	0	2.88	2.13	0	0	0	0	0	0	21.88
SG	SSS	2.25	2.50	2.75	0	0	0	0	0	0	0	0	0	0	0	0	0	7.50
	EVS	1.50	1.75	1.63	0	0	0	0	0	0	0	0	0	0	0	0	0	4.88
	SES	1.38	1.38	1.38	0	0	0	0	0	0	0	0	0	0	0	0	0	4.13
Sum		37.00	37.13	37.13	23.13	23.00	25.38	20.88	18.63	25.50	22.88	13.88	13.00	15.13	17.75	16.50	16.13	

Table 12.6 The total influence matrix of 16 principal-PPIs

		CA			SA			FP		US		TSCI			SG						
		OPC	PDC	LTC	HCS	OCS	ICS	PFF	LSF	SFU	SCU	ITST	VAST	ICIT	SSS	EVS	SES	R_i	pr_i^+	Normalised pr_i^+	pr_i^-
CA	OPC	0.17	0.24	0.23	0.17	0.18	0.19	0.18	0.16	0.19	0.19	0.14	0.12	0.14	0.15	0.13	0.13	2.70	5.47	9.42%	−0.07
	PDC	0.26	0.18	0.26	0.19	0.19	0.20	0.19	0.16	0.20	0.20	0.15	0.13	0.15	0.14	0.14	0.13	2.86	5.63	9.70%	0.09
	LTC	0.25	0.26	0.18	0.18	0.18	0.20	0.18	0.16	0.22	0.19	0.15	0.12	0.16	0.15	0.14	0.12	2.83	5.59	9.64%	0.07
SA	HCS	0.21	0.22	0.21	0.10	0.10	0.11	0.15	0.14	0.17	0.16	0.11	0.12	0.13	0.14	0.12	0.12	2.32	4.14	7.13%	0.50
	OCS	0.21	0.21	0.20	0.10	0.10	0.11	0.16	0.13	0.18	0.15	0.12	0.11	0.13	0.13	0.12	0.12	2.27	4.08	7.04%	0.45
	ICS	0.23	0.23	0.24	0.11	0.11	0.12	0.16	0.14	0.18	0.16	0.13	0.12	0.14	0.14	0.12	0.11	2.43	4.41	7.60%	0.46
FP	PFF	0.14	0.14	0.14	0.11	0.11	0.11	0.06	0.05	0.07	0.06	0.05	0.04	0.05	0.11	0.10	0.13	1.45	3.23	5.57%	−0.34
	LSF	0.14	0.13	0.13	0.11	0.11	0.11	0.06	0.05	0.07	0.06	0.05	0.04	0.05	0.11	0.12	0.11	1.45	3.04	5.24%	−0.13
US	SFU	0.20	0.20	0.19	0.15	0.16	0.16	0.15	0.14	0.10	0.09	0.07	0.06	0.07	0.08	0.07	0.07	1.95	3.97	6.84%	−0.07
	SCU	0.18	0.18	0.17	0.11	0.11	0.12	0.15	0.14	0.08	0.07	0.05	0.05	0.06	0.07	0.06	0.06	1.68	3.52	6.06%	−0.16
TSCI	ITST	0.19	0.18	0.19	0.13	0.13	0.16	0.09	0.08	0.16	0.13	0.06	0.06	0.07	0.07	0.06	0.06	1.80	3.04	5.24%	0.56
	VAST	0.17	0.16	0.16	0.12	0.12	0.13	0.08	0.07	0.15	0.14	0.05	0.05	0.06	0.06	0.05	0.05	1.62	2.75	4.75%	0.49
	ICIT	0.20	0.20	0.20	0.15	0.15	0.18	0.10	0.09	0.17	0.15	0.07	0.06	0.07	0.07	0.07	0.06	1.98	3.31	5.70%	0.65
SG	SSS	0.11	0.11	0.12	0.04	0.04	0.04	0.04	0.03	0.04	0.04	0.03	0.02	0.03	0.03	0.03	0.03	0.76	2.24	3.86%	−0.72
	EVS	0.07	0.08	0.07	0.02	0.02	0.03	0.02	0.02	0.03	0.03	0.02	0.02	0.02	0.02	0.02	0.02	0.49	1.86	3.21%	−0.88
	SES	0.06	0.06	0.06	0.02	0.02	0.02	0.02	0.02	0.02	0.02	0.02	0.01	0.02	0.02	0.01	0.01	0.42	1.74	3.00%	−0.91
C_j		2.77	2.77	2.76	1.82	1.82	1.97	1.79	1.58	2.02	1.84	1.24	1.13	1.33	1.48	1.37	1.32				

(Panayides and Song 2009); financial performance yielded by a company's core business operations and any earning from the company's investment (SU et al. 2003); customer satisfaction measured by the perceived service qualities (Brooks and Schellinck 2013); an appropriate safety and security scheme for port efficiency and competitiveness (Woo et al. 2011). Consequently, the principal-PPIs of OPC (output), PFF (profitability), LSF (liquidity and solvency) and SFU (service fulfilment) are obviously classified in an effect factor. Therefore, the DEMATEL model is verified by both contents and technical validity.

4.2 AHP and Weights of 60 PPIs

AHP is used to obtain weights of 60 PPIs mainly because that (1) interdependency among the PPIs is partially modelled through their parent level 16 principal PPIs; (2) the influence of the interdependency is moderate due to their location at the bottom level of the hierarchy; and (3) using ANP and DEMATEL to analyse the weights of 60 PPIs is a time consuming job requiring a high amount of data to be collected which is very difficult, although it is not impossible, in reality. The judgements of five among the eight evaluators have verified with the CR of 0.10 or less. Generally, the value of CR is greater than 0.1 and the evaluators need to revise their pairwise judgements. However, five judgements presenting consistent input data, which are sufficient to provide a reasonable AHP outcome (Büyüközkan et al. 2012) are used to derive the weights of 60 PPIs at the bottom level. It is noteworthy that the weights obtained are local weights at the same level (6 dimensions, 16 principal-PPIs and 60 PPIs respectively). Further computation has been conducted to obtain global weights of the bottom level PPIs by multiplying their local weights with the ones of their associated upper level criteria (i.e. 16 principal PPIs and 6 dimensions). For instance, the global weight of 'throughput growth' can be obtained as 0.012 (=0.1861 (the local weight of core activities) × 0.0942(the local weight of output) × 0.696 (the local weight of throughput)). Consequently, the local weights of all criteria, the global weights of 60 PPIs and their normalised global weights are shown in Table 12.7. Derived from the results of DEMATEL and AHP, throughput growth (OPC1) is the most important PPI, which has a relative importance value of 7.16%, followed by vessel turnaround (LTC1, 6.34%), crane productivity (PDC4, 3.65%), overall service cost (SCU1, 3.35%), vessel call size growth (OPC2, 3.13%), IT systems (ICS1, 2.82%), networks (ICS3, 2.6%), training and education (HCS3, 2.57%), overall service reliability (SFU1, 2.48%) and teamwork (OCS4, 2.37%), being the top 10 highest scores in Table 12.7. In the contrast, disclose of information (0.4%), total water consumption (0.42%), waste recycling (0.43%) and carbon footprint (0.45%) under sustainable growth (SG) are the least important PPIs. The weights obtained by DEMATEL and AHP can be synthesised with the evaluations of each terminal against all PPIs using ER and its associated intelligent decision system (IDS) software (Yang and Xu 2000).

Table 12.7 PPIs' relative weights

PPIs	LW	GW	NGW
Core activities (CA)	0.1861		
Output (OPC)	0.0942		
Throughput growth (OPC1)	0.696	0.012	7.16%
Vessel call size growth (OPC2)	0.304	0.005	3.13%
Productivity (PDC)	0.097		
Ship load rate (PDC1)	0.158	0.003	1.67%
Berth utilization (PDC2)	0.132	0.002	1.40%
Berth occupancy (PDC3)	0.107	0.002	1.13%
Crane productivity (PDC4)	0.345	0.006	3.65%
Yard utilization (PDC5)	0.103	0.002	1.09%
Labour productivity (PDC6)	0.155	0.003	1.64%
Lead time (LTC)	0.0964		
Vessel turnaround (LTC1)	0.602	0.011	6.34%
Truck turnaround (LTC2)	0.185	0.003	1.95%
Container dwell time (LTC3)	0.213	0.004	2.24%
Support activities (SA)	0.1738		
Human capital (HCS)	0.0713		
Knowledge and skills (HCS1)	0.246	0.003	1.79%
Capabilities (HCS2)	0.243	0.003	1.77%
Training and education (HCS3)	0.354	0.004	2.57%
Commitment and loyalty (HCS4)	0.157	0.002	1.14%
Organisation capital (OCS)	0.0704		
Culture (OCS1)	0.175	0.002	1.26%
Leadership (OCS2)	0.296	0.004	2.12%
Alignment (OCS3)	0.198	0.002	1.42%
Teamwork (OCS4)	0.33	0.004	2.37%
Information capital (ICS)	0.076		
IT systems (ICS1)	0.364	0.005	2.82%
Database (ICS2)	0.301	0.004	2.33%
Networks (ICS3)	0.335	0.004	2.60%
Financial strength (FS)	0.1628		
Profitability (PFF)	0.0557		
Revenue growth (PFF1)	0.318	0.003	1.69%
EBIT(operating profit) margin (PFF2)	0.328	0.003	1.74%
Net profit margin (PFF3)	0.354	0.003	1.88%
Liquidity & Solvency (LSF)	0.0524		
Current ratio (LSF1)	0.342	0.003	1.71%
Debt to total asset (LSF2)	0.349	0.003	1.75%
Debt to equity (LSF3)	0.309	0.003	1.55%
Users' satisfaction (US)	0.1714		
Service fulfilment (SFU)	0.0684		
Overall service reliability (SFU1)	0.361	0.004	2.48%

(continued)

Table 12.7 (continued)

PPIs	LW	GW	NGW
Responsiveness to special requests (SFU2)	0.147	0.002	1.01%
Accuracy of documents & information (SFU3)	0.134	0.002	0.92%
Incidence of cargo damage (SFU4)	0.188	0.002	1.29%
Incidence of service delay (SFU5)	0.17	0.002	1.17%
Service costs (SCU)	0.0606		
Overall service cost (SCU1)	0.549	0.006	3.35%
Cargo handling charges (SCU2)	0.315	0.003	1.92%
Cost of terminal ancillary services (SCU3)	0.137	0.001	0.83%
Terminal supply chain integration (TSCI)	0.1531		
Intermodal transport systems (ITST)	0.0524		
Sea-side connectivity (ITST1)	0.466	0.004	2.19%
Land-side connectivity (ITST2)	0.159	0.001	0.75%
Reliability of multimodal operations (ITST3)	0.197	0.002	0.93%
Efficiency of multimodal operations (ITST4)	0.178	0.001	0.84%
Value-added services (VAST)	0.0475		
Facilities to add value to cargos (VAST1)	0.369	0.003	1.57%
Service adaptation to customers (VAST2)	0.172	0.001	0.73%
Capacity to handle different types of cargo (VAST3)	0.262	0.002	1.12%
Tailored services to customers (VAST4)	0.197	0.001	0.84%
Information/communication integration (ICIT)	0.057		
Integrated EDI for communication (ICIT1)	0.291	0.003	1.49%
Integrated IT to share data (ICIT2)	0.261	0.002	1.34%
Collaborate with Channel members (ICIT3)	0.232	0.002	1.19%
Latest port IT systems (ICIT4)	0.216	0.002	1.11%
Sustainable growth (SG)	0.1527		
Safety and Security (SSS)	0.0386		
Identifying restricted areas and access control (SSS1)	0.298	0.002	1.03%
Formal safety and security training practices (SSS2)	0.206	0.001	0.71%
Adequate monitoring and threat awareness (SSS3)	0.231	0.001	0.80%
Safety and security officers and facilities (SSS4)	0.265	0.002	0.92%
Environment (EVS)	0.0321		
Carbon footprint (EVS1)	0.158	0.001	0.45%
Water consumption (EVS2)	0.145	0.001	0.42%
Energy consumption (EVS3)	0.248	0.001	0.71%
Waste recycling (EVS4)	0.149	0.001	0.43%
Environment management programmes (EVS5)	0.3	0.001	0.86%
Social engagement (SES)	0.03		
Employment (SES1)	0.578	0.003	1.55%
Regional GDP (SES2)	0.272	0.001	0.73%
Disclose of information (SES3)	0.15	0.001	0.40%

Note: *LW* local weight, *GW* global weight, *NGW* normalised global weight

Table 12.8 Examples of assessment grades of the PPIs

PPIs	Assessment grades				
Output (OPC)					
Throughput volume growth	leq 0%	5%	10%	15%	20%	geq 25%
Vessel call size growth	leq 0%	5%	10%	15%	geq 20%	
Human capital (HCS)						
Knowledge and skills	Very poor	Poor	Medium	Good	Very good	
Capability	Very poor	Poor	Medium	Good	Very good	

Note: *leq* less than or equal to, *geq* great than or equal to

5 Evaluation of Port Performance in Korean Container Terminals

5.1 Setting Assessment Grades to Each PPI

Assessment grades are set to all PPIs which are of either qualitative or quantitative based on their features. For assessing qualitative PPIs, different sets of measurement grades (linguistic terms) defined by domain experts can be used (Yang 2001). For example, in order to measure the "Knowledge and skills of port employees", a set of the fuzzy linguistic terms {very poor, poor, medium, good, very good} are used. If PPI is of quantitative nature, it can be assessed using numerical grades (Yang 2001) based on various data (i.e. consulting reports, journal papers and internal data of terminal operators). From this perspective, a set of quantitative grades, for example, {leq 0%, 5%, 10%, 15%, geq 20%} for "vessel call size growth (the number of vessel calls × the size of vessels)" are developed based on the publicly available data from ports around the globe. In a similar manner, assessment grades for all PPIs can be developed. Again, the defined assessment grades for both quantitative and qualitative PPIs are justified by the experts. The examples of assessment grades allocated to both quantitative (i.e. output) and qualitative PPIs (i.e. human capital) are shown in Table 12.8.

5.2 Data Collection

The empirical study in this chapter needs to collect both primary and secondary data for the various types of quantitative and qualitative PPIs. The secondary data of quantitative PPIs were directly collected from terminal operating companies and information systems/databases managed by port authorities and Korean government. The qualitative PPIs were collected using questionnaires from three groups of operators, users and administrators to assess their own associated PPIs to measure each terminal container terminal performance. The survey was conducted through an online survey tool as well as distributed by emails during 2014–2015. The detailed responses of the survey are listed in Table 12.9.

Table 12.9 Response details

Terminal	Stakeholder	Total distributed	Email	Online	Usable response	Judgement on:
T1	TO	25	0	12 (11)	11	SA, TSCI, SSS, EVS
	PU	200	38 (31)	20(12)	43	US, TSCI
	AD	40	0	9 (6)	6	SG
T2	TO	25	0	9 (8)	8	SA, TSCI, SSS, EVS
	PU	200	38 (30)	20(12)	42	US, TSCI
	AD	40	0	9 (6)	6	SG
T3	TO	25	2 (2)	12 (10)	12	SA, TSCI, SSS, EVS
	PU	200	38 (30)	20(12)	42	US, TSCI
	AD	40	0	9 (6)	6	SG
T4	TO	25	1 (1)	6 (6)	7	SA, TSCI, SSS, EVS
	PU	200	38 (30)	20(12)	42	US, TSCI
	AD	40	0	9 (6)	6	SG
T5	TO	25	4 (4)	13 (10)	14	SA, TSCI, SSS, EVS
	PU	200	38 (30)	20(12)	42	US, TSCI
	AD	40	0	9 (6)	6	SG
T6	TO	25	0	7 (7)	7	SA, TSCI, SSS, EVS
	PU	200	38 (30)	20 (12)	42	US, TSCI
	AD	40	0	9 (6)	6	SG

5.3 Evaluations of Quantitative and Qualitative PPIs at the Bottom Level

In this chapter, each PPI can be assessed using a belief degree represented by judgements (Yang 2001; Yeo et al. 2014). The judgements can be presented by degree of beliefs (DoBs) which belong to either linguistic terms (for the qualitative PPIs) or numerical values (for the quantitative PPIs). The former can be directly obtained by expert judgements and the latter needs to be calculated through various location measurement techniques (Yang et al. 2009). Using location measurement techniques, the degree of belief associated to numerical grades can be transformed by Eq. 12.18. Any quantitative number $h_{j,i}$ (with an evaluation grade H_j) is

Table 12.10 Throughput growth (2012–2013)

Terminal	2012	2013	Growth ('12-'13)
T6	1,988,675	2,391,890	20.28%

evaluated between $h_{j-1,i}$ (with an evaluation grade H_{j-1}) and $h_{j+1,i}$ (with an evaluation grade H_{j+1}).

$$\textit{If } h_{j-1,i} < h_{j,i} < h_{j+1,i} \textit{ then } B_{j+1,i} = \frac{h_{j,i} - h_{j-1,i}}{h_{j+1,i} - h_{j-1,i}}, B_{j-1,i} = 1 - h_{j+1,i} \quad (12.18)$$

where $B_{j+1,i}$ represents the degree of belief associated quantitative number with the grade $H_{j+1}$and $B_{j-1,i}$ represents the degree of belief associated quantitative number with the grade H_{j-1}. For example, a set of quantitative grades $H = \{leq\ 0\,\%\ (H_1), 5\,\%\ (H_2), 10\,\%\ (H_3), 15\,\%\ (H_4), 20\,\%\ (H_5), \text{geq}\ \ 25\,\%\ (H_6)\}$ for "throughput growth" is already defined. If the assessment of the throughput growth in terminal 6 (T6) is 20.28% (Table 12.10), thus, $B_{j+1,i} = \frac{20.28-20}{25-20} = 0.056$ DoB with H_6 and $B_{j-1,i} = 1 - 0.056 = 0.944$ DoB with H_5. Therefore, the throughput growth (TG) set in T6 is assessed as follows:

$$H^{TG} = \{(leq\ 0\%, 0), (5\%, 0), (10\%, 0), (15\%, 0), (20\%, 0.944), (geq\ 25\%, 0.056)\}$$

The DoB obtained is used as an input data for mapping process in Sect. 5.3. In a similar way, the bottom level PPIs' sets of all ports can be obtained.

5.4 *Mapping Process –Transform the Evaluation from the Bottom Level PPIs to Top Level PPI*

This chapter defined various and different assessment grades in terms of the terms (i.e. linguistic terms for qualitative; numerical terms for quantitative PPIs) and numbers (from 2 to 7 scales) at the lower-level PPIs and their associated upper-level PPIs. The defined grades, thus, need to be interpreted and transformed into a unified format for assessment of the associated upper level PPIs (Yeo et al. 2014). This can be done using a fuzzy IF-THEN rule based belief structure. Yang (2001) developed the rule based utility techniques that can be applied for transforming qualitative and quantitative data in this study. Furthermore, the mapping techniques to convert DoBs of the bottom-level PPIs to their associated upper level principal-PPIs can be soundly conducted in a unified manner (Yang 2001; Yang et al. 2009; Yeo et al. 2014). The core of this technique is a fuzzy mapping technique to transform fuzzy inputs to fuzzy outputs. As shown in Fig. 12.2, I^i ($\sum_{i=1}^{n} I^i \leq 1$) indicates the fuzzy input associated with a lower-level PPI and O^j ($O^j = \sum_{i=1}^{n} I^i \beta_i^j$) represents the fuzzy output transformed from I^i. B_i^j ($\sum_{j=1}^{n} \beta_i^j = 1$) denotes the DoBs assigned by experts for presenting the relationship between assessment grades of different

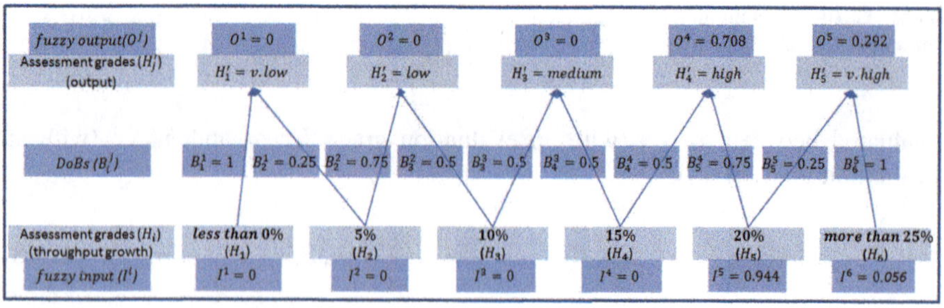

Fig. 12.2 Fuzzy mapping process

Table 12.11 Fuzzy rule base belief structure

Throughput growth (TG) to output (OPC)	R^1: *IF terminal operator's "throughput growth (TG)" is "more than 25% (TG6)", then "output (OPC)" is "very high (OPC5)" with 100% DoB. This can be simplified and presented by symbols as* R^1: *If "TG" is "TG6", then "OPC" is "100% OPC5"* *Similarly,* R^2: *If "TG" is "TG5", then "OPC" is "25% OPC5" and "75% OPC4"* R^3: *If "TG" is "TG4", then "OPC" is "50% OPC4" and "50% OPC3"* R^4: *If "TG" is "TG3", then "OPC" is "50% OPC3" and "50% OPC2"* R^5: *If "TG" is "TG2", then "OPC" is "75% OPC2" and "25% OPC1"* R^6: *If "TG" is "TG1", then "OPC" is "100% OPC1"*

levels. For example, the upper level PPI "output (OPC)" can be expressed using linguistic terms as "very low (OPC1)", "low (OPC2)", "medium (OPC3)", "high (OPC4)" and "very high (OPC5)". The numerical grades used to assess the lowest level PPI "throughput growth (TG)" can be expressed "leq 0% (TG1)", "5% (TG2)", "10% (TG3), "15% (TG4)", "20% (TG5)" and geq 25% (TG6). The decision makers have assigned the fuzzy rules for mapping the transformation from throughput growth to output (Table 12.11). It is noteworthy that a throughput growth of the "leq0%" means that the output is said to be equivalent to a grade "very low" using fuzzy rules. Based on $R^1$and R^2, it can be transformed into 70.8% OPC4 ($O^4 = 0.944 \times 0.75$) and 29.2% OPC5 ($O^5 = (0.944 \times 0.25) + (0.056 \times 1)$) respectively. The TG output set in T6 is assessed as follows:

$$H^{TG\ OPC} = \{(very\ low, 0), (low, 0), (medium, 0), (high, 0.708), (very\ high, 0.292)\}$$

This mapping process can be conducted from the lowest level PPIs to the top level goal in a similar manner (Fig. 12.3).

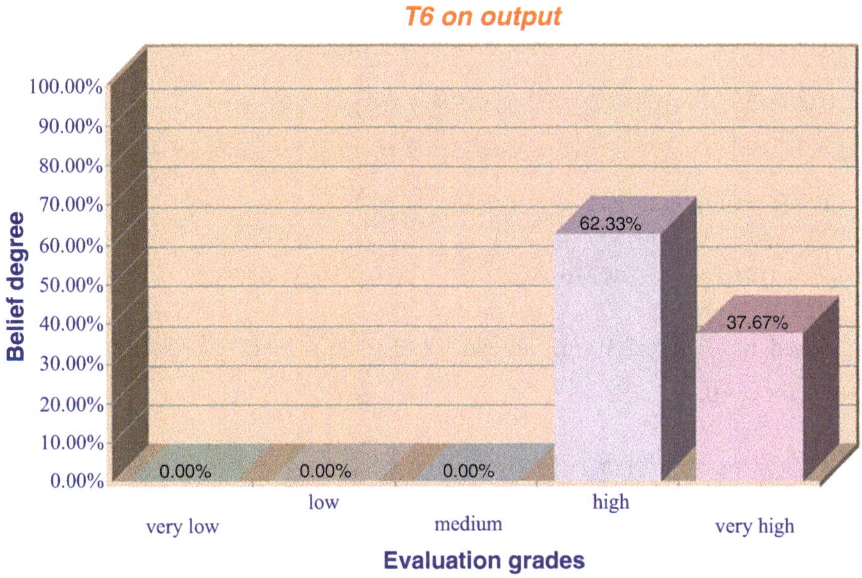

Fig. 12.3 Mapping result of output in T6 (IDS software)

Table 12.12 Mapping results and relative weights and aggregation (output)

Output	Very low	Low	Medium	High	Very high	Weight
Throughput growth	0	0	0	0.708	0.292	0.696
Vessel call size growth	0	0	0	0.294	0.706	0.304
Aggregation results	0	0	0	**0.62330**	**0.37669**	

5.5 Synthesis of DoBs and Weights of PPIs Using ER Algorithm

The DOBs transformed results from the lowest level PPIs to the top level goal and their interdependent weights can be synthesised by IDS software incorporating the ER algorithm (Eqs. 12.6, 12.7, 12.8, 12.9, 12.10, 12.11, 12.12, 12.13, 12.14 and 12.15) and utility technique (Eqs. 12.16 and 12.17). In this section, based on the results from mapping process and relative weights (see Table 12.12 and Fig. 12.3), aggregation of the bottom level PPIs (i.e. throughput growth and vessel call size growth) under principal-PPI (i.e. output) is demonstrated as an example, which is presented as follows:

Based on Eqs. (12.6, 12.7, 12.8 and 12.9) and Table 12.12, $m_1^1 = w_1\beta_1^1 = 0, m_2^1 = w_1\beta_2^1 = 0, m_3^1 = w_1\beta_3^1 = 0, m_4^1 = w_1\beta_4^1 = 0.696 \times 0.708 = 0.49276, m_5^1 = w_1\beta_5^1 = 0.696 \times 0.292 = 0.20303, \bar{m}_D^1 = 1 - w_1 = 1 - 0.696 = 0.304, \tilde{m}_D^1 = w_1 \left(1 - \sum_{j=1}^{5} \beta_j^1\right) = 0.696(1 - 1) = 0.$

$m_1^2 = w_2\beta_1^2 = 0, m_2^2 = w_2\beta_2^2 = 0, m_3^2 = w_2\beta_3^2 = 0, m_4^2 = w_2\beta_4^2 = 0.304 \times 0.294 = 0.08938, m_5^2 = w_2\beta_5^2 = 0.304 \times 0.706 = 0.21462, \bar{m}_D^2 = 1 - w_2 = 1 - 0.304 = 0.696, \tilde{m}_D^2 = w_2\left(1 - \sum_{j=1}^{5}\beta_j^2\right) = 0.304(1-1) = 0.$

Based on Eq. (12.13), $K_{c(2)} = \left[1 - \sum_{j=1}^{5}\sum_{\substack{t=1 \\ j \neq t}}^{5} m_j^{c(1)} m_t^2\right]^{-1} = [1 - (0 + 0 + 0 + 0.10576\,(= 0.49276 \times 0.21462) + 0.01816\,(= 0.20323 \times 0.08938)]^{-1} = 1.14145.$

Based on Eqs. (12.10, 12.11 and 12.12), $m_1^{c(2)} = K_{c(2)}\left(m_1^{c(1)} m_1^2 + m_1^{c(1)} m_D^2 + m_D^{c(1)} m_1^2\right) = 0,$

$$m_2^{c(2)} = K_{c(2)}\left(m_2^{c(1)} m_2^2 + m_2^{c(1)} m_D^2 + m_D^{c(1)} m_2^2\right) = 0,$$

$$m_3^{c(2)} = K_{c(2)}\left(m_3^{c(1)} m_3^2 + m_3^{c(1)} m_D^2 + m_D^{c(1)} m_3^2\right) = 0,$$

$$\begin{aligned} m_4^{c(2)} &= K_{c(2)}\left(m_4^{c(1)} m_4^2 + m_4^{c(1)} m_D^2 + m_D^{c(1)} m_4^2\right) \\ &= 1.14145(0.49276 \times 0.08938 + 0.49276 \times 0.696 + 0.08938 \times 0.304) \\ &= 0.47276, \end{aligned}$$

$$\begin{aligned} m_5^{c(2)} &= K_{c(2)}\left(m_5^{c(1)} m_5^2 + m_5^{c(1)} m_D^2 + m_D^{c(1)} m_5^2\right) \\ &= 1.14145(0.20323 \times 0.21462 + 0.20323 \times 0.696 + 0.21462 \times 0.304) \\ &= 0.28572, \end{aligned}$$

$$\tilde{m}_D^{c(2)} = K_{c(2)}\left(\tilde{m}_D^{c(1)} \tilde{m}_D^2 + \tilde{m}_D^{c(1)} \bar{m}_D^2 + \bar{m}_D^{c(1)} \tilde{m}_D^2\right) = 0,$$

$$\bar{m}_D^{c(2)} = K_{c(2)}\left(\bar{m}_D^{c(1)} \bar{m}_D^2\right) = 1.14145(0.304 \times 0.696) = 0.24151.$$

Based on Eq. (12.14), $\{D_j\} : \beta_j = \dfrac{m_j^{c(L)}}{1 - \bar{m}_D^{c(L)}} = 0, \ j = 1, 2, 3,$

$$\{D_j\} : \beta_4 = \frac{m_4^{c(2)}}{1 - \bar{m}_D^{c(2)}} = \frac{0.47276}{1 - 0.24151} = 0.62330,$$

$$\{D_j\} : \beta_5 = \frac{m_5^{c(2)}}{1 - \bar{m}_D^{c(2)}} = \frac{0.28572}{1 - 0.24151} = 0.37669.$$

Table 12.13 Aggregation of 6 dimensions (goal)

Goal (port performance)	Very poor	Poor	Medium	Good	Very good	Weight
Core activities	0.0201	0.0675	0.0336	0.1907	0.6880	0.1861
Support activities	0.0490	0.1031	0.0602	0.4113	0.3764	0.1738
Financial strength	0.2770	0.1945	0.0834	0.1657	0.2794	0.1628
Users' satisfaction	0.1238	0.1809	0.0746	0.3358	0.2849	0.1714
Terminal supply chain integration	0.0674	0.1183	0.0588	0.3640	0.3915	0.1531
Sustainable growth	0.0216	0.0730	0.0472	0.1799	0.6782	0.1527
Aggregation results	**0.1087**	**0.0943**	**0.0253**	**0.2069**	**0.5649**	

Table 12.14 Calculation of port performance (utility techniques)

	Very poor	Poor	Medium	Good	Very good
Preference value	1	2	3	4	5
U_j	$\frac{1-1}{5-1}=0$	$\frac{2-1}{5-1}=0.25$	$\frac{3-1}{5-1}=0.5$	$\frac{4-1}{5-1}=0.75$	$\frac{5-1}{5-1}=1$
β_j	0.1087	0.0943	0.0253	0.2069	0.5649
$\beta_j U_j$	0.0000	0.0236	0.0127	0.1552	0.5649
R_C	$R_C=\sum_{j=1}^{5}\beta_j U_j=0.7563$				

Similarly, the aggregations of the bottom level PPIs can be obtained, then the transformed results from the lowest level PPIs to the top level goal and their relative weights can be synthesised. By help of the IDS software, DoBs of six dimensions are aggregated as shown in Table 12.13. The fuzzy set for the T6's performance can be expressed as follows:

$$H^{PP} = \{(very\ poor, 0.1087), (poor, 0.0943), (medium, 0.0253), (good, 0.2069), (very\ good, 0.5649)\}$$

However, it is not straightforward to use the overall aggregated result (DOBs) obtained using ER to rank each candidate terminal. It needs to be transformed into a single crisp value for the top-level goal (port performance) of T6. Utility techniques can be used to obtain a single crisp value. Based on Eqs. (12.16 and 12.17), the performance of T6 can be calculated as shown in Table 12.14. The overall performance of T6 is evaluated with 0.7563. In a similar way, the performance scores of the six alternative container terminals are obtained and presented in Table 12.15.

Table 12.15 Performance score of each container terminal

	Performance	Ranking index	Ranking
T 1	VP 0.2389; P 0.919; M 0.0282; G 0.2157; VG 0.4253	0.6241	5
T 2	VP 0.2807; P 0.1513; M 0.0350; G 0.1971; VG 0.3353	0.5391	6
T 3	VP 0.1910; P 0.1127; M 0.0322; G 0.2257; VG 0.4384	0.6519	4
T 4	VP 0.1250; P 0.0694; M 0.0222; G 0.1971; VG 0.5863	0.7626	1
T 5	VP 0.1778; P 0.0906; M 0.0296; G 0.2478; VG 0.4542	0.6775	3
T 6	VP 0.1087; P 0.0943; M 0.0253; G 0.2069; VG 0.5649	0.7563	2

5.6 *Results*

Based on the proposed framework, the performance scores of each terminal can be easily compared so that decision makers can straightforwardly identify the strengths and weaknesses of their terminals. The results derived from IDS software are shown in Tables 12.15, 12.16, 12.17 and 12.18. This software provides a tailored information with respect to criteria in different level. Table 12.15 shows the overall performance score of each terminal in terms of performance ranking index. The results suggest that T4 shows the best results (0.7626), followed by T6 with 0.7563. T2 is assessed to be the least competitive terminal with lowest performance score (0.5391).

However, the overall performance score only provides port managers with ranking index among the six terminals but does not offer diagnostic instruments to decision makers in identifying the particular areas for improvement. The performance of individual PPIs with respect to the alternative terminals is shown in Table 12.16. The information is very straightforward and port managers can easily identify and interpret their strength and weakness in terms of each port activity. For example, both the container throughput growth and vessel call size growth PPIs in T6 are shown as the highest performance with 0.9004 and 0.9587 respectively[2]. This leads to the remarkable performance result of all profitability PPIs (i.e. revenue growth (PFF1) with 1.000, operating profit margin (PFF2) with 0.9175, and net profit margin (PFF3) with 0.5489) in 2013. This is a very natural result due to the highest performance on the container throughput and vessel call PPI that generates revenues for terminal operators. This relationship is also evidenced by individual DEMATEL pr_i^- value (−0.04) between output (OPC, 0.14) and profitability (PFF,

[2]The container throughput in T6 increased dramatically from 1,988,675 TEU in 2012 to 2,391,890 TEU in 2013 (20.28%), and the vessel capacity growth (total gross tonnage (GT) of the vessels divided by total number of vessel calls) increased radically from 40,684 GT/ship in 2012 to 47,981 GT/ship in 2013(18.04%).

Table 12.16 Performance score on 60 PPIs

PPIs	T1	T2	T3	T4	T5	T6
OPC1	0.0000	0.1579	0.0000	0.0124	0.0000	0.9004
OPC2	0.0000	0.0000	0.5884	0.0057	0.0019	0.9587
PDC1	0.9256	1.0000	0.2056	0.4393	0.3983	0.2466
PDC2	0.6355	0.3731	0.7156	0.9736	1.0000	1.0000
PDC3	0.0422	0.1844	1.0000	0.0000	0.4211	1.0000
PDC4	0.6078	0.6078	0.6797	0.8516	0.5359	0.3922
PDC5	0.0105	0.0791	0.6381	0.5592	0.6775	1.0000
PDC6	0.4321	0.4518	0.2786	0.3104	0.8619	0.9097
LTC1	1.0000	1.0000	1.0000	1.0000	1.0000	1.0000
LTC2	0.9283	0.9198	0.7613	1.0000	1.0000	0.9641
LTC3	0.9367	0.9104	0.9684	0.9420	0.9473	0.9631
HCS1	0.8945	0.6933	0.8443	0.9399	0.8117	0.9251
HCS2	0.6815	0.5000	0.7373	0.7249	0.6973	0.6696
HCS3	0.5050	0.4223	0.6481	0.6696	0.5868	0.6292
HCS4	0.6565	0.4439	0.8307	0.8096	0.7046	0.7249
OCS1	0.7370	0.3798	0.7728	0.7396	0.7801	0.7288
OCS2	0.7465	0.5000	0.8623	0.8402	0.7599	0.7396
OCS3	0.6470	0.5434	0.8623	0.7801	0.6696	0.6844
OCS4	0.7170	0.4587	0.8623	0.8096	0.6696	0.7544
ICS1	0.8520	0.6420	0.7959	0.3856	0.7096	0.7299
ICS2	0.6870	0.5473	0.7643	0.7801	0.5868	0.6844
ICS3	0.7360	0.5257	0.7288	0.6844	0.5868	0.7801
PFF1	0.0000	0.0000	0.0000	0.0000	0.0000	1.0000
PFF2	0.0000	0.0000	0.0153	0.9175	0.8962	0.9175
PFF3	0.0000	0.0000	0.1158	0.9437	0.7175	0.5489
LSF1	0.1953	1.0000	0.8047	0.8047	0.8047	0.1953
LSF2	0.1953	0.1953	0.8398	0.8398	0.1953	0.1953
LSF3	0.7549	0.0000	1.0000	1.0000	0.0000	0.0000
SFU1	0.8080	0.6765	0.6018	0.7099	0.7102	0.6557
SFU2	0.7131	0.5881	0.5700	0.6126	0.6400	0.5929
SFU3	0.7081	0.6455	0.6828	0.7246	0.7883	0.7410
SFU4	0.7312	0.7173	0.5652	0.7678	0.7707	0.7381
SFU5	0.5931	0.5605	0.5495	0.6899	0.6902	0.6284
SCU1	0.6497	0.6079	0.5673	0.6100	0.6007	0.5550
SCU2	0.6297	0.5871	0.5197	0.5523	0.5573	0.5205
SCU3	0.6060	0.5573	0.5395	0.5355	0.5021	0.4703
ITST1	0.6944	0.6268	0.6326	0.7210	0.7207	0.7370
ITST2	0.7318	0.6939	0.6920	0.7289	0.7207	0.7070
ITST3	0.6965	0.6941	0.6920	0.7715	0.7586	0.7270
ITST4	0.7060	0.6620	0.6215	0.7265	0.7418	0.7112
VAST1	0.5965	0.5473	0.5710	0.6134	0.6886	0.6570
VAST2	0.6550	0.6052	0.5931	0.6955	0.7857	0.7070

(continued)

Table 12.16 (continued)

PPIs	T1	T2	T3	T4	T5	T6
VAST3	0.6786	0.5737	0.6250	0.7052	0.7578	0.6265
VAST4	0.6292	0.5831	0.5831	0.6831	0.7286	0.6649
ICIT1	0.7754	0.6899	0.6870	0.7581	0.7718	0.7686
ICIT2	0.7339	0.6415	0.6694	0.7341	0.7507	0.7428
ICIT3	0.6923	0.6394	0.6641	0.7118	0.6973	0.6923
ICIT4	0.6407	0.5773	0.6136	0.7110	0.7202	0.6923
SSS1	0.9299	0.7668	0.9289	0.9736	0.8954	0.9757
SSS2	0.8886	0.7318	0.8923	0.9652	0.8338	0.9673
SSS3	0.9036	0.7594	0.8739	0.9821	0.8444	0.9842
SSS4	0.9299	0.7826	0.7170	1.0000	0.9254	0.9842
EVS1	0.4616	0.4424	0.3398	0.7286	0.6131	0.6150
EVS2	0.6873	0.5450	0.8489	0.7907	0.8007	0.8849
EVS3	0.8159	0.6820	0.8859	0.9209	0.8733	0.9515
EVS4	0.7036	0.6544	0.7836	0.7191	0.7510	0.7407
EVS5	0.6400	0.5326	0.4919	0.6921	0.5303	0.6826
SES1	0.7823	0.6302	0.5631	0.9652	0.7152	0.6471
SES2	0.8623	0.6302	0.5800	0.9473	0.9473	0.8623
SES3	0.5631	0.5039	0.5631	0.1725	0.5631	0.4369

0.18) in Table 12.6. The same relationship (pr_i^- value with −0.02) between output (OPC, 0.14) and liquidity and solvency (LSF, 0.16) can be found in Table 12.6, but performance of LSF PPIs is less competitive because this is analysed that T6 started up its operation since 2011 with a rather recent heavy initial capital spending raising from financial institutions for terminal superstructure, state-of-the-art systems and equipment.

Furthermore, port managers can analyse performance at the upper level criteria (i.e. 6 dimensions and 16 principal-PPIs). Table 12.17 demonstrates the performance scores of the sixteen principal-PPIs, which is derived from the transformed values through the mapping process from the lowest level PPIs to their associated principal-PPIs, and the lowest level PPIs' weights. The results can lead to performance scores for the six dimensions (Table 12.18). From the results, T4 outperforms the other terminals in terms of human capital, liquidity and solvency, intermodal transport systems, safety and security. On the other hand, T6 shows the highest performance on output, productivity, profit and environment but is less competitive at the level of two principal-PPIs such as liquidity & solvency and service costs. Another striking feature of terminals demonstrates that they relatively outperform on lead-time and safety and security, while they show relatively poor performance on output, profitability and liquidity and solvency.

Table 12.17 Performance score on 16 principal-PPIs

	T 1	T 2	T 3	T 4	T 5	T 6	Ranking
OPC	0.0000	0.1240	0.0943	0.0087	0.0002	**0.9123**	T6 > T2 > T3 > T4 > T5 > T1
PDC	0.5320	0.5175	0.5820	0.6274	0.6190	**0.6406**	T6 > T4 > T5 > T3 > T1 > T2
LTC	0.9856	0.9806	0.9731	0.9942	**0.9947**	0.9930	T5 > T4 > T6 > T1 > T2 > T3
HCS	0.6771	0.5155	0.7543	**0.7789**	0.6981	0.7360	T4 > T3 > T6 > T5 > T1 > T2
OCS	0.7301	0.4714	**0.8592**	0.8137	0.7288	0.7447	T3 > T4 > T6 > T1 > T5 > T2
ICS	0.7791	0.5872	**0.7804**	0.6051	0.6435	0.7504	T3 > T1 > T6 > T5 > T4 > T2
PFF	0.0000	0.0000	0.0395	0.6652	0.5741	**0.8211**	T6 > T4 > T5 > T3 > T5 = T1
LSF	0.3535	0.3995	**0.8873**	**0.8873**	0.3364	0.1296	T3 = T4 > T2 > T1 > T5 > T6
SFU	**0.7480**	0.6566	0.5990	0.7180	0.7304	0.6803	T1 > T5 > T4 > T6 > T2 > T3
SCU	**0.6459**	0.6024	0.5546	0.5934	0.5857	0.5422	T1 > T2 > T4 > T5 > T3 > T6
ITST	0.7124	0.6604	0.6585	**0.7444**	0.7438	0.7415	T4 > T5 > T6 > T1 > T2 > T3
VAST	0.6428	0.5746	0.5970	0.6748	**0.7424**	0.6728	T5 > T4 > T6 > T1 > T3 > T2
ICIT	0.7294	0.6498	0.6731	0.7439	**0.7515**	0.7416	T5 > T4 > T6 > T1 > T3 > T2
SSS	0.9255	0.7825	0.8650	**0.9861**	0.8909	0.9837	T4 > T6 > T1 > T5 > T3 > T2
EVS	0.6851	0.5832	0.7012	0.7882	0.7091	**0.7950**	T6 > T4 > T5 > T3 > T1 > T2
SES	0.7869	0.6235	0.5685	**0.9003**	0.7594	0.6777	T4 > T3 > T5 > T6 > T2 > T3

Table 12.18 Performance score on 6 dimensions

	T 1	T 2	T 3	T 4	T 5	T 6	Ranking
CA	0.5096	0.5469	0.5645	0.5587	0.5569	0.8724	T6 > T3 > T4 > T5 > T2 > T1
SA	0.7468	0.5310	0.8119	0.7432	0.7020	0.7600	T3 > T6 > T1 > T4 > T5 > T2
FS	0.1468	0.1530	0.4367	0.7998	0.4517	0.4922	T4 > T6 > T5 > T3 > T2 > T1
US	0.7153	0.6410	0.5850	0.6750	0.6789	0.6289	T1 > T5 > T4 > T2 > T6 > T3
TSCI	0.7141	0.6420	0.6580	0.7405	0.7632	0.7389	T5 > T4 > T6 > T1 > T3 > T2
SG	0.8298	0.6893	0.7479	0.9206	0.8155	0.8617	T4 > T6 > T1 > T5 > T3 > T2

6 Conclusions

Existing studies on port performance lack a systematic approach capable of incorporating concerns from multi-stakeholders. In addition, they mainly focus on seaside operational performance using operational PPIs (i.e. CA) and treat the associated PPIs as independent from each other. Port performance measurement (PPM) is a typical MCDM problem under uncertainty (i.e. uncertain and incomplete data) and complexity (i.e. multi-stakeholder environment). In the MCDM applications, the evaluations of PPIs and their importance should be conducted separately and then incorporated into a single value for each terminal with respect to each PPI in a different hierarchy to rank the best performance from the alternative terminals or to diagnose their own situations. To this end, this chapter applies a hybrid PPM model that deals with PPIs' interdependency instead of PPIs' independency and evaluates them in a quantitative manner. This is achieved by the combination of DEMATEL and AHP and incorporating them into ER algorithm and utility technique. The proposed framework has been successfully implemented in dealing with both objective data and subjective data in a unified manner to incorporate multiple objectives of key stakeholders, which is validated through the case study of six container terminals in South Korea. From the case study results, decision makers in the terminals can identify the particular areas for improvement to enhance their competitiveness based on any necessary comparisons. In consequence, the hybrid methodology has proven to be a sound approach in dealing with MCDM problems under uncertainty which the previous studies have done little with on PPM.

The hybrid approach using DEMATEL and AHP incorporating FER is in particular useful in dealing with the following issues for PPM.

First, we use a combined method of DEMATEL and AHP for PPIs' relative weights instead of absolute weights to determine interdependent weights of the PPIs, making trade-off against each PPI. The selected PPIs are not equally important, which tends to be traded-off among the PPIs. The combined relative weighting method using pairwise comparisons is useful to identify traded-off among the PPIs.

Secondly, in the FER applications, this chapter used a belief structure (i.e. degrees of belief) to offer judgement flexibility to assessors by assessing on either one grade or even more instead of assessing only on one grade to avoid uncertainties in subjective judgement. On top of that we permitted incomplete judgements (i.e. the sum of DoB is less than 1) when assessors are not able to conduct a precise judgement due to inadequacy of information, which can be assigned to unknown scales. In this regard, we could minimise the missing data problems, which have been well recognised by researchers. But it has not well been tackled yet in the literature.

Thirdly, regardless of the number of assessment grades, the mapping techniques to convert DoBs of the bottom-level PPIs to their associated upper level principal-PPIs can be conducted in a unified manner. Through this technique, port managers look at the performance for each individual PPI, principal-PPI and dimension. This is another novelty of the approach, which has not been used in port performance studies.

Fourthly, DoBs in a belief structure can be assigned to an interval including several grades instead of a single grade (Xu et al. 2006). This chapter utilised a number of quantitative data that are confidential and sensitive for terminal operators. From real observations, they were reluctant to provide the data, hence there was a need to develop a powerful assessment tool capable of dealing with the inherent data uncertainties. Where there is no precise data available, using interval grades can be a second best solution in order to collect the required data.

However, it is noteworthy that this chapter discusses each applied methodology and their applications to PPM rather than empirical investigation, therefore the empirical results are only drawn from Korean container terminal cases and further empirical studies in different regions/areas need to be conducted.

Acknowledgements This research was supported by an EU Marie Curie grant (ENRICH – 612546).

References

Alavi M, Kayworth TR, Leidner DE (2006) An empirical examination of the influence of organizational culture on knowledge management practices. J Manag Inf Syst 22(3):191–224

Albadvi A, Keramati A, Razmi J (2007) Assessing the impact of information technology on firm performance considering the role of intervening variables: organizational infrastructures and business processes reengineering. Int J Prod Res 45(12):2697–2734

Bagozzi RP, Yi Y, Phillips LW (1991) Assessing construct validity in organizational research. Adm Sci Q 36:421–458

Barney J (1991) Firm resources and sustained competitive advantage. J Manag 17(1):99–120

Bichou K (2006) Chapter 24 Review of Port Performance Approaches and a Supply Chain Framework to Port Performance Benchmarking. Res Transp Econ 17(0):567–598

Bichou K, Gray R (2004) A logistics and supply chain management approach to port performance measurement. Marit Policy Manag 31(1):47–67

Brooks MR (2006) Issues in measuring port devolution program performance: a managerial perspective. Res Transp Econ 17:599–629

Brooks MR, Schellinck T (2013) Measuring port effectiveness in user service delivery: What really determines users' evaluations of port service delivery? Res Transp Bus Manag 8:87–96

Buyukozkan C, Cifci G (2012) A novel hybrid MCDM approach based on fuzzy DEMATEL, fuzzy ANP and fuzzy TOPSIS to evaluate green suppliers. Expert Syst Appl 39:3000–3011

Büyüközkan G, Arsenyan J, Ruan D (2012) Logistics tool selection with two-phase fuzzy multi criteria decision making: A case study for personal digital assistant selection. Expert Syst Appl 39(1):142–153

Cullinane K, Song D-W, Gray R (2002) A stochastic frontier model of the efficiency of major container terminals in Asia: assessing the influence of administrative and ownership structures. Transp Res Part A: Policy and Pract 36(8):743–762

Cullinane K, Wang T-F, Song D-W, Ji P (2006) The technical efficiency of container ports: Comparing data envelopment analysis and stochastic frontier analysis. Transp Res Part A: Policy and Pract 40:354–374

Darbra RM, Pittam N, Royston KA, Darbra JP, Journee H (2009) Survey on environmental monitoring requirements of European ports. J Environ Manag 90(3):1396–1403

De Langen P (2002) Clustering and performance: the case of maritime clustering in The Netherlands. Marit Policy Manag 29(3):209–221

De Monie G (1987) Measuring and evaluating port performance and productivity. United Nations, New York

Dooms M, Verbeke A (2007) Stakeholder management in ports: a conceptual framework integrating insights from research in strategy, Corporate social responsibility and port management. Int Assoc Marit Econ (IAME), Athens, Greece, 4–6 July 2007

ESPO (2010) Port Performance Indicators Selection and Measurement Work Package 1(WP1): Pre-Selection of an initial set of indicators. TREN/ 09/ SUB/ G2/ 170.2009/ S12.552637. European Sea Ports Organization, 26 October 2010

Gabus, A, Fontela, E (1973) Perceptions of the world problematique: communication procedure, communicating with those bearing collective responsibility. DEMATEL Report No. 1, Battelle Geneva Research Centre, Geneva, Switzerland

IMO (2002) ISPS code conference of contracting governments to the international convention for the SOLAS 1974: SOLAS/ CONF.5/34. IMO: London, 17 December 2002

Kaplan RS, Norton DP (2004) Measuring the strategic readiness of intangible assets. Harv Bus Rev 82(2):52–63

Lee PTW, Wu JZ, Hu KC, Flynn M (2013) Applying analytic network process (ANP) to rank critical success factors of waterfront redevelopment. Int J Shipping Transp Logist 5 (4):390–411

Liou JJ, Tzeng G-H, Chang H-C (2007) Airline safety measurement using a hybrid model. J Air Transp Manag 13:243–249

Lirn TC, Thanopoulou HA, Beynon MJ, Beresford AKC (2004) An application of AHP on transhipment port selection: a global perspective. Marit Econ Logist 6(1):70–91

Marlow PB, Paixão Casaca AC (2003) Measuring lean ports performance. Int J Transp Manag 1 (4):189–202

Panayides PM, Song D-W (2009) Port integration in global supply chains: measures and implications for maritime logistics. Int J Logist: Res Appl 12(2):133–145

Peris-Mora E, Orejas JD, Subirats A, Ibáñez S, Alvarez P (2005) Development of a system of indicators for sustainable port management. Mar Pollut Bull 50(12):1649–1660

Roll Y, Hayuth Y (1993) Port performance comparison applying data envelopment analysis (DEA). Marit Policy Manag 20:153–161

Saaty TL (1980) The analytic hierarchy process: Planning, priority setting, resources allocation. McGraw, New York

Saaty TL (2004) Decision making—the analytic hierarchy and network processes (AHP/ANP). J Syst Sci Syst Eng 13(1):1–35

Song DW, Panayides PM (2008) Global supply chain and port/terminal: integration and competitiveness. Marit Policy Manag 35(1):73–87

Su Y, Liang G-S, Chin-Feng L, Tsung-Yu C (2003) A study on integrated port performance comparison based on the concept of balanced scorecard. J East Asia Soc Transp Stud 5:609–624

Tongzon JL (1995) Determinants of port performance and efficiency. Transp Res Part A Policy Pract 29(3):245–252

UNCTAD (1976) Port performance indicators, TD/B/C.4/131/Supp.1/ Rev.1, UNCTAD, New York

Woo S-H, Pettit S, Beresford AK (2011) Port evolution and performance in changing logistics environments. Marit Econ Logist 13(3):250–277

Woo S-H, Pettit SJ, Beresford AK (2013) An assessment of the integration of seaports into supply chains using a structural equation model. Supply Chain Manag Int J 18(3):235–252

Xu DL, Yang JB, Wang YM (2006) The ER approach for multi-attribute decision analysis underinterval uncertainties. Eur J Oper Res 174(3):1914–1943

Yang J-B (2001) Rule and utility based evidential reasoning approach for multi-attribute decision analysis under uncertainties. Eur J Oper Res 131(1):31–61

Yang J, Xu D (2000) An introduction to an intelligent decision system: IDS.' In: Proceedings of the 42nd Annual Conference of the UK Operational Research Society, Swansea, UK, 12–14 September 2000

Yang J-B, Xu D-L (2002) On the evidential reasoning algorithm for multiple attribute decision analysis under uncertainty. Syst Man Cyb Part A Syst Hum, IEEE Trans 32(3):289–304

Yang Z, Bonsall S, Wang J (2009) Use of hybrid multiple uncertain attribute decision making techniques in safety management. Expert Syst Appl 36(2):1569–1586

Yang ZL, Ng AKY, Wang J (2016) Towards robust management of maritime risk and security. In: Lee P, Cullinane K (eds) Dynamic shipping and port development in the globalized economy. Palgrave, Houndmills

Yeo G-T, Ng AK, Lee PT-W, Yang Z (2014) Modelling port choice in an uncertain environment. Marit Policy Manag 41(3):251–267

Index

P.T.-W. Lee, Z. Yang (eds.), *Multi-Criteria Decision Making in Maritime Studies and Logistics*, International Series in Operations Research & Management Science 260,
DOI 10.1007/978-3-319-62338-2

MIX
Papier aus verantwortungsvollen Quellen
Paper from responsible sources
FSC® C105338

Printed by Books on Demand, Germany